Technology, Peace and Security | Technologie, Frieden und Sicherheit

Series Editor

Christian Reuter, Darmstadt, Germany

Technology plays a crucial role in various aspects of peace and security. This book series explores the intersection of computer science with peace and security studies. With a focus on cyber security and privacy, human-computer interaction as well as peace and conflict studies, it addresses topics such as peace informatics and technical peace research (cyber war, peace, arms control, and dual use), crisis informatics and information warfare (social media and collaborative technologies in conflict and crisis situations, misinformation and opinion manipulation), as well as usable safety, security, and privacy (resilient digital infrastructures, security and privacy enhancing technologies).

Technologie spielt eine entscheidende Rolle in verschiedenen Aspekten von Frieden und Sicherheit. Diese Buchreihe befasst sich mit Fragen an der Schnittstelle der Informatik mit der Friedens- und Sicherheitsforschung. Mit einem Fokus auf Cybersicherheit und Privatheit, Mensch-Computer-Interaktion sowie Friedens- und Konfliktforschung werden Themen wie Friedensinformatik und technische Friedensforschung (Cyberkrieg, Frieden, Rüstungskontrolle und Dual-Use), Kriseninformatik und Informationskrieg (soziale Medien und kollaborative Technologien in Konflikt- und Krisensituationen, Desinformation und Meinungsmanipulation) sowie nutzbare Sicherheit und Privatheit (resiliente digitale Infrastrukturen, Technologien zur Verbesserung der Sicherheit und des Datenschutzes) behandelt.

Katrin Hartwig

Navigating Misinformation

User-Centered Design and Evaluation of Indicator-Based Digital Interventions

Springer Vieweg

Katrin Hartwig
Informatik
Technische Universität Darmstadt
Walldorf, Germany

Dissertation approved by the Department of Computer Science of the Technical University of Darmstadt for the award of the academic degree Doktor-Ingenieurin (Dr.-Ing.) submitted by Katrin Hartwig, M.Sc. (born in Kempen)
First examiner: Prof. Dr. Dr. Christian Reuter
Second examiner: Prof. Dr. Amanda Lee Hughes
Date of submission: 11.07.2024
Date of disputation: 28.08.2024

ISSN 3004-9318 ISSN 3004-9326 (electronic)
Technology, Peace and Security | Technologie, Frieden und Sicherheit
ISBN 978-3-658-47884-1 ISBN 978-3-658-47885-8 (eBook)
https://doi.org/10.1007/978-3-658-47885-8

This Springer Vieweg imprint is published by the registered company Springer Fachmedien Wiesbaden GmbH, part of Springer Nature.
The registered company address is: Abraham-Lincoln-Str. 46, 65189 Wiesbaden, Germany

Foreword

Navigating the complex digital information landscape is increasingly challenging due to the overabundance of both reliable and false information, especially on social media. Technologies to help users navigate within this information have the potential to have a positive impact on users' abilities to handle the situation. Katrin Hartwig's PhD thesis focuses on user-centered digital misinformation interventions, such as smartphone apps and web apps, to help users manage misinformation in various forms. She employs a human-centered design approach to analyze and categorize existing interventions systematically, as well as to design new interventions and evaluate their perception and effectiveness through user studies. Key findings and important contributions include the identification of valuable and comprehensible indicators for different content modalities (text, audio, video), designing and testing a simulated TikTok app, and recognizing both the potential and limitations of these interventions, such as users' blind trust in them. The research also offers insights into the transferability of these findings to cybersecurity contexts and makes an important contribution.

The studies included in her PhD thesis have been published as eight peer-reviewed papers. In total, she has been involved in 19 publications. Her first publication began during the first semester of the PEASEC chair's existence in 2017 when Katrin (a master's student at that time) attended a seminar with us. She contributed to a study on the perception of fake news and misinformation in Germany, leading to her first peer-reviewed scientific publication and conference participation and presentation. She luckily joined our group for her PhD. After working in the field of cybersecurity, she contributed to acquiring and later coordinating the BMBF-funded NEBULA project (user-centered AI-based detection of fake news and misinformation), and she dedicated her dissertation to this topic. Additionally, she has been actively involved in the lecture "Safety-Critical Human-Computer Interaction" as well as in our seminars.

Katrin Hartwig has proven that she is capable of independent scientific work. Thus, in August 2024, her dissertation was accepted by the Department of Computer Science at the Technical University of Darmstadt for the degree of *Dr.-Ing.* as the seventh PhD thesis in our research group PEASEC. I would like to see a further focus on topics of such high importance. Katrin, thank you for your contribution and for allowing me to accompany you on your way to your PhD. I wish you all the best and every success for the future.

Prof. Dr. Dr. Christian Reuter
Professor for Science and Technology for
Peace and Security (PEASEC) and Dean
of the Department of Computer Science at
Technical University of Darmstadt
Darmstadt, Germany

Acknowledgements

I would like to express my sincere gratitude to the many people who have contributed to the success of this dissertation. Firstly, I would like to thank my doctoral supervisor, *Prof. Dr. Dr. Christian Reuter*, who has provided me with continuous guidance, support, and motivation. Without his trust and advice, this work could not have been possible. Furthermore, I wish to thank my co-examiner *Prof. Dr. Amanda Lee Hughes*, as well as *Prof. Dr. Justus Thies*, and *Prof. Dr. Martin Steinebach* who kindly join the doctoral committee.

I have been fortunate to work with the PEASEC team at the Technical University of Darmstadt, with wonderful colleagues whose support and team spirit made this journey so enjoyable. Working alongside my colleagues not only enriched my academic experience but also led to lasting friendships. I am truly thankful for your insightful conversations, fun team events, and encouragement throughout the years. In particular, I want to acknowledge the invaluable contributions of my co-authors Tom Biselli, Stefka Schmid, and Franziska Schneider, with whom I worked closely on significant research projects that played a critical role in this dissertation. I also extend my gratitude to Helene Pleil, Frederic Döll, Ruslan Sandler, Robert Cieslinski, Atlas Englisch, and Jan Pelle Thomson whose collaborative efforts as students have been greatly appreciated. Additionally, I would like to thank the former student assistants who provided invaluable support in software development, research, and proofreading of various publications.

This work was supported by the German Federal Ministry for Education and Research (BMBF) in the project NEBULA (13N16361—User-centered AI-based detection of misinformation and disinformation), which aims to automatically detect misinformation and disinformation for civil safety and security and subsequently provides user-centered presentations of the detection results for end users to promote media literacy on various social media platforms. In addition, the dissertation was supported by the German Federal Ministry of Education and Research (BMBF) within the project CYLENCE (13N16636), by the BMBF and the Hessian Ministry of Higher Education, Research, Science and the Arts within their joint support of the National Research Center for Applied Cybersecurity ATHENE, and by the Deutsche Forschungsgemeinschaft (DFG, German Research Foundation) within the Collaborative Research Center CROSSING (SFB 1119–236615297).

Finally, I want to give my deepest thanks to my family—particularly my parents –, my partner, and friends. Without their support and patience, this dissertation would not have been possible.

Abstract

Informed navigation of the complex digital information realm constitutes a major challenge for society and individuals, which is constantly growing due to the over-abundance of reliable and false information. In particular, this includes dealing with misinformation on social media, a central source of information procurement and exchange, in everyday life and in times of crisis. Complementing efforts like educational measures in classrooms or critical journalism, the research field of *Human-Computer Interaction* (HCI) suggests digital misinformation interventions as a user-centered countermeasure. Researchers and practitioners may benefit from the findings of this dissertation, which contributes to the design, implementation, and evaluation of interventions in different forms and degrees of transparency. It facilitates social media users' informed navigation within the complex digital information landscape encompassing multiple modalities, such as text, audio, and video.

This dissertation provides (1) a structured understanding of the state-of-the-art of misinformation interventions within a developed taxonomy. This encompasses diverse intervention designs (e.g., correcting misinformation in the comments, showing indicators, or reducing the visibility of misinformation content), active and passive user interaction types that do or do not require an engagement of the user with the intervention (e.g., confirmation before sharing), and various timings (e.g., taking place before or after a confrontation with misinformation). The dissertation demonstrates that the current research is mostly U.S.-based, addresses higher-educated participants, and particularly targets platforms like Twitter/X and Facebook. There is a lack of research regarding trending platforms that focus particularly on multimodal content and are highly relevant for specific user groups. It further highlights trends toward transparent and comprehensible interventions in contrast to top-down approaches (e.g., deleting or binary flagging of misinformation).

The findings informed (2) the design, implementation, and evaluation of a simulated smartphone app for TikTok, an intervention for misleading voice messages, and a web app for Twitter/X content as indicator-based interventions along with additional contrasting concepts and artifacts (e.g., nudges to reduce problematic sharing intentions). Therefore, (3) the dissertation identified user-centered misinformation indicators for various modalities (e.g., text, audio, and video) that were perceived as useful and comprehensible by adult and adolescent participants in empirical studies. Among these are emotion-evoking sounds and music in TikTok videos, critical discussions in the comment sections of social media posts, and rhetorical characteristics like claims to absolute truth.

The dissertation empirically demonstrates that (4) indicator-based interventions are positively received overall, particularly due to their transparency in contrast to less comprehensible top-down approaches. However, they also come with severe limitations and challenges, such as users' blind trust and lack of realistic assessments of potential errors or biases.

This research contributes to user-centered and informed navigation of misinformation in a complex digital information environment by triangulating findings from diverse data collection techniques. It outlines opportunities, limitations, and implications for future research on misinformation interventions.

Author's Publications

In sum, 18 publications have been published in the context of the author's work.

The following 8 publications are published or in submission and are included as chapters in part II of this thesis:

1. Katrin Hartwig, Frederic Doell, and Christian Reuter. "The Landscape of User-centered Misinformation Interventions—A Systematic Literature Review". In: *ACM Computing Surveys (CSUR)* 56.11 (2024), pp. 1–36. https://doi.org/10.1145/3674724 (CORE-A*)
2. Katrin Hartwig, Stefka Schmid, Tom Biselli, Helene Pleil, and Christian Reuter. "Misleading Information in Crises: Exploring Content-Specific Indicators on Twitter from a User Perspective". In: *Behaviour & Information Technology* (2024). https://doi.org/10.1080/0144929X.2024.2373166 (Impact Factor 3.7)
3. Stefka Schmid, Katrin Hartwig, Robert Cieslinski, and Christian Reuter. "Digital Resilience in Dealing with Misinformation on Social Media during COVID-19 A Web Application to Assist Users in Crises". In: Information Systems Frontiers 26.2 (2024), pp. 477–499. https://doi.org/10.1007/s10796-022-10347-5 (Impact Factor 5.9)
4. Katrin Hartwig, Ruslan Sandler, and Christian Reuter. "Navigating Misinformation in Voice Messages: Identification of User-Centered Features for Digital Interventions". In: Risk, Hazards & Crisis in Public Policy 15.2 (2024), pp. 203–235. https://doi.org/10.1002/rhc3.12296 (Impact Factor 4.38)

5. Katrin Hartwig, Tom Biselli, Franziska Schneider, and Christian Reuter. "From Adolescents' Eyes: Assessing an Indicator-Based Intervention to Combat Misinformation on TikTok". In: Proceedings of the CHI Conference on Human Factors in Computing Systems. ACM, 2024, pp. 1–20. https://doi.org/10.1145/3613904.3642264 (CORE-A*)
6. Tom Biselli, Katrin Hartwig, and Christian Reuter. "Mitigating Misinformation Sharing on Social Media through Personalised Nudging". In: Proceedings of the ACM: Human-Computer Interaction (PACM): Computer-Supported Cooperative Work and Social Computing (CSCW). ACM, 2025. https://doi.org/10.1145/3711034 (CORE-A)
7. Katrin Hartwig and Christian Reuter. "Nudge or Restraint: How Do People Assess Nudging in Cybersecurity—A Representative Study in Germany". In: Proceedings of the 2021 European Symposium on Usable Security (EuroUSEC). Association for Computing Machinery, 2021, pp. 141–150. doi: https://doi.org/10.1145/3481357.3481514
8. Katrin Hartwig, Atlas Englisch, Jan Pelle Thomson, and Christian Reuter. "Finding Secret Treasure? Improving Memorized Secrets Through Gamification". In: Proceedings of the 2021 European Symposium on Usable Security (EuroUSEC). ACM, 2021, pp. 105–117. doi: https://doi.org/10.1145/3481357.3481509

The following 10 papers are not included in the thesis, although their findings are supplementary to it:

9. Katrin Hartwig and Christian Reuter. "Nudging Users Towards Better Security Decisions in Password Creation Using Whitebox-based Multidimensional Visualizations". In: Behaviour & Information Technology (BIT) 41.7 (2021), pp. 1357–1380. https://doi.org/10.1080/0144929X.2021.1876167
10. Anjuli Franz, Verena Zimmermann, Gregor Albrecht, Katrin Hartwig, Christian Reuter, Alexander Benlian, and Joachim Vogt. "SoK: Still Plenty of Phish in the Sea—A Taxonomy of User-Oriented Phishing Interventions and Avenues for Future Research". In: Proceedings of the Seventeenth Symposium on Usable Privacy and Security (SOUPS 2021). USENIX, 2021, pp. 339–358. url: https://www.usenix.org/conference/soups2021/presentation/franz
11. Elise Özalp, Katrin Hartwig, and Christian Reuter. "Trends in Explainable Artificial Intelligence for Non-Experts". In: Artificial Intelligence—Limits and Prospects. Ed. by Peter Klimczak and Christer Petersen. 4. Transcript Verlag, 2023, pp. 223–243. url: https://doi.org/10.14361/9783839457320

12. Katrin Hartwig and Christian Reuter. "Fake News technisch begegnen: Detektions-und Behandlungsansätze zur Unterstützung von NutzerInnen". In: Wahrheit und Fake News im postfaktischen Zeitalter. Ed. by Peter Klimczak and Thomas Zoglauer. Vol. 3. ars digitalis. Springer Vieweg, 2021, pp. 133–150. url: https://link.springer.com/chapter/10.1007/978-3-658-32957-0_7
13. Katrin Hartwig and Christian Reuter. "Transparenz Im Technischen Umgang Mit Fake News". In: Technik & Mensch 2 (2021), pp. 9–11. url: https://tuprints.ulb.tu-darmstadt.de/19267/1/VDI_TuM_2.2021.pdf
14. Katrin Hartwig and Christian Reuter. "TrustyTweet: An Indicator-Based Browser-Plugin to Assist Users in Dealing with Fake News on Twitter". In: Proceedings of the International Conference on Wirtschaftsinformatik (WI). Vol. 14. AIS, 2019, pp. 1844–1855. url: https://aisel.aisnet.org/wi2019/specialtrack01/papers/5/
15. Katrin Hartwig and Christian Reuter. "Fighting Misinformation on Twitter: The Plugin Based Approach TrustyTweet". In: SCIENCE PEACE SECURITY'19—Proceedings of the Interdisciplinary Conference on Technical Peace and Security Research. Ed. by Christian Reuter, Jürgen Altmann, Malte Göttsche, and Mirko Himmel. TUprints, 2019, pp. 67–69. url: https://publications.rwth-aachen.de/record/781544/files/781544.pdf#page=67
16. Christian Reuter, Katrin Hartwig, Jan Kirchner, and Noah Schlegel. "Fake News Perception in Germany: A Representative Study of People's Attitudes and Approaches to Counteract Disinformation". In: Proceedings of the International Conference on Wirtschaftsinformatik (WI). AIS, 2019, pp. 1–15. url: https://aisel.aisnet.org/wi2019/track09/papers/5/
17. Cristina Laura Oyarzun, Katrin Hartwig, Alexander Distergoft, Tim Hoffmann, Kathrin Scheckenbach, Melanie Brüsseler, and Stefan Wesarg. "Automatic Segmentation of the Structures in the Nasal Cavity and the Ethmoidal Sinus for the Quantification of Nasal Septal Deviations". In: Medical Imaging 2021: Computer-Aided Diagnosis. Vol. 11597. SPIE, 2021, pp. 582–591. https://doi.org/10.1117/12.2582251
18. Cristina Laura Oyarzun, Katrin Hartwig, Anna-Sophie Hertlein, Florian Jung, Jan Burmeister, Jörn Kohlhammer, Stefan Wesarg, and Guido Sauter. "Web-Based Prostate Visualization Tool". In: Current Directions in Biomedical Engineering 6.3 (2020), pp. 563–566. https://doi.org/10.1515/cdbme-2020-3144

Contents

List of Figures

List of Tables

Part I
Synopsis

Introduction 1

1.1 Motivation and Problem Statement

In our contemporary world that is closely tied to the digital information realm, the emergence of complex global crises such as the COVID-19 pandemic and ongoing conflicts and wars in several parts of the world particularly demonstrates an overabundance of both accurate and misleading information. Social media platforms play an essential and indispensable role in exchanging information, both in everyday life and crises [537]. However, social media platforms are further characterized by attributes like high speed of dissemination, low cost, and (alleged) anonymity, which collectively create an environment conducive to the dissemination of misleading information [539]. Indeed, various forms of misleading information can potentially cause significant harm in multiple dimensions of society, politics, and on individual levels. In certain contexts, they can manipulate elections [376], lead to polarization [325], and even cause serious health consequences, including fatalities [343, 422]. The volume and significant effects of misleading information pose a crucial challenge for society and individuals, emphasizing the need for user-centered measures that facilitate informed navigation within the digital information landscape. Misleading information, whether spread intentionally (i.e., 'disinformation' or 'fake news') or unintentionally (i.e., 'misinformation'), can cause great harm. In some cases, this can even include satirical content that aims to present topics ironically but can still be misleading if the recipients lack understanding or context information [225]. This dissertation follows the suggestion of other research and uses the umbrella term 'misinformation' for better readability, encompassing both intentionally and unintentionally misleading or false content [15, 97, 597].

K. Hartwig, *Navigating Misinformation*, Technology, Peace and Security | Technologie, Frieden und Sicherheit,
https://doi.org/10.1007/978-3-658-47885-8_1

Interdisciplinary research has started to investigate how to tackle the challenges of misinformation that individuals and society face. This encompasses perspectives from computer science, psychology, social sciences, and communication science among others, and, for instance, covers efforts for automatic detection of misinformation, often using machine-learning [253], decisions for user feedback post-detection that can be automatic, expert-based (e.g., by official fact-checkers), or crowd-sourced, or aspects that are detached from detection efforts at all, such as motivations for misinformation dissemination or correction efforts. HCI research, in particular, focuses on how technology can support people with such questions and problems in a user-centered way. In doing so, efforts have been made to design, implement, and evaluate user interventions to encourage a specific behavior or perception in various forms and with various targets of behavior change. This includes improved credibility assessment, reduced sharing of misinformation, successful distinguishing of misinformation and credible content, or decreased overall flow of misinformation on social media [214]—to this point with controversial findings and contentious debates about efficacy, appropriateness, and potential backfire effects, and a definitive solution is yet to be found [225].

This dissertation aims to contribute to the user-centered design, implementation, and evaluation of digital misinformation interventions that equip users with the capabilities they need to navigate the digital information landscape in an informed manner. Thus, it depicts efforts with qualitative and quantitative findings on interventions as one of multiple building blocks for dealing with misinformation from an HCI perspective. The dissertation understands solutions to the misinformation challenges as a strongly interdisciplinary field that requires the combination of multiple efforts from diverse perspectives. This includes, among others, professional journalism and fact-checking, media literacy training at school, technology-driven approaches for detection, and user-centered digital interventions as a complementing piece of the puzzle that comes with opportunities and limitations.

Previous research suggests user preferences for transparency and comprehensibility of interventions, i.e., users tend to have more trust in interventions that can explain comprehensibly, for instance, why a piece of information should be classified as misinformation or should be examined critically and prefer them over interventions that rather give top-down feedback without explanations or lead to censorship [188, 224, 299]. Efforts in explainable artificial intelligence have started to look into how detection approaches can be made transparent in terms of highlighting pieces of content that have been particularly considered by the decision of the detection algorithm [621]. Other approaches are detached from a preceding automatic detection and look into indicators for misinformation from a particularly user-centered perspective. By deriving comprehensible and (easily) visible indicators for content credibility and displaying them to the end user, indicator-based misinformation interventions aim to address the preference for comprehensibility to increase trust and overcome unintended backfire effects such as reactance. While research indicates promising advantages of indicator-based interventions [224] such as reduced uncertainty [188], alignment with pre-existing mental models and practices of users [513], beneficial perceived utility [188], and facilitated autonomous assessment skills [491], challenges persist [224] and are controversially discussed with interventions being perceived as patronizing [226] or biased [188].

This dissertation delves into the opportunities and limitations of indicator-based misinformation interventions, considering a diversity of information modalities. While most of current related research focuses on text, this dissertation considers the variety of modalities depicted in contemporary social media platforms. Thus, it addresses diverse platforms that each come with unique attributes and potential to mislead, including *Video-Sharing Platforms* (VSP), messengers facilitating voice messages, and mainly text- or image-based platforms. Further, the dissertation considers diverse target groups, following the assumption of related research that not all users are equally vulnerable to misinformation. Individuals vary regarding susceptibility to misinformation [224] in terms of decision-making styles [34], cognitive abilities [124], demographic factors [273], and language skills [473]. While the majority of current research focuses on 'the average user' or adult college students as user study participants [214], this dissertation complements their findings by applying specific foci not only on adults but also on adolescents as a unique and highly relevant user group on social media [224, 226], and on potentials to apply personalization of interventions via psychometric differences. Furthermore, due to its timing, this dissertation was shaped by global crises such as the COVID-19 pandemic and ongoing conflicts and wars, which is depicted in specific crisis-related case studies. Addressing the research gaps in HCI, the dissertation offers a user-centered and technology-driven contribution to the challenges of multimodal misinformation and digital interventions to encourage informed navigation within the complex digital information realm.

The methodology of the dissertation follows an iterative human-centered design and evaluation approach, that covers literature-based in-depth insights and systematization of state-of-the-art research on misinformation interventions to form a knowledge base, qualitative user studies (e.g., focus group discussions and individual think-aloud studies) to gain rich insights into users' needs and preferences, as well as (representative) quantitative online experiments and surveys to derive statistically significant findings. The triangulation of data collection techniques allowed for an iterative user-centered design and evaluation process of digital artifacts, deriving opportunities, limitations, and implications for future research.

1.2 Aim and Research Question

This dissertation aims to contribute to the user-centered design, implementation, and evaluation of digital misinformation interventions as a measure to facilitate informed navigation in the complex digital information landscape, complementing interdisciplinary efforts such as professional fact-checking, media literacy training, and automatic misinformation detection approaches.

The main research question of this dissertation is:

RQ How can user-centered digital interventions assist users in navigating misinformation across modalities?

This research question can be divided into four sub-questions:

RQ1 How does research from different fields inform the design and evaluation of user-centered misinformation interventions?

In an in-depth systematic literature review, the dissertation derives an understanding of the current interdisciplinary research landscape of user-centered misinformation interventions, developing a taxonomy, and revealing trends and opportunities that strongly influenced the subsequent design contributions of the dissertation, motivating an indicator-based approach. Answers to this question can be found in Chap. 6.

RQ2 What indicators of misinformation can be considered user-centered?

Motivated by the derived insights into the current research landscape, the dissertation provides a basis for designing indicator-based interventions. Thus, to contribute to understanding individuals' information processing and interaction, this dissertation derives characteristics of misinformation and evaluates how they are (not) applied autonomously and how they are (not) perceived as useful or comprehensible in diverse modalities and by diverse user groups. These indicators will subsequently be applied to the design of indicator-based interventions. Answers to this question can be found for mainly text-based content and for images in Chap. 7, for voice messages in Chap. 9, and for VSPs in Chap. 10.

RQ3 How can indicator-based interventions be designed and evaluated to assist users in dealing with misinformation online?

The interdisciplinary research landscape on digital interventions to combat misinformation is confronted with various approaches. This dissertation suggests complementing indicator-based interventions for diverse modalities (i.e., mainly text-based social media posts, short-videos, and voice messages). It addresses users' preferences, particularly for transparency and comprehensibility, and derives design implications. Answers to this question can be found for a conceptual design and evaluation in Chap. 7, and for the development of artifacts (e.g., a web app and a smartphone app) as low-fidelity prototypes, high-fidelity prototypes, and fully functional applications in Chaps. 8, 10, and 9. Chapter 11 complements the findings on indicator-based interventions with the design and evaluation of various contrasting nudges.

RQ4 What are the opportunities and limitations of indicator-based interventions to assist users in dealing with misinformation?

While indicator-based interventions aim at addressing users' preferences for transparency, they come with both opportunities and challenges. This dissertation derives opportunities and limitations from diverse perspectives and for diverse contexts and discusses influencing factors and mitigation options (e.g., personalization) on perception and efficacy. It further gives a first outlook on how our findings on misinformation interventions can be applied to other contexts with sensitive information, using the example of (hybrid) nudging in cybersecurity, and takes a closer look at user preferences, attitudes, and influencing factors. Answers to this question can be found based on literature in Chap. 6, based on empirical studies in the context of misinformation in Chaps. 7, 9, 10, and 11, and regarding the outlook on applicability to the cybersecurity context in Chaps. 12 and 13.

1.3 Content and Structure of the Dissertation and Contributions of the Authors

This dissertation consists of two parts: a synopsis (I) and publications (II).

Part I: Synopsis
The synopsis consists of the presentation of theoretical foundations for this dissertation and a systematic summary of related work within this area of research. Further, this first part of the dissertation presents the research design and the findings of the empirical contributions, which are further discussed, leading to a conclusion that highlights limitations and potential for future research.

Chapter 1 *(Introduction)* comprises the motivation, aims and objectives, and the structure of the work.
Chapter 2 *(Related Work)* outlines the relevance of different forms of misleading information on social media, gives an overview on the diverse landscape of digital interventions to assist users in navigating misinformation, and presents recent research on indicator-based interventions. Relevant terms related to the research area of this dissertation are introduced, and research gaps are emphasized.
Chapter 3 *(Research Design)* suggests the research approach and methodology of this dissertation, including the design of user-centered artifacts as well as the quantitative and qualitative evaluations.
Chapter 4 *(Findings and Discussion)* highlights and discusses the findings of this dissertation in relation to implications for digital interventions to assist users in navigating misinformation online. This comprises (1) insights into the interdisciplinary research landscape of user-centered misinformation interventions regarding intervention design, user interaction, and timing, (2) characteristics of misinformation that are perceived as useful and comprehensible in diverse modalities and by diverse user groups, (3) the design, implementation, and evaluation of indicator-based misinformation interventions as well as nudging interventions, and (4) derived implications for opportunities and limitations of indicator-based misinformation interventions including influencing factors and mitigation options, as well as a first outlook on how the findings on misinformation interventions can be applied to other contexts with sensible information such as cybersecurity.
Chapter 5 *(Conclusion)* summarizes the main contributions, emphasizes the overall conclusions and presents perspectives on future research on the design,

implementation, and evaluation of digital user interventions in the context of misinformation.

Part II: Publications
The second part of the dissertation consists of the publications that provide the empirical and theoretical findings to answer the research questions. They have been previously published as articles (journal articles, conference papers, and book chapters). Major parts of this dissertation are written by several authors. The purpose of this section is to list the independent academic contributions that have been approved by all authors. For all but two papers that are part of this dissertation, I was the lead and corresponding author, but the contributions of the co-authors, which I gratefully acknowledge, are also very important, as I will describe below.

Chapter 6 *(The Landscape of User-centered Misinformation Interventions—A Systematic Literature Review)* systematically reviews the diverse landscape of user-centered misinformation interventions to build a knowledge base of the state-of-the-art. Based on 172 scholarly publications, a taxonomy was developed to cluster interventions regarding their design, type of user interaction, and timing. This chapter was originally published in *ACM Computing Surveys* (CORE-A*; Impact Factor 23.8) [214] (Paper A).

AUTHOR STATEMENT It constitutes a joint work of Katrin Hartwig, Frederic Döll, and Christian Reuter. As the corresponding and leading author, Katrin led and conducted the paper's overall research design, management, and writing process of the final paper. Frederic contributed the initial subset of the literature review with N = 43 scientific papers based on his Bachelor thesis in 2022. For the final version of the paper, Katrin extended the review to an overall number of 172 papers to cover a broader perspective on misinformation interventions and the current literature until 2024 and adapted the research questions and overall framing of the review to the state-of-the-art research landscape. This includes rewriting the overall sections of the paper to encompass the novel focus and the novel results based on the final set of papers. Frederic's thesis was used as a basis for the derived taxonomy, providing valuable examples of interventions within the tables. Christian was a general advisor to this work, contributed to the idea generation and framing of the research, and gave continuous feedback during all phases of the paper writing process.

Chapter 7 *(Misleading Information in Crises: Exploring Content-specific Indicators on Twitter from a User Perspective)* presents findings of a threefold evaluation to derive topical, formal, and rhetorical characteristics of misleading tweets on Twitter/X during crises, evaluate autonomous assessment strategies of users, and evaluate the perceived usefulness and comprehensibility of identified misinformation indicators. Implications for user assistance in dealing with misinformation during crises online are derived. This chapter was originally published in *Behaviour & Information Technology* (Impact Factor 3.7) [225] (Paper B).

AUTHOR STATEMENT It constitutes a joint work of Katrin Hartwig, Stefka Schmid, Tom Biselli, Helene Pleil, and Christian Reuter. As corresponding and leading author, Katrin led the paper's overall research design, data collection, management, and writing process in thorough consultation with Stefka regarding steps 1 and 2, and in thorough consultation with Tom regarding step 3. Katrin and Stefka developed the general methodological approach of the study to equal parts. Tom contributed to the paper after steps 1 and 2 were conducted and evaluated, focusing on step 3. Katrin and Stefka planned and conducted steps 1 (quantitative Twitter analysis) and 2 (online survey) together. In general, Stefka was primarily responsible for the qualitative insights (e.g., including citations of participants) and an in-depth reflection of content-specific themes of conspiracy theories, while Katrin was primarily responsible for quantitative insights and automatically retrievable characteristics. Step 1 was led by Katrin and evaluated by Katrin, Stefka, and Helene. Step 2 was led and evaluated by Stefka. Helene was especially involved in the coding of tweets and provided relevant literature based on a thorough literature search. Katrin and Tom planned, conducted, and evaluated step 3 (think-aloud sessions) together. The methods and results sections of the paper were primarily written by Katrin (steps 1 and 3), Stefka (step 2), and Tom (step 3) accordingly. Katrin was responsible for writing the final introduction, based on input from Stefka, and the related work sections. Katrin finalized the paper's sections and addressed reviewers' comments during revisions, with continuous input and feedback from Stefka, Tom, and Christian. Christian was a general advisor of this work and contributed to idea generation and framing of the research, as well as continuous feedback during all phases of the paper writing process.

Chapter 8 *(Digital Resilience in Dealing with Misinformation on Social Media during COVID-19)* proposes a web app based on *Social Network Analysis* (SNA) to provide an overview of potentially misleading vs. non-misleading content on Twitter/X, which can be explored by users to enable foundational learning. The artifact was evaluated within a two-fold think-aloud study. This chapter was originally published in *Information Systems Frontiers* (Impact Factor 5.9) [491] (Paper C).

AUTHOR STATEMENT It constitutes a joint work of Stefka Schmid, Katrin Hartwig, Robert Cieslinski, and Christian Reuter. In collaboration, Katrin and Stefka formulated the overall research idea of developing a web app based on SNA (Introduction). As the corresponding and leading author, Stefka led the research design process, including the identification and presentation of related and conceptual work on resilient information systems (Related Work). Robert developed and implemented the web app and carried out the iterative evaluation process (Concept and Evaluation). Katrin supported the research process by presenting the design science approach including think aloud studies as a method for validation (Methodology). She also prepared data and gave input on presenting technical information (Concept) and conducting the iterative evaluation (Evaluation). Stefka built on the design science approach and organized the process and presentation of findings (Concept, Evaluation) and discussion (Discussion) according to the overall interest of problem-solving and improvement of an information system. Christian supervised the research and supported it by giving constructive feedback on the methodological approach (Methodology) and information systems perspective (Discussion, Conclusion).

Chapter 9 *(Navigating Misinformation in Voice Messages: Identification of User-Centered Features for Digital Interventions)* investigates how users can be assisted in dealing with misinformation in voice messages. Features on various levels (e.g., word, audio, creator) were automatically extracted and evaluated in terms of perceived usefulness and comprehensibility. This chapter was originally published in *Risk, Hazards & Crisis in Public Policy* (Impact Factor 4.38) [222] (Paper D).

AUTHOR STATEMENT It constitutes a joint work of Katrin Hartwig, Ruslan Sandler, and Christian Reuter. As corresponding and leading author, Katrin led the overall research design, management, and writing process of the paper. Ruslan contributed to the complete technical implementation of indicator detection and a knowledge base on indicators based on scientific literature. The thesis of Ruslan was used as a basis for the subsection on features of misinformation in text and audio and the section on automatic detection of features. Katrin designed, conducted, and analyzed a novel user study and reframed the paper as a contribution to public policy. This includes rewriting the introduction, adding related work excluding Sect. 2.3, Chap. 4 regarding the novel user study, the discussion, and the conclusion. Christian was a general advisor of this work and contributed to idea generation and framing of the research, as well as continuous feedback during all phases of the paper writing process.

Chapter 10 *(From Adolescents' Eyes: Assessing an Indicator-Based Intervention to Combat Misinformation on TikTok)* evaluates comprehensible indicators for misinformation on TikTok, confirming indicators from other (text-based) platforms and identifying indicators unique to VSPs. Based on that, a simulated smartphone app was developed and evaluated with adolescents. This chapter was originally published in the *Proceedings of the CHI Conference on Human Factors in Computing Systems* (CORE-A*) [224] (Paper E).

AUTHOR STATEMENT It constitutes a joint work of Katrin Hartwig, Tom Biselli, Franziska Schneider, and Christian Reuter. As the corresponding and leading author, Katrin led the paper's overall research design, data collection, management, and writing process. Katrin developed the general methodological approach of the study in consultation with Tom. Franziska identified the stimuli for misleading and credible TikTok videos. Katrin, Tom, and Franziska conducted the focus groups, think-aloud sessions, and the coding of qualitative results together, based on Katrin's planned and prepared analysis design. Katrin applied the coded results to the corresponding sections and wrote the results section. Franziska contributed to the section on autonomous assessment strategies, inserted the quantitative results of the survey items, and prepared pictures and a table for the paper. Tom was responsible for the writing part of the analysis within the methodology section, based on an initial version by Katrin, contributed to the description of the study procedure and recruitment, and wrote the subsection on adolescents as vulnerable users. Katrin took the lead in writing the discussion, which Tom supplemented with related literature. Tom and Katrin contributed equal parts to the derived design implications. Katrin wrote the paper's other sections with continuous input and feedback from Tom, Franziska, and Christian. Christian was a general advisor of this work and contributed to idea generation and framing of the research, as well as continuous feedback during all phases of the paper writing process.

Chapter 11 *(Mitigating Misinformation Sharing on Social Media through Personalized Nudging)* investigates how personalized nudges as a contrasting and less transparent intervention can assist users in sharing less misinformation online. Therefore, the efficacy of various nudges was evaluated in an online experiment. This chapter was originally published in the *Proceedings of the ACM Human-Computer Interaction (PACM): Computer-Supported Cooperative Work and Social Computing (CSCW)* (CORE-A) [54] (Paper F).

AUTHOR STATEMENT As the corresponding and leading author, Tom led the paper's overall research design, data collection, management, and writing process. Tom developed the study's general methodological approach and the stimuli in consultation with Katrin. Katrin was responsible for the theoretical foundation, writing the related work Sects. 2.1 and 2.2 alone and writing the related work Sect. 2.3 together with Tom. Tom wrote the introduction and method section, conducted the statistical analysis, and wrote the corresponding results section. Tom was also mainly responsible for writing the discussion, with suggestions from and in consultation with Katrin, especially in Sect. 5.2. Christian was a general advisor of this work and contributed to idea generation and framing of the research, as well as continuous feedback during all phases of the paper writing process.

Chapter 12 *(Nudge or Restraint: How do People Assess Nudging in Cybersecurity—A Representative Study in Germany)* constitutes a first outlook into how the findings on misinformation interventions can be applied to an additional context with sensible information. It investigates the attitude towards nudging in cybersecurity by conducting a representative study in Germany, asking German citizens about their attitude towards nudging in that specific context. The study considers aspects of transparency and personalization. The chapter was originally published in the *European Symposium on Usable Security* (EuroUSEC) (CORE-B) [217] (Paper G).

AUTHOR STATEMENT It constitutes a joint work of Katrin Hartwig, and Christian Reuter. Katrin planned and led the general study design, performing and analyzing the online survey. Christian was a general advisor of this work and contributed with continuous feedback during all phases of the paper writing process.

Chapter 13 *(Finding Secret Treasure? Improving Memorized Secrets Through Gamification)* suggests further initial insights on how misinformation findings can be transferred to the context of cybersecurity, focusing on comprehensibility and transparent information. This chapter evaluates the efficacy, usability, and effects on password memorability of multiple nudging implementations, including a novel nudge with gamification elements in two user studies ($N_1 = 16, N_2 = 1,000$). The chapter was originally published in the *European Symposium on Usable Security* (EuroUSEC) (CORE-B) [223] (Paper H).

AUTHOR STATEMENT It constitutes a joint work of Katrin Hartwig, Atlas Englisch, Jan Pelle Thomason, and Christian Reuter. Katrin contributed by giving the general idea on nudges for stronger passwords building on her previous research, planning and leading the general study design, re-framing the paper for the final submission at EuroUsec, performing the literature review for Chap. 2, and discussing the results in Chap. 6. She further contributed with continuous feedback on the implementation, which was conducted by

Atlas and Jan Pelle. Atlas and Jan Pelle contributed equally to the paper by giving concrete ideas for the evaluated novel nudges. Further, they took key responsibilities for the implementation of the nudges and the provision of their evaluation. Atlas and Jan Pelle further contributed to the writing process by providing an initial version of Chaps. 3, 4, 5, and 7, which were then refined by Katrin. Christian was a general advisor of this work and contributed with continuous feedback during all phases of the paper writing process.

Summary—Introduction

- Misinformation on social media poses a *significant threat to society and individuals*, in crises and everyday life.
- Interdisciplinary research suggests controversial findings and contentious debates about digital misinformation interventions.
- This dissertation contributes to *designing, implementing, and evaluating indicator-based misinformation interventions* for content with various modalities to facilitate informed navigation of the complex digital information realm.

Related Work

2

2.1 Misinformation on Social Media from an HCI Perspective

In the contemporary digital age, social media platforms have become indispensable for exchanging information. Providers such as TikTok, Facebook, Telegram, and X (formerly Twitter), among others, offer tailored options for receiving, creating, or interacting with information in various modalities. The landscape of social media platforms is in a state of constant evolution, adapting to the advancing technological possibilities, providers' intentions, and users' preferences. Research suggests beneficial effects of social media platforms like connecting communities (e.g., for social support in mental health crises: [369, 487]), or facilitating social interaction and creativity via collaborative content creation [393]. For example, TikTok offers a duet feature that allows users to collaboratively edit and re-share videos in multiple iterations, resulting in diverse creative artifacts that can be assigned to music generation, satire, or experience reports, among others [393].

Social media allows for communication and entertainment not only in everyday life but also for vital information channels during disruptive situations [195] both for individuals and official authorities [102]. In this respect, the platforms enable the rapid dissemination of information (e.g., by the government or other authorities) in the event of a disaster, as well as the rapid acquisition of information by individuals [460]. Furthermore, crisis-specific safety-check features [321] enable users to indicate their safety status during disruptive situations (e.g., a shooting or earthquake), thereby providing reassurance to family and friends of their well-being.

K. Hartwig, *Navigating Misinformation*, Technology, Peace and Security | Technologie, Frieden und Sicherheit,
https://doi.org/10.1007/978-3-658-47885-8_2

Although social media platforms are an integral part of our everyday lives and offer numerous benefits, recent crises such as the Russian war of aggression against Ukraine and the global pandemic of COVID-19 have once again highlighted the challenges and risks associated with these platforms [288]. Research has shown that attributes of social media platforms such as high speed of dissemination, low cost, and (alleged) anonymity facilitate an environment for misleading information [539]. Indeed, in an era of multiple complex crises, we are confronted with an overabundance of credible and misleading information [611]. Individuals and society as a whole struggle to manage the large volume of information and distinguish misinformation from factual content. The risk of this should not be underestimated, as misinformation can have severe or even fatal consequences [422], lead to polarization or manipulated elections [283, 395], and, thus, pose a threat to democracy. Research has highlighted the threat and impact on individuals [41] but also on governments worldwide that are confronted with disinformation campaigns [71] that may manipulate diplomatic relations and policies.

The diverse landscape of social media platforms, with its various features and modalities, entails a unique potential to mislead. The vast majority of research has focused on mainly textual content. However, misinformation is commonly spread within formats like voice messages [458] or short videos [16] as well. Indeed, current popular platforms like TikTok or other VSPs offer various layers of content [387] that each may be able to mislead. In addition, misinformation in voice messages sent via popular messenger applications such as Telegram or WhatsApp has emerged as a concerning phenomenon [351, 384] both in everyday life and in uncertain crisis situations, where individuals seek information in public messenger groups. While some platforms or modalities are mainly used by a broad variety of users, others are particularly relevant for specific demographic groups. For instance, VSPs like TikTok constitute a central platform, especially for adolescents. However, content on VSPs for younger users tends to be shared on other platforms with a certain degree of timely latency and, thus, reaches an even more diverse audience there, including older users on Facebook. Studies indicate that while everyone may be vulnerable to the effects of misinformation to a certain degree, some attributes particularly impact vulnerability. For instance, there are controversial discussions on adolescents being very capable digital natives on the one hand [157, 580], but they are considered as particularly vulnerable by others, due to limited critical thinking skills [408] or low self-efficacy [401]. Research discusses other potential vulnerability indicators such as individiual attributes [34] or language skills [473], suggesting the importance of considering these differences in misinformation research.

The phenomenon of misinformation on social media has been tackled by research in various disciplines, such as HCI, psychology, communication science, journalism, and others. Human factors play a significant role to consider, as users' behavior, perceptions, and reactions are strongly impacted by misinformation and impact misinformation dissemination vice versa. Research in HCI addresses this human-centeredness from various angles, focusing on socio-technological challenges. For example, related research investigates motivations and intentions for creating misinformation to derive implications on misinformation spread [97], evaluates sharing behavior on diverse social media platforms and under diverse conditions to understand who shares misinformation, how, and why [336], looks into the perception and handling of misinformation by users [97, 462], or investigates how technology can be part of the solution to combat misinformation. User-centered research of the misinformation phenomenon is considered a crucial concern of *Computer-Supported Cooperative Work* (CSCW) as it *"function[s] to undermine the integrity of the information space and reduce[s] human agency by overwhelming our capacity to make sense of information"* [p. 2 533], resulting in extensive studies within that area. Furthermore, the research field of *Crisis Informatics* that encompasses perspectives on how computing and social science knowledge are used during crises and in the aftermath of disruptive situations [404] includes a focus on misinformation, suggesting particular differences during crises compared to everyday life [428, 573].

Research in HCI and related fields highlights that misinformation and its effects strongly depend on the surrounding cultural and social contexts [473]. While most misinformation studies focus on the U.S., some studies have contributed insights into misinformation perspectives from the Global South [545, 575], allowing for transnational discourses that still need to be extended. In addition, the current research landscape in HCI is dominated by a focus on platforms such as X (formerly Twitter) and Facebook, disregarding the growing relevance of often multi-modal content on VSPs, image-based platforms, or voice messages. The always-evolving social media landscape, with its adaption to novel technological possibilities, changing user preferences and intentions or expectations towards platform providers, implicates the necessity of continuous interdisciplinary research.

2.2 The Landscape of Digital Misinformation Interventions

With severe consequences of misinformation and their effects on individuals and society worldwide being indisputable, finding adequate countermeasures has become a significant research interest from multiple interdisciplinary perspectives. Research highlights that fighting the effects of misinformation requires a multidisciplinary approach, whereby a range of different efforts are integrated to construct a comprehensive solution—comparable to assembling multiple pieces of a puzzle. Among others, this encompasses education in classrooms, policy implementations, and critical journalism. The computer science research community contributes to these efforts by suggesting technology-driven solutions as *digital misinformation interventions* that may comprise automatic detection approaches to distinguish credible from misleading information [253], particularly applying machine learning techniques and implementations of interventions that take place after manual or automatic detection or detached from it.

HCI research takes a human-centered perspective on technological efforts by suggesting the design, implementation, and evaluation of *user-centered digital misinformation interventions* that complement purely algorithmic solutions (e.g., for automatic detecting and deleting content) and have a direct influence on end users via information withholding or presentation [214]. These interventions can take various forms and vary widely in their primary goal. Some interventions aim to reduce sharing of misinformation or generally decrease the dissemination of misinformation overall, for instance by addressing the subconscious of users via nudges [467]. Others aim to address individual users in fostering their critical reflection and facilitating informed credibility assessment [31]. With their diverse targets of behavior change, user-centered misinformation interventions come in various forms, including corrections by expert fact-checkers or the community [136, 263], labels to flag misinformation in varying granularity [40, 292], reducing the visibility of misinformation [199], and a wide collection of other heterogeneous approaches. Research has started to systematically look into the diverse landscape of interventions, for instance, with a focus on applied communication elements [97] or driving factors [7], leaving an in-depth review of design, implementation, and evaluation attributes of misinformation intervention studies from an HCI perspective for future work.

Although promising interventions have been suggested by research, the field is still far from a decisive solution. Researchers are confronted with challenges and debates about censorship and the responsibility of platform providers. Further, there are controversial findings about the limited effects of digital interventions under specific circumstances [35], potential backfire effects [610], and users feeling patronized and developing feelings of reactance [391]. Research suggests that users prefer interventions that provide some degree of transparency and comprehensibility to foster trust in interventions [188, 299], i.e., interventions that provide an understanding of how they come to their conclusions of credibility judgment (e.g., via explainable artificial intelligence efforts for detection) or measures to comprehend that a content is misleading and should be critically examined. While efforts to make automatic detection approaches transparent are a significant research field, they still struggle to provide users with easily comprehensible explanations. Other approaches that aim to address users' preference for comprehensibility refrain from making detection efforts transparent but rather focus on facilitating users' own assessment skills. Interventions have been proposed that guide autonomous critical thinking and media literacy. Research defines *media literacy* as the ability to decode, evaluate, analyze, and produce print and electronic media to gain a sense of critical autonomy in dealing with all media [28]. The term entails controversial discussions that, on the one hand, optimistically point out the effect of facilitated media literacy on critical thinking [280, 530, 607], and on the other hand criticize an over-simplification of the complex information space [75, 226], and relying too much on rationality [62]. Nevertheless, drawing on the optimistic view on interventions fostering media literacy, researchers have begun exploring indicator-based misinformation interventions.

2.3 Indicator-Based Misinformation Interventions

In response to users' preference for comprehensible explanations, studies have started to design and evaluate indicator-based misinformation interventions. In this dissertation, *indicator-based misinformation intervention* is defined as "a digital countermeasure that uses identifiable characteristics (indicators) of misinformation to assess and convey the credibility of online content" [224]. Research suggests that indicator-based interventions come with reduced uncertainty [188], align with pre-existing mental models and practices [513], and positively affect trust [51, 299] and perceived utility [188].

Indicators of misinformation can appear in a variety of forms, encompassing information about the author [51], the source of information [513], propaganda techniques such as loaded language or exaggeration [350, 607], or the application of voice changers in videos [165] among others. Often, interventions apply indicators by displaying them to users involving highlighting design, for instance, using color-code [350], guiding autonomous credibility assessment and promoting critical thinking and trust [51].

Indicators for misinformation can largely differ, depending on the corresponding modality of content or social media platform. Platforms provide multiple components of content that each brings along different misleading potential. For instance, content on VSPs provides layers of information that include the video, audio material, interactions with the video, and textual descriptions [387], and voice messages come with unique misleading potential due to their emotion-evoking attributes [351]. Primary research has mainly focused on text-based indicators, particularly with a focus on their applicability for training misinformation detection algorithms [518, 613]. Some text-based indicators may be transferable to other modalities (e.g., loaded language). However, a fuller picture of user-centered indicators in different modalities is still to be formed. This applies also to varying contexts and user groups. Research indicates that misinformation indicators might differ in crises. For instance, [428] suggest reviewing a text's grammatical correctness is still an effective indicator of trustworthiness when time is limited during disruptive situations. Potential differences might also apply to diverse user groups with different requirements and preferences, suggesting a need for tailored interventions that have proven to be promising in various contexts [17, 146]. For example, research indicates adolescents assess content of platforms like TikTok or Facebook differently than adults [157, 226, 408]—with controversial findings on varying levels of proficiency.

To enable the design of indicator-based misinformation interventions, it is necessary to form a knowledge base on indicators of misinformation that are considered comprehensible and useful by users. Primary research has started to investigate which indicators are user-centered, for example, from the perspective of journalists and news consumers [188, 626]. Complementing research targeting the user perspective within specific contexts, i.e., diverse user groups, platforms, and modalities, as well as crisis and everyday situations, is still to be addressed more exhaustively. While indicator-based misinformation interventions have shown promising results, they entail challenges and limitations as well. Future research might complement initial insights on whether users still feel patronized [226] or reject approaches as they are not sufficiently comprehensible and appear biased in their implementation [188].

2.4 Research Gap

This dissertation advances HCI research on user-centered misinformation interventions through the design, implementation, and evaluation of indicator-based interventions and contrasting types of interventions applying less transparent approaches. The existing body of research presents several noticeable gaps that require further exploration.

1st gap: Strong U.S. perspective. The vast majority of research takes a U.S. perspective, complemented by initial insights from the global south [545, 575]. Misinformation and its effects strongly depend on the surrounding social and cultural contents [473], highlighting the need to expand these insights to include transnational discourses. This dissertation offers a vantage point, addressing the German context.

2nd gap: Heterogeneous interventions. The landscape of user-centered interventions as one piece of the puzzle to tackle the challenge of misinformation is highly heterogeneous in various dimensions [7, 97, 467], including intervention design, addressed user groups, and evaluation methods. This necessitates a systematic overview from an HCI perspective to form a knowledge base and allow for subsequent complementing contributions.

3rd gap: Focus on text-based content. There is a predominant perspective on misinformation research regarding mainly text-based content and corresponding platforms [171, 511, e.g.,]. However, recent crises have shown that platforms like TikTok [16] and messenger groups fostering voice messages [351] are a crucial breeding ground for misinformation as well. Current platforms come with unique misleading potentials due to their features and modalities, and particularly vulnerable user groups (e.g., adolescents) [408]. Addressing these conditions is necessary to keep on tackling the evolving challenges.

4th gap: Risks and limitations. As there are controversial discussions about the applicability of indicator-based interventions and other approaches addressing various levels of comprehensibility [62, 75], this necessitates perspectives of various user groups on the opportunities, risks, and limitations of the interventions. It encompasses deriving implications to prevent unintended effects like blind trust and overestimating or underestimating the interventions' capabilities.

5th gap: User preferences and needs. Existing research indicates that susceptibility to misinformation varies among users. Individual attributes like decision-making style [34] and cognitive ability [124], and demographic factors like age [273] and language skills [473] have an impact. Research suggests the beneficial effects of personalized interventions that address particular user preferences and needs instead of one-size-fits-all interventions [13, 120]. In the context of misinformation, there is a need to complement studies with further research on varying preferences regarding the transparency of interventions to address preferences for autonomy and feelings of information overload.

Summary—Related Work

- *Social media platforms* can be *vital during disruptive situations*, however, they *foster the dissemination of misinformation.*
- *Research in HCI* explores, e.g., misinformation creation, sharing intentions, and technological countermeasures.
- Various starting points exist to tackle misinformation, including *user-centered digital misinformation interventions* (e.g., corrections in the comment section, warning labels, visibility reduction).
- Research suggests no definitive solution yet; challenges include avoiding user reactance.
- Indicator-based misinformation interventions can reduce uncertainty, align with mental models, and enhance trust and perceived utility.
- Effective design requires an understanding which indicators work for *different modalities and user groups.*

Research Design 3

The following chapter places the dissertation in its research context (see Sect. 3.1) and introduces the overarching research method of *Design Science Research* (DSR) in Sect. 3.2. It shows how various methods were used in the dissertation to contribute to the perspective on user-centered digital misinformation interventions (see Sect. 3.3).

3.1 Research Context

This dissertation was particularly embedded in the research project NEBULA (User-centered AI-based detection of misinformation and disinformation), funded by the German Federal Ministry of Education and Research (BMBF). NEBULA has a duration of three years (2022–2025), covering most of my dissertation's time, and combines perspectives from HCI, machine learning, and social and communication sciences. Its interdisciplinary research goals are the automatic detection of misinformation and disinformation for civil safety and security and a subsequent user-centered presentation of the detection results for end users to promote media literacy on various social media platforms. Furthermore, the project addresses two crucial user groups: First, users who are vulnerable due to age or language barriers, and second, authorities and organizations with security tasks who need to monitor misinformation on social media to react quickly and in a user-centered way. Within NEBULA, the research network investigated automatic approaches for the detection of misinformation using knowledge graphs, explored users' perceptions, needs, and requirements regarding the navigation of digital information, and contributed to the development and evaluation of user-centered digital misinformation interventions.

K. Hartwig, *Navigating Misinformation*, Technology, Peace and Security I Technologie, Frieden und Sicherheit,
https://doi.org/10.1007/978-3-658-47885-8_3

Sharing the focus on user-centered handling of misleading and sensitive information, the research context was further embedded within the projects CYLENCE (Strategies and tools against cyberbullying and hate speech), funded by the BMBF, the National Research Center for Applied Cybersecurity ATHENE, funded by the BMBF and the Hessian Ministry of Higher Education, Research, Science and the Arts, as well as the Collaborative Research Center CROSSING, funded by the German Research Foundation. While CYLENCE comes with a research context highly related to misinformation, ATHENE and CROSSING allowed for a perspective on cybersecurity as an additional context with sensible information.

The continuous exchange with contacts from all research projects provided insights into different interdisciplinary perspectives and strongly informed my dissertation.

3.2 Research Approach

This dissertation examines the phenomenon of misinformation from the intersection of computer science and psychology. It is primarily located in the research field of HCI, which deals with the interplay of computers and people [436], and in the research field of CSCW in particular, which constitutes interdisciplinary research and development of technology to support collaborative activities and the understanding of social interactions within [191]. Additionally, the focus on crisis-specific scenarios of parts of the dissertation allows for an embedding in the context of Crisis Informatics [404].

This dissertation integrates various user-centered research methods across individual publications, following the overarching DSR methodology, which corresponds to an iterative, participatory approach common in HCI research. It can be broken down into six iterative steps [417]: (1) the identification and motivation of the research problem, (2) the definition of the objectives of a solution to outline the scope, (3) the design and development of an artifact, (4) the demonstration of the artifact as a solution to the problem within a suitable context, (5) the evaluation of the effectiveness and efficacy of the artifact, and (6) the communication of gained insights to inform the research discipline. In the following, I give an overview of how my publications contribute to each step of DSR.

3.3 Methods

To shed light into how user-centered digital interventions can assist users in navigating misinformation across modalities, a variety of complementary studies with various methods have been conducted. An overview can be found in Table 3.1. From an overarching perspective, they follow the iterative and human-centered DSR process in six steps [417].

Table 3.1 Applied evaluation methods of this dissertation

Method	N	Paper
Systematic literature review	108	A: Hartwig, Doell, and Reuter [214]
Quantitative content analysis	2,382	B: Hartwig et al. [225]
Think-aloud study	22	B: Hartwig et al. [225]
	9	C: Schmid et al. [491]
	7	C: Schmid et al. [491]
	20	D: Hartwig, Sandler, and Reuter [222]
	18	E: Hartwig et al. [224]
	16	H: Hartwig et al. [223]
Focus group	21	E: Hartwig et al. [224]
Online survey	44	B: Hartwig et al. [225]
	1,012	G: Hartwig and Reuter [217]
Online experiment	396	F: Biselli, Hartwig, and Reuter [54]
	1,000	H: Hartwig et al. [223]

3.3.1 Step 1: Problem Identification and Motivation

The dissemination of misinformation has been identified as a significant challenge worldwide, and finding adequate ways to support users online is crucial. This dissertation is motivated by how users are confronted with an overabundance of information on social media and struggle with its complexity in various modalities. It emphasizes digital misinformation interventions as one piece of the puzzle contributing to an informed navigation of the complex information landscape. Chap. 7 highlights real-world examples of social media content's potentially misleading nature, focusing on the COVID-19 pandemic as a particularly disruptive crisis. It presents in-depth findings of a manual content analysis of 2,382 German tweets on Twitter/X to derive topical, formal, and rhetorical characteristics and demonstrate the variety of misleading potentials. A systematic literature review following the PRISMA flow [403] contributes to the problem identification and motivation by providing a comprehensive overview of the state-of-the-art misinformation interventions, as well as trends and challenges for future research, which are addressed in this dissertation (see Chap. 6). The literature review screened over 6,000 scholarly publications and systematically reviewed 172 articles that address user-centered misinformation interventions from an interdisciplinary perspective in diverse forms, such as corrections, removal, or showing indicators. A taxonomy was derived to cluster interventions according to their design, user interaction, and timing. In addition, methodological characteristics of misinformation studies were collected to inform this dissertation. Considering various social media platforms (e.g., TikTok, Twitter/X, Telegram) facilitates a user-centered perspective on multimodal content with high relevance for misinformation. A representative online survey ($N = 1,012$) was conducted in Germany to gain insights into users' attitudes towards different intervention types, particularly focusing on nudging and hybrid nudging in the context of cybersecurity as an additional context with sensitive information (see Chap. 12). This allowed for a look beyond the horizon of misinformation research, tackling connecting phenomena like transparency, reactance, and manipulation. Participants were recruited via the panel provider Respondi and were representative of the German adult population aged 18 to 74. In addition to the developed items, the survey contained two standardized psychometric scales.

3.3.2 Step 2: Objectives of a Solution

Building on the knowledge base informed by Step 1, quantitative and qualitative user studies were conducted to evaluate users' experiences in dealing with misinformation, and their needs, preferences, and requirements for interventions were derived. Various demographic groups, such as adolescents and adults of various ages, participated in individual think-aloud sessions, focus group discussions, and (representative) online surveys and experiments and motivated an indicator-based approach as a user-centered misinformation intervention to address calls for comprehensibility and transparency. Qualitative and quantitative evaluations enabled the derivation of comprehensible indicators for misinformation in social media posts containing text, images, voice messages, and short-videos that were perceived as useful for assessing online content.

In Chap. 7, a qualitative online survey ($N = 44$) was conducted with adult participants indicating which characteristics they already autonomously apply when assessing the credibility of tweets regarding the COVID-19 pandemic. Subsequently, individual think-aloud sessions ($N = 23$) were conducted, presenting indicators derived from the previous quantitative content analysis and the qualitative online survey to evaluate their perceived usefulness and comprehensibility. In this part of the study, the data set was extended to include tweets regarding the Russian war of aggression against Ukraine as an additional significant crisis with a more security-related nature. A comparable methodological approach was conducted for the modality of voice messages in Chap. 9 with 20 individual think-aloud sessions, and for content on VSPs like TikTok in Chap. 10 with focus group discussions in combination with a paper-based survey ($N = 21$). Hereby, Chap. 10 particularly focuses on adolescents aged 13 to 16, which comes with numerous specific methodological requirements. The young participants were recruited in cooperation with a local media competence center via schools and youth centers. The study procedure and stimuli selection required a thorough involvement of the university's ethics committee and coordination with teachers and parents. For the informed consent, stimuli, and intervention design, age-appropriate language was crucial and was achieved by iterative feedback loops with the media competence center and teachers.

3.3.3 Step 3: Design and Development

While Steps 1 and 2 motivated and defined an indicator-based approach as one piece of a puzzle to facilitate informed online navigation, Step 3 designs and develops the approach from a perspective on different platforms and modalities that each come with unique requirements, for instance, regarding the main target group. This involves artifacts on a conceptual level (e.g., designing the overall concept of an indicator-based intervention using a low-fidelity prototype), high-fidelity prototypes (e.g., a simulated smartphone app with realistic UI elements), and fully functional implementations. On a more conceptual level, Chap. 7 designs the overall indicator-based approach by highlighting indicators in tweets containing text and images using color code in an experimental setting not involving technical implementations but rather a low-fidelity prototype. Similarly, Chap. 11 designs low-fidelity UI elements as nudges to reduce misinformation sharing intention. Chap. 9 expands the conceptual level by implementing an automatic extraction of features of misinformation from voice messages, including automatic transcription using WHISPER [446], text preprocessing (e.g., tokenization), and storing human-readable features in a .JSON file. In addition, Chap. 8 implements a web app based on SNA that—based on indicators—facilitates exploratory navigation of content on Twitter/X to identify patterns of misleading content. The implementations for textual content and voice messages were complemented by Chap. 10 that addresses the highly relevant context of VSPs and expands previous user studies on autonomous assessment strategies of adolescents and the perceived usefulness and comprehensibility of misinformation indicators on TikTok by simulating a smartphone app based on indicators using Figma[1]. As in the previous steps of the DSR process, this dissertation contains an outlook on the context of cybersecurity, designing and developing low-fidelity and high-fidelity prototypes of user interventions in the form of nudges for stronger passwords.

3.3.4 Step 4: Demonstration

An integral part of the DSR process is to demonstrate that the developed artifacts *generally work* to solve the defined problem [417]. In the case of the dissertation, Step 4 aims to demonstrate that the developed misinformation interventions help to combat misinformation. This can be interpreted in multiple ways, for instance, by reducing the spread of misinformation, reducing interaction with it, improving credibility assessment, and encouraging critical reflection, among other aspects. Chap. 11 simulates multiple misinformation interventions as low-fidelity prototypes in an online experiment and quantitatively evaluates their efficacy with 396 adult participants, involving intervention types that differ in their level of transparency. The study particularly focuses on the potential of personalization to increase efficacy. In this context, efficacy is measured by reduced sharing intention of misinformation—a measure that is commonly applied in misinformation research. Similarly, Chap. 13 quantitatively evaluates the efficacy of interventions in cybersecurity through online experiments ($N = 1,012$, see Chaps. 12 and $N = 1,000$, see Chap. 13). While these parts of the dissertation constitute an obvious mechanism to demonstrate efficacy, qualitative studies can contribute to that demonstration as well. In various parts of this dissertation, qualitative

[1] https://figma.com

insights during think-aloud sessions and focus group discussions revealed a potential for indicator-based interventions to lead to a learning effect and corresponding transferability of gained knowledge to novel content (e.g., see Chap. 10).

3.3.5 Step 5: Evaluation

In addition to the interventions generally being effective in combating misinformation as quantitatively measured in Step 4, Step 5 of the DSR process aims to evaluate *how well* the interventions work [417]. Therefore, various user studies have been conducted based on the designed and developed artifacts. In Chaps. 7 and 9 the overall concept of the indicator-based intervention is evaluated in terms of its applicability and suitability in think-aloud sessions with 23 and 20 participants, focusing on the unique characteristics of modalities (e.g., challenges when highlighting indicators in voice messages). Further, Chap. 10 evaluates the intervention's suitability in 18 individual think-aloud sessions, including a particular perspective on the challenges and risks of the indicator-based approach from the perspective of adolescents. From a methodological perspective, think-aloud sessions play a particular role during the design and development phase as they create in-depth insights into the participants' impressions while the intervention is actually carried out on them [163], allowing for iterative deriving of needs and perceptions and subsequent improvements of the intervention.

3.3.6 Step 6: Communication

Comprehensibly communicating the scientific contribution of the gained knowledge to relevant audiences is an important step to facilitate the impact of the research. In conjunction with the aforementioned publications, this dissertation facilitates the dissemination of knowledge through scientific channels. Four of the nine embedded papers are conference papers that have been presented in research presentations digitally due to COVID-19 restrictions or in person. Additional participation at symposia and networking events with research project-related stakeholders facilitated the dissemination of findings and valuable critical discussions beyond the 'research bubble'.

Summary—Research Design

- The dissertation was particularly embedded within the *research project NEBULA* which combines perspectives of machine learning, HCI, and social and communication sciences to counter misinformation online.
- It is primarily located in the research field of *HCI*, taking on perspectives from CSCW and Crisis Informatics.
- The overarching methodology aligns with the *iterative and human-centered DSR process* in six steps.
- *Various methods* have been applied within the DSR steps, including a systematic literature review, think-aloud sessions, focus group discussions, and (representative) online experiments and surveys.

4 Findings and Discussion

The digital information landscape confronts people worldwide with an overabundance of both accurate and false or misleading content. This has become evident, particularly in crises. While social media platforms are an essential means for information exchange in various modalities, they facilitate an environment of misinformation dissemination, causing significant harm on individual, political, and societal levels. This dissertation investigates indicator-based misinformation interventions as one piece of the puzzle to tackle the challenges of misinformation in a user-centered way. It aims to answer the following main research question: **How can user-centered digital interventions assist users in navigating misinformation across modalities?** This section highlights the essential findings of my contribution. An illustration of the answered research questions and the corresponding scientific papers can be found in Fig. 4.1.

4.1 User-centered Digital Misinformation Interventions

Establishing a foundational knowledge base on the state-of-the-art is a crucial first step in answering my main research question. Consequently, Chap. 6 systematizes the current interdisciplinary research landscape of user-centered misinformation interventions, aiming to answer the question RQ1: **How does research from different fields inform the design and evaluation of user-centered misinformation interventions?**

K. Hartwig, *Navigating Misinformation*, Technology, Peace and Security | Technologie, Frieden und Sicherheit,
https://doi.org/10.1007/978-3-658-47885-8_4

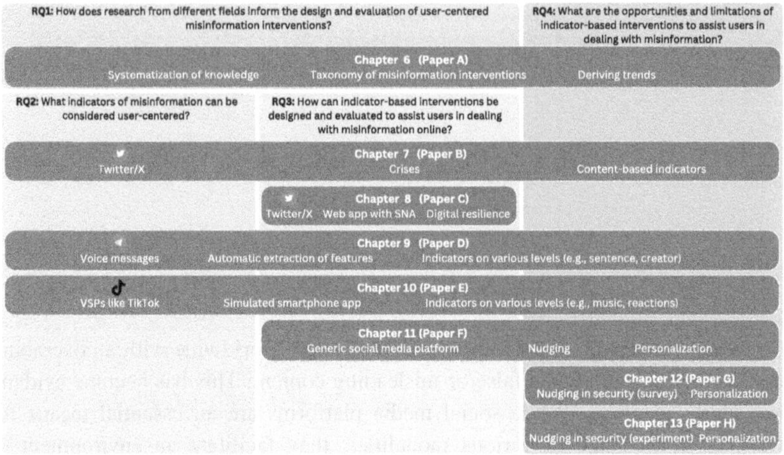

Fig. 4.1 Contributions of eight scientific papers to the research questions. Icons used in this figure are in the public domain (CC0 1.0)

While there are multiple approaches that aim to tackle the challenge of misinformation, including critical journalism and media literacy training at school, *user-centered misinformation interventions* can be considered one piece of the puzzle. They constitute digital countermeasures that go beyond a purely algorithmic back-end solution while exerting a direct influence on end users in the form of information presentation or information withholding. Based on a systematic literature review of 172 scientific articles following the PRISMA process [403], the chapter analyzes methodological characteristics of evaluations of misinformation interventions and clusters the derived interventions within a taxonomy distinguishing multiple intervention designs, timings, and types of user interaction. Further, the chapter discusses the trends, perceptions, and impacts of the reviewed misinformation interventions.

The landscape of user-centered misinformation interventions constitutes a variety of interventions shaped by interdisciplinary research encompassing perspectives from HCI, psychology, information systems, communication sciences, and medical research, among others. This entails a wide range of terms which the individual publications refer to. The vast majority of publications ($N = 149$) use the terms *misinformation, disinformation, fake news,* or *misleading information.* The terms *rumors, news credibility*, or *myths* are also used in this context. The form in which interventions are evaluated is as diverse as the interdisciplinary composition of the publications and covers empirical studies from small groups of participants (e.g., <20) to large-scale (representative) samples with over 1,000 participants. Studies collect empirical data mostly in online experiments; this is understandable due to the digital nature of social media usage and corresponding interventions. Online experiments allow researchers to feasibly collect a large amount of data and perform statistical analysis, for instance, to detect significant effects in behavior changes in a controlled environment. However, in addition to the advantages of this form of evaluation, its controlled setting also involves the disadvantage that a completely realistic picture of social media usage behavior cannot be achieved. Field studies offer a complementary opportunity to observe interventions in natural user behavior within a real-world scenario—a form of evaluation covered by very few publications ($N = 9$). Taking a closer look at participants in misinformation intervention studies, we can see that these are often clearly biased, with the vast majority being U.S.-based adults and college students. Isolated publications have investigated specific (vulnerable) target groups like adolescents, marginalized communities, or low-vision social media users. Misinformation is relevant on all social media platforms, that all have unique potential to mislead due to particular features (e.g., sending voice messages) or modalities (e.g., text, audio, video). Most of the current research landscape focuses on Twitter/X ($N = 33$) or Facebook ($N = 36$) and often does not cover videos or audio content, which can be explained by the already established data access and easy linkage to existing related literature. This leaves room for future investigations on highly relevant multimodal social media platforms, including VSPs like TikTok.

User-centered misinformation interventions come in various forms and aim to protect users from misinformation with diverse starting points. This encompasses, among others, deleting misinformation after identification, warning about its existence or effects, or strengthening users' media literacy to facilitate autonomous critical reflection of content. In the context of my research, a taxonomy was developed (see Fig. 4.2) to gain a structured overview of the state-of-the-art user-centered misinformation interventions in their variety. The 237 interventions within 172 scientific articles were clustered accordingly. The taxonomy differentiates between nine types of *intervention designs*: warnings (e.g., as labels or stop signs; "This post was disputed"), correction/debunking (e.g., naturally or artificially occurring user comments or official corrections), showing indicators (e.g., displaying relevant characteristics of misinformation within content, infographics), (binary) labels (e.g., tagging posts as true or false, traffic light symbols), highlighting design (e.g., color-coding content according to its accuracy, highlighting propaganda techniques), visibility reduction (e.g., reducing opacity or size), removal of misinformation overall, complicate sharing (e.g., additional confirmation barrier before sharing), and specific visualizations (e.g., visualizing sentiment scores, SNA visualization). It is striking that a majority of research focuses on corrections/debunking, which is examined in great detail and from multiple perspectives such as timing and repetition [112, 125] or regarding corrections by layperson versus experts or officials [210, 263, 583]. This observation corresponds to related reviews that have identified corrections as the most common type of intervention strategy for corrective communication [97]. While corrections can create some learning effects when perceived and reflected, other intervention designs aim to encourage critical reflection and increase media literacy more explicitly. This includes interventions that display linguistic or content-related indicators as characteristics of misinformation or credible content, which can come in various forms. For instance, some approaches highlight propaganda techniques using color-code [350] or emphasize the original publication date of a video [513]. Research has demonstrated that users tend to prefer this kind of transparency and opportunity for comprehensible interventions [299], and there are even studies that suggest tendencies of transparency can reduce feelings of reactance, censorship, paternalism, or lack of control [299, 391]. In particular, indicator-based interventions tend to reduce uncertainty [188] and align with pre-existing mental models and practices [513]. This is discussed in more detail in Chap. 10.

Nevertheless, other types of intervention designs take a more rigorous approach to reducing the spread of misinformation, for instance, by deleting content immediately after detection or by reducing its visibility. Indeed, this is beneficial when considering that users tend to remember misinformation content itself and not whether it was potentially displayed with a correction or warning [187], and demonstrates the relevance of the interventions' *timing*. When reviewing the diverse interventions more closely, their largely different timing becomes evident. Most of the interventions ($N = 120$) occur during exposure to misinformation, for instance, when displaying corrections below a post or highlighting indicators within the content. On the one hand, this results in the content already being seen by users and (in the worst case) being remembered as correct. On the other hand, this allows for the opportunity to immediately demonstrate false statements or misleading characteristics with the example at hand.

The taxonomy in Chap. 6 further differentiates between active and passive *user interactions*, where the majority of interventions are passive (i.e., users do not have to interact with the intervention during social media usage actively but have the choice to ignore it) and fewer interventions require an active engagement to keep on using the social media platform (e.g., an additional confirmation before liking or sharing misinformation). This can potentially impact subsequent user behavior, perceptions of interventions, and—again—feelings of reactance or paternalism.

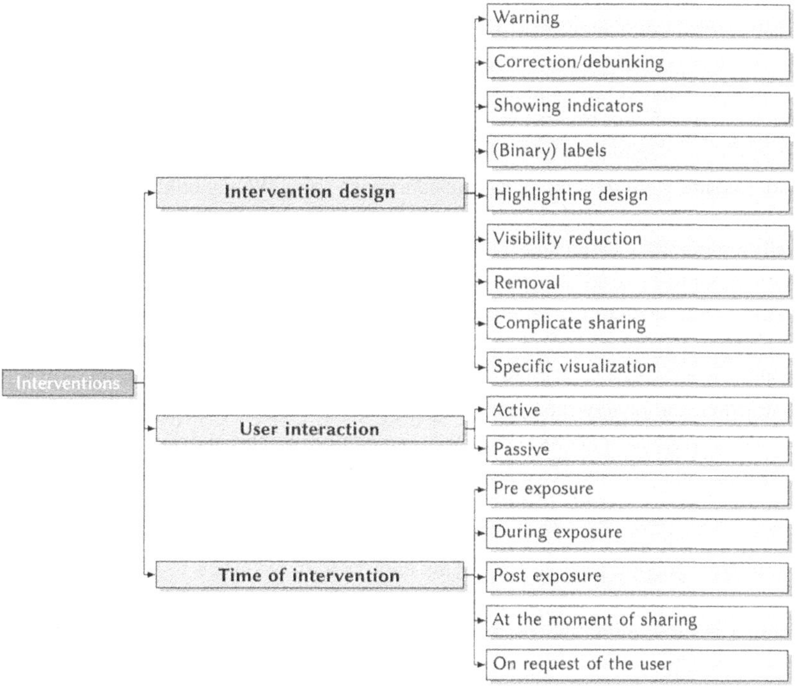

Fig. 4.2 A taxonomy for user-centered misinformation interventions. (Figure adapted from [214])

The systematic review of current literature on misinformation interventions revealed particular trends, opportunities, and challenges, raising questions for future research. Building on these insights, parts of these questions have been tackled in the following chapters of this dissertation.

(1) *Are approaches for specific platforms transferable to other new platforms?* and (2) *How can collected findings and technical approaches for text-based interventions be applied to emerging video- and image-based misinformation?* The review highlighted a strong research focus on Facebook and Twitter/X and on text-based content, and likewise, trending types of platforms (e.g., VSPs) with multimodal content have not yet received substantial research attention at the beginning of my dissertation. Each platform is characterized by unique features that impact the potential to mislead. On the other hand, there are common characteristics across platforms inviting transferability of gained knowledge. I tackle these questions in Chap. 9 for voice messages on platforms like Telegram and in Chap. 10 for short-videos on TikTok.

(3) *How can opportunities of digital misinformation interventions be effectively combined with the advantages of human experts?* and (4) *How can automatic detection be combined with user-centered feedback?* Finding ways to combine the advantages of human experts and algorithms (e.g., for automatic detection) is a major challenge requiring many considerations (e.g., Who should be the arbiter of content credibility?). While Chap. 6 did not systematically analyze the arbiters for each intervention within the review, it gives common examples and emphasizes open challenges. Additionally, this dissertation suggests an approach combining an automatic detection as pre-filtering of misinformation (*simulated* for instance in Chaps. 10 and 9) and a subsequent display of user-centered indicators for misinformation that is explicitly designed for human comprehensibility.

Further, the review raises the following questions: (5) *How can media literacy be encouraged without exposing users to misinformation?* and (6) *Can vulnerable people profit from the general findings of participants with high media literacy? How can we reach vulnerable people with official tools?* The main focus of this dissertation is on indicator-based interventions that are typically displayed alongside the misinformation they address. This approach allows users to directly compare the misinformation with the indicators, making it easier to understand why the content is misleading and promoting a learning effect to identify misinformation autonomously; nevertheless, there are opposing ideas that aim to refrain from exposure at all and rather hide or delete content. These intervention types have not explicitly been investigated in this dissertation. However, I provide insights that argue for a more transparent, informed, and empowering approach to prevent feelings of reactance or paternalism. This has been tackled not only for adult social media users of various demographic backgrounds (e.g., education, gender, and age) but also for adolescents as potentially more vulnerable users [233, 408]. I address the user group of adolescents in Chap. 10, discussing the controversial state of research regarding their vulnerability to misinformation.

The findings from Chap. 6 highlight the diverse landscape of user-centered misinformation interventions and systematically cluster them within a taxonomy. Insights from current research indicate user preferences for transparent and comprehensible interventions that foster autonomy instead of top-down approaches. In the following section, I elaborate on this, where I turn to the specific individuals' credibility assessment strategies and perceptions of misinformation indicators for various modalities and platforms as a foundation for digital interventions.

4.2 User-centered Misinformation Indicators

After building a knowledge base on the state-of-the-art of user-centered misinformation interventions and revealing potentials for indicator-based interventions as one piece of the puzzle toward comprehensible and empowering countermeasures, Chaps. 7, 9, and 10 aim to answer the question RQ2: **What indicators of misinformation can be considered user-centered?** With a foundational understanding of interdisciplinary contributions, these chapters now contribute to an understanding of credibility assessment strategies that are already autonomously applied and evaluate how indicators of misinformation are (not) perceived as useful and comprehensible by various user groups and for various modalities and platforms.

Multiple empirical studies of this dissertation evaluated which indicators users already autonomously apply when assessing the credibility of social media content. For social media posts on Twitter/X with a particular focus on text, Chap. 7 presents a qualitative online survey ($N = 44$) with twelve tweets as stimuli. This study looks at misinformation specifically in the context of crises, particularly in the COVID-19 pandemic and the Russian war of aggression against Ukraine. The adult participants of different socio-demographic backgrounds were asked whether they thought tweets to be true or false and how they reached their respective conclusions. Contributing to related research on assessment strategies [513], the findings reveal that participants tend to focus on simple and easily visible indicators (e.g., presence or absence of references) while less explicit indicators (e.g., rhetorical indicators) were mostly not considered. The study focuses on content-specific indicators as one of the multiple relevant dimensions, differentiating topical (e.g., 5G), formal (e.g., links), and rhetorical (e.g., sarcasm) characteristics of misinformation. Regarding topical indicators, participants autonomously applied well-known buzzwords such as 'mainstream media' for credibility assessment and went for the general appearance of conspiracy theories rather than reflecting on specific conspiracy theories (e.g., regarding science denialism) or their combined appearance. For formal characteristics of tweets, participants were very aware of links to (not) credible sources, referred to hashtags (e.g., '#DeepState') that were easily identifiable and considered a lack of punctuation or excessive punctuation as indicative of misinformation. Rhetorical characteristics of misinformation (e.g., relativizations or claims to absolute truth) did not regularly appear to be autonomously and consciously applied. In general, participants focused on easily visible indicators without dedicating much effort and time when assessing tweets' credibility.

While the study in Chap. 7 particularly focused on content-specific indicators, there are many more levels on which indicators can be applied. This has been addressed in Chap. 9 for voice messages on a micro-level (i.e., for misinformation indicators on a character, word, sentence, audio, and creator level), and even more comprehensively by Chap. 10 for TikTok videos, differentiating multiple levels of short-videos (e.g., spoken word, sound and music, and reactions). In Chap. 9, an online think-aloud study ($N = 20$) with adults from diverse social-demographic backgrounds was conducted that, among others, aimed to reveal autonomous assessment strategies when listening to voice messages forwarded within platforms such as Telegram. The findings demonstrate that when assessing voice messages, users mostly rely on their knowledge and only partly draw on characteristics such as tonality. This contrasts the findings on mostly textual content from Chap. 7, where the visual representation of the content invites users to look for additional cues. For voice messages, the study reveals that all participants, at least to some extent, used their existing knowledge or gut feeling, and only a minority further considered additional characteristics (e.g., usage of swear words, angry or very emotional tonality). When specific buzzwords were mentioned within the voice message (e.g., 'BRD GmbH' as a common conspiracy theory in Germany), this impacted credibility assessment, complementing findings on the usage of buzzwords in tweets.

VSPs like TikTok offer content on various levels (e.g., the video itself, the video description, underlying music or sound, and the comment section), each entailing unique misleading potential. Thus, for Chap. 10, I aimed to grasp a more comprehensive view of indicators on these different levels. To evaluate autonomous assessment strategies, three parallel focus group discussions with integrated individual surveys were conducted at a local school in Germany with adolescents aged 13 to 16. Involving adolescents in the research design was crucial, as VSPs are highly relevant for that age group, and assessment strategies and indicator perceptions may vary strongly between adults and adolescents. Real-world TikTok videos containing misinformation of various forms and a credible video were presented in succession, and the adolescents were asked to individually indicate their credibility assessment, giving detailed reasons in a paper-based survey. Afterward, the adolescents discussed their perceptions in a focus group setting. Assessment skills strongly differed between individual adolescents. It was striking that many participants could successfully identify specific indicators for artificially generated or manipulated videos (e.g., looking for additional fingers or the skin color of the face not matching the rest of the body). On the other hand, they were often not aware of the emotionalizing characteristics of videos (e.g., emotion-evoking music in the background or facial expressions) as indicators for misinformation. Many adolescents autonomously paid attention to the content of the video description, interactions with the video, and the general appearance of the video. This indicates that some adolescents routinely navigate the social media platform, opening the comments section or a creator's profile, for example, to search specifically for credibility indicators. Naturally, this behavior refers to the theoretical assessment skills when required in an experimental setting and does not represent a natural social media usage behavior, where users tend just to scroll away rather than perform an in-depth credibility assessment.

A closer look at the autonomous assessment skills of users of various socio-demographic groups and on various social media platforms allowed us to understand potential misinformation indicators on multiple levels. These insights extended the content analysis of Chap. 7 where a large number of tweets were manually analyzed and indicators were derived, and the literature-based collection of misinformation indicators for various modalities (see the Related Work and Methodology sections of the publications within my dissertation, e.g., in Chaps. 9 and 10). The individual empirical studies (i.e., particularly think-aloud studies with integrated surveys) evaluated the perceived usefulness and comprehensibility of indicators for the specific user groups and platforms, building a knowledge base on which indicators are user-centered and should be integrated within an indicator-based misinformation intervention. This encompasses reflections about indicators' granularity or level of detail to be comprehensible enough without being too overwhelming.

For Chap. 7, with a focus on content-specific indicators of tweets during crises (particularly the COVID-19 pandemic and the Russian war of aggression against Ukraine), the majority of the adult participants rated the proposed topical, formal, and rhetorical indicators as useful and comprehensible (i.e., they were capable of explaining what they meant correctly). Topical (e.g., 'science denialism') and rhetorical (e.g., usage of negative phrases) were assessed as slightly more useful than formal indicators (e.g., wrong punctuation). The participants were very aware of the limited value of spelling and punctuation as indicators when considering non-native speakers and people with dyslexia. It was interesting to observe that while users did not autonomously apply rhetorical indicators consciously without assistance, they perceived them as very useful and comprehensible when presented during the experiment. Some participants further demonstrated sophisticated interpretations of indicators when differentiating whether they appeared within a quote or the original tweet. Including adults with various socio-demographic backgrounds, Chap. 7 further reveals insights into how factors such as age, education, and political orientation impact the perceived usefulness and comprehensibility of indicators with various complexity and granularity. For example, the topical indicator 'restriction of liberty' was fully comprehensible for all participants with a high level of education but not fully comprehensible for a subgroup of lower-educated participants.

These findings were complemented in Chap. 9, focusing on voice messages with different potentials to mislead due to their modality. The adult participants rated indicators on word, sentence, audio, and creator level as rather useful and could often correctly describe what they referred to. However, this strongly depended on the level of abstraction. Suspicious conspiracy theory keywords, emotionally loaded words, and words referring to the speaker's belief (e.g., "I think that [...]") were particularly comprehensible and useful. Contrarily, participants struggled with interpreting abstract features (e.g., number of words, percentage of adjectives and adverbs) as indicators for credibility assessment. For example, only a few isolated participants could comprehend the number of first-person pronouns to be an indicator for creators of voice messages falsely presenting themselves as experts or eye-witnesses—an indicator inspired by [351]. In summary, the findings of Chap. 9 underline the necessity to combine multiple features from various levels to enable users to gain a full understanding.

We pick up on the aim to create a fuller picture by considering various levels of content in Chap. 10, where we assessed the perceived usefulness and comprehensibility of indicators for TikTok videos with adolescents. In doing so, we clustered indicators within the following levels: video, spoken words, sound and music, layout, content and message, profile, reactions and interactions, and overall impression. We found that indicators on various levels expand and confirm the adolescents' perceptions. Some indicators' usefulness must be considered limited due to their perceived obviousness (e.g., regarding manipulated videos that were already autonomously identified by many participants). Similarly, the usage of filters was referred to as a common feature on TikTok, and the adolescents suggested this indicator as rather useful for older users who are easily deceived. While the young participants did not autonomously apply emotionalizing characteristics of the videos during credibility assessment, they deemed them comprehensible and useful when presented as potential indicators. Similar to our user study with adults regarding tweets on Twitter/X, the adolescents critically evaluated indicators that referred to grammar or spelling due to the issue of dyslexia, sharing personal experiences that shaped that perception. Our findings extend and support current research on assessment strategies of adolescents [226], for example, when confirming that adolescents consider content being refuted in the comment section a useful and comprehensible indicator. In addition to objective indicators, we included the reflection of the overall impression of a video's credibility as a subjective indicator. This was inspired by feedback from a local media competence center and was emphasized during observations in the focus group discussions. Instead of automatically detecting and highlighting the indicator, as is the case for many others, the idea of this indicator is to be displayed as a default nudge for critical reflection. The participants rated its usefulness positively, especially when phrased more generically. In sum, our findings extend existing research on trust heuristics for VSPs [226] and emphasize the high relevance of comment sections to check for sources or social cues.

The insights on user-centered indicators of misinformation for various modalities pave the way for designing and evaluating indicator-based misinformation interventions. In the following, I discuss the implementation of these artifacts.

4.3 Design and Evaluation of Indicator-based Misinformation Interventions

Based on the derived user-centered misinformation indicators, this dissertation aims to answer RQ 3: **How can indicator-based interventions be designed and evaluated to assist users in dealing with misinformation online?** This encompasses the development of artifacts on a conceptual level or low-fidelity prototypes regarding the overall concept of indicator-based interventions (see Chap. 7) and contrasting nudging interventions (see Chap. 11). Chapter 9 expands the conceptual level by implementing an automatic extraction of misinformation features and storing them within a human-readable .JSON file. Further, Chaps. 8 and 10 present high-fidelity prototypes (i.e., as a simulated smartphone app) and fully functional artifacts (i.e., as a web app). Artifacts have been developed for various modalities and platforms, addressing adults of different socio-demographic backgrounds as well as adolescents as a particularly relevant user group. The empirical findings result in further implications for designing indicator-based misinformation interventions, building a foundation for the iterative DSR process.

Chapter 7 presents the general concept of an indicator-based misinformation intervention, developing a low-fidelity prototype that—within an experimental setting of an online think-aloud study ($N = 23$)—displays tweets on Twitter/X and simulates that (1) a fact-checker (i.e., a human expert, an organization, or an algorithm) has detected this tweet as misleading and in addition (2) a set of indicators have been identified which should help the users comprehend that decision. These indicators were manually highlighted using color-code within the tweet. Within the think-aloud study, participants rated the approach's overall comprehensibility and perceived usefulness (very) high. In addition, there were several suggestions for improvement and interesting criticism regarding the naming of indicators, habituation effects when imagining being confronted with the intervention regularly, and drawing attention to problematic parts of the tweet by highlighting them. Participants further suggested concrete steps toward fully implementing the conceptual approach, including features for more detailed information about indicators and their sources using pop-ups or a mouse-over effect—these suggestions motivated our next steps for subsequent studies. The positive assessment of the indicator-based approach is consistent with related studies that highlight the potential for fostering media literacy [51, 299]. Interestingly, the approach was often perceived as suitable "Not for me, rather for others." We observed this in our subsequent user studies with different user groups (e.g., adolescents) as well [224]. Indeed, it aligns with findings of related research and, to some extent, reflects tendencies toward overconfidence in credibility assessment [75, 390].

In light of the chapter's conceptual design and evaluation, we derive four design implications that strongly motivate our subsequent work and give orientation for future interventions: (1) *Consider comprehensive indicator-based interventions to address the need for transparency.* The transparency of the approach was particularly mentioned as decisive for the approach's positive perception, and research indicates opportunities to avoid counterproductive socio-psychological effects such as reactance and backfire effects [391]. (2) *Keep indicators rather general and simple by default.* To avoid information overload and foster productive processes of sense-making [244, 533], and in line with the impressions of our user studies in Chap. 7, we suggest keeping indicators within the intervention rather general per default but allow for more detailed information when preferred, to foster a reflection on more complex indicators (e.g., underlying legitimization strategies [129] or rhetorical tactics [131]) derived from related literature. Nevertheless, our findings suggest focusing on comprehensible indicators at first sight, which also includes a (3) *focus on visible indicators* as users prefer not to spend too much time and effort during the assessment. During our user studies, we found that preferences among participants strongly differed, partly impacted by different socio-demographic backgrounds but also by individual attitudes and needs. This indicates a need for (4) *personalization of indicator-based interventions* instead of one-size-fits-all solutions, consistent with research that emphasizes personalization as promising means [146, 494].

The promising possibility of personalization was taken up in Chap. 11. Contrasting the transparent indicator-based approach, various nudges were examined here for their effectiveness in reducing the sharing intention of misinformation in generic social media posts. A nudge is defined as an intervention that *"alters people's behavior in a predictable way without significantly changing their economic incentives"* [554, p.6] and can take on various forms. The evaluated nudges include (a) hiding the social media post content behind a blurred panel until the user actively confirms to deblur it, (b) a just-in-time prompt triggered upon user interaction that confronts potential misinformation sharing and requires confirmation, (c) a social comparison nudge that displays the user's contacts avoiding the sharing of potential misinformation, (d) colored backgrounds to indicate a reliable or unreliable account according to its posting behavior, and (e) presenting fact-checked articles with different viewpoints to instill doubt. Nudge (e) constitutes the only nudge that can be considered 'hybrid'; the term 'hybrid nudge' (also called 'enriched nudge') has been proposed to refer to nudges that are enriched with additional, comprehensive information [454]. Indeed, an indicator-based intervention could also be understood as a hybrid nudge. Within an online experiment ($N = 399$), we investigated how personalization based on psychometric dimensions (i.e., general decision-making style, consideration of future consequences, and need for cognition) can increase the nudges' efficacy instead of one-size-fits-all nudges. We quantitatively evaluated the statistically significant effect of various nudges to reduce misinformation-sharing intentions, revealing a positive impact of personalization on nudging efficacy; however, with limited scope and remaining (ethical) concerns regarding the collection of private data, that is necessary for the segmentation of user groups (e.g., applying psychometric tests). While this chapter provides an outlook on less transparent interventions in the form of nudges, this approach will be picked up on again in Sect. 4.4, where its application to cybersecurity as an additional sensitive context is investigated.

Coming back to the more transparent approach of indicator-based interventions, Chap. 8 extended the initial conceptual ideas of Chap. 7 with the implementation of a web app based on SNA for tweets on Twitter/X as an artifact that encompasses a full implementation of features for user interaction. The web app was evaluated within a two-step think-aloud study ($N = 16$). This study particularly focuses on digital resilience as a goal that allows social media users to *resiliently* encounter misinformation during crises, requiring transparency, learning, diversity and flexibility, equality, and rapidity. The indicator-based artifact constitutes the visualization of a network of tweets, where nodes represent a tweet, and its author and edges represent thematic connections. Additionally, the tweets are automatically classified as probably misleading or not, using the output of a BERT classifier, and indicators (e.g., excessive capitalization, frequent or conspicuous emojis, or frequent hashtags) are displayed to the user. The findings reveal design implications concerning the previously listed requirements. This encompasses, among other things, the slowing down of information processing if users want to pay attention to tweets, even if this causes higher costs, as a complement to interventions focusing on efficient information retrieval and assessment. This approach for intensive data exploration contrasts our additional artifacts that aim to facilitate a more efficient credibility assessment. Our findings further emphasize the necessity to balance between detailed information and features to enhance transparency and a potentially resulting information overload. Again, the findings argue for personalization to distinguish different user assistance needs and requirements.

While the previous chapters focus on mainly text-based social media posts on Twitter/X, Chap. 9 complements the findings with a focus on voice messages as a highly relevant context for misinformation that comes with unique potential to mislead [351]. It presents an artifact that encompasses the automatic processing of a voice message to derive and store features of various granularity within a human-readable .JSON file as potential user-centered indicators for misinformation. These features have been evaluated within a think-aloud study, highlighting them within the Telegram messenger's user interface. The chapter demonstrates how interventions for voice messages, in contrast to interventions for text, image, or even video, come with the unique attribute of not having any visual anchors that users could relate to when reflecting on indicators—despite information such as the name of the author or additional text sent via the message. As a result, we found that users cannot remember or verify indicators when receiving their feedback after listening to a voice message. The chapter highlights this challenge for future HCI research.

Chapter 10 further extends the development and evaluation of artifacts to the context of short-videos in TikTok, which is particularly relevant for adolescents. The overall applicability of the indicator-based approach was evaluated within individual think-aloud sessions ($N = 18$) where adolescents used a simulated smartphone app for TikTok that was created using Figma. The prototype simulated a realistic 'ForYou page' on TikTok, where participants could share videos with the misinformation app over the share button, which other apps commonly use to check its credibility. The participants were then forwarded to the simulated assessment result that states: *"A fact-checker (algorithm or human/organization) has determined that this is most likely false or misleading information. The following clues should help identify this for yourself: [list of indicators]"* (see Fig. 4.3b). The participants could then swipe through detailed indicator descriptions. We evaluated the app's overall usability with the System Usability Scale (SUS) adapted for adolescents [442] and received a positive score of 81.7. This aligns with the positive qualitative feedback we gained from our participants, who stressed the familiar and intuitive features, positive social effects of being well-informed, and beneficial transparency. Consistent with our previous studies on indicator-based interventions, the participants strongly differed regarding their needs for compact versus detailed information. A few adolescents criticized the app for being too complex, as they found the feedback overwhelming and unsuitable for people who dislike reading or are illiterate. This emphasizes the adolescents' diverse experiences and ability to empathize with potential other users. As potential mitigation, some participants suggested a tailored version of the app. As we observed a tendency of participants to find the approach helpful for others but not for themselves in our previous studies in other contexts and with adult participants, we were interested in how this observation occurs for adolescents. Indeed, while our participants saw value for themselves, they stated that older or very young users would benefit the most from the app and reported about personal experiences, where they take on the role of fact-checkers for older relatives—especially regarding video content. Adolescents are often considered digital natives with sophisticated navigation capabilities within social media [580]. On the other hand, research indicates they are particularly vulnerable regarding misinformation due to limited critical thinking skills or overconfidence [408, 635]. Our results confirm that misinformation on VSPs can be highly complex and not easy to distinguish as it can mislead in multiple dimensions—this was also the case in our user studies.

(a) Video Pets (misinformation)

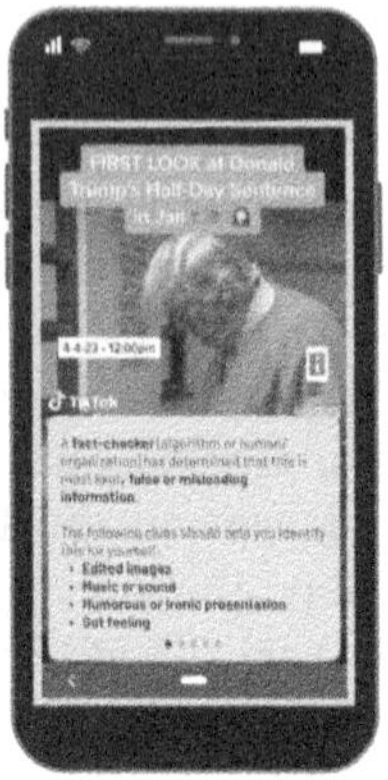

(b) Video Trump (deep fake)

(c) Video Clouds Part I (satire)

(d) Video Clouds Part II (satire)

Fig. 4.3 Overview on the stimuli videos, demonstrating several views of the smartphone app's UI within the Figma Prototype. Faces of private persons were pixelated for publication. UI elements were translated from German to English. (Figure adapted from Hartwig et al. [224])

Our findings extend and confirm existing research on design implications for indicator-based misinformation interventions, including the previous studies of this dissertation on other platforms and for other user groups. For user-centered countermeasures targeting adolescents, five design implications have been derived. (1) *Promote self-reflection*: The findings suggest a necessity to encourage users' critical evaluation of content based on their own gut feeling and more rational and objective indicators. Self-reflection can be fostered by including default prompts like "What does your gut say?" in pre-filtered misinformation content. Further, this study, in line with our previous findings, emphasizes the (2) *usage of comprehensive and transparent misinformation indicators* across multiple components of video content (e.g., sound, layout, interactions) as one option among others to allow for a well-founded and informed assessment. Potential users' diverging needs and preferences suggest a need for (3) *personalization* to customize interventions and allow for adaption to individual circumstances and abilities. Recognizing that adolescents may be overconfident in their credibility assessment skills, interventions may be designed that (4) *promote self-awareness for overconfidence mitigation*. Further, users' (5) *blind trust in algorithms must be addressed*, establishing a clear understanding of limitations and potential errors. This observation was part of this dissertation's RQ4, which will be answered in detail in Sect. 4.4. The design and evaluation of concepts and artifacts as indicator-based misinformation interventions reveal crucial perspectives on their practical application. This leads us to examine their overall opportunities and limitations in assisting users in navigating misinformation.

4.4 Opportunities and Limitations of Indicator-based Interventions

This dissertation provides an in-depth design and evaluation of indicator-based misinformation interventions. Just like all other types of interventions, they come with both opportunities and limitations. With a detailed understanding of the design and evaluation of indicator-based interventions, I now explore their broader opportunities and limitations. Accordingly, I aim to answer RQ4: **What are the opportunities and limitations of indicator-based interventions to assist users in dealing with misinformation?** Answers to this question can be found based on literature in Chap. 6 and based on empirical studies in the context of misinformation in Chaps. 10, 11, 9, and 7. An additional outlook on the applicability and resulting opportunities and limitations for the context of cybersecurity can be found in Chaps. 12, and 13.

To make valid assumptions, it is necessary to consider indicator-based interventions in light of the overall context of misinformation interventions. Chapter 6 highlights how interventions and their according studies strongly differ in their target of behavior change (e.g., improving credibility assessment, reducing the sharing of misinformation, or decreasing the overall flow of misinformation), their applied evaluation method (including qualitative and quantitative perspectives), specific contexts in which interventions were evaluated (e.g., social media platforms, user groups etc.), and applied evaluation conditions (e.g., in comparison to a control group or state-of-the-art interventions of platforms). When taking a closer look at studies that quantitatively evaluated the efficacy of an intervention, there are (sometimes contradictory) findings about potential influencing factors such as timing of corrections that sometimes but not always matters [115, 464], intervention mechanisms or source types (e.g., social versus algorithmic corrections or warnings by citizens versus news agencies or experts) [60, 207, 245, 340, 344], or the modality of interventions (e.g., images, audio, or videos) [412, 528, 560, 605, 627]. In sum, comparability of effects is often not possible due to the very diverse settings and contexts in which individual studies take place, and finding a research consensus for efficacy measurements of misinformation interventions has been highlighted as an important step toward more successful countermeasures [193]. The review reveals not only statistically significant or qualitatively promising findings (for the majority of studies) but also for fewer studies what did not work [25, 86, 297, 506] or what unintended or counterproductive effects were discovered. This encompasses, for instance, spillover effects on accurate content [156, 206], over-reliance on interventions [344], increased belief in misinformation [29, 429, 550] (e.g., when repeatedly exposed to it), or increased distrust in authentic content [572]. Limitations and risks are very individual for interventions—some of them might be mitigated by a different type of intervention.

After having shed light on the positive and negative impacts of diverse misinformation interventions in Chap. 6, the subsequent Chapters with their empirical findings particularly highlight the opportunities and limitations of indicator-based interventions as observed and reported with our participants of different sociodemographic backgrounds and regarding interventions for various modalities. Most of these studies included in the dissertation focus on in-depth user perspectives and perceptions, depicting the impacts of the interventions mostly qualitatively. Still, the empirical findings reveal outlooks on the expected effects of the indicator-based approach. For example, Chap. 7 assessed *how credibility ratings changed* after the adult participants were confronted with indicators for crisis-related tweets on Twitter/X. For all nine tweets, there were changes for the better (e.g., from (rather) credible to not credible), and we particularly observed reinforcing (i.e., reduced uncertainty) of the pre-exposure presumptions.

A potential learning effect of applying indicators autonomously is up to future research. However, we provide initial insights into *how participants learned to apply* and autonomously name indicators when being confronted with novel tweets (see Chap. 7) or novel videos on TikTok (see Chap. 10) within the short-term experimental settings. It is striking, however, that this is true for credible content as well, where participants looked for indicators and expected content to be misinformation when a reliable social media post contained, for instance, a rhetorical question. This is an implication that needs to be thoroughly considered when designing indicator-based interventions and ties in thematically with research by [572], who found that interventions sometimes tend to *increase distrust in authentic content*. Our empirical findings revealed very specific limitations that our participants outlined. For example, they were very aware of indicators not being valid within a quote or indicators, drawing even more attention to misinformation when highlighting them (see Chap. 7). Perceived opportunities and limitations have particularly been assessed within individual think-aloud sessions with adolescents within the context of our smartphone app for TikTok (see Chap. 10). Extending related research on over-reliance [344], we found that the young participants tended to *blindly trust* the simulated algorithm on whether a TikTok video is misleading or not. Interestingly, a reason for that was sometimes a general belief that computers were more intelligent than humans and, thus, were trustworthy. We further made the worrisome observation that the adolescents *struggled to envision the consequences* of the app making mistakes. While the majority was, in general, aware of the app, algorithms, or general AI systems making mistakes, four out of 18 participants were sure that the specific app would never make mistakes and work a lot more precisely than humans. Indeed, all our participants were confident that the app making mistakes would not have any severe consequences. However, it is important to notice that the participants' trust in the intervention was partly influenced by how stimuli and detection results were presented (i.e., there were no false positives). The lack of skepticism toward automated systems was discussed by related research as well [23, 312], and our complementing findings suggest a potentially high level of vulnerability, making it all the more important to use transparent approaches when assisting adolescent users in dealing with misinformation instead of relying solely on top-down feedback without explanations. Displaying indicators while transparently establishing a clear understanding of an algorithm's limitations and potential errors can—to some extent—mitigate the risk by encouraging critical reflection of content.

Presenting indicators comprehensibly was crucial for building trust and giving a sense of autonomy in the adolescents' assessments. Many participants viewed the intervention as a supplementary aid to support autonomous critical thinking, demonstrating the positive potential of the indicator-based approach. Only one participant explicitly did not favor displaying indicators as additional information but rather preferred a binary label to reduce effort and not be too text-heavy. Approaches fostering media literacy have been criticized in parts of related literature as well, as it is only one of many components of the complex information space [75, 226], relies on rationality [62], and creates a false sense of confidence [75]. This prompts calls to "rethinking media literacy in the age of platforms" [75, p. 17]. In this dissertation, I draw on a rather optimistic view that indicates media literacy education promotes critical thinking and alters behavior for the better [530], particularly among younger people [607]. The opportunities and criticism of media literacy are specifically discussed in Chap. 10.

Indicator-based misinformation interventions entail both opportunities and limitations and, as other digital interventions, can only be considered one piece of the puzzle to counter misinformation. Due to their transparent nature and their goal to foster informed navigation of social media content, this intervention type has the potential to tackle undesired effects such as reactance. Positive perceptions and impacts have been demonstrated within this dissertation, and related literature has been extended and confirmed. This dissertation demonstrates the applicability of the intervention type for various modalities, scenarios, and user groups. This encompasses short-text on Twitter/X, videos with additional information (e.g., comment sections) on VSPs, and voice messages. It was demonstrated with adults of multiple socio-demographic backgrounds and with adolescents aged 13 to 16 as an additional highly relevant user group. Further, the intervention was applied with real-world stimuli of everyday content and within acute crisis situations. The dissertation highlights how indicator-based interventions have the potential to be applied by laypersons using social media on an individual level (see e.g., Chaps. 10 and 7) but also by people working in crisis management or governmental agencies to automatically analyze large amounts of data and quickly identify problematic trending misinformation for support in monitoring and analysis (see Chap. 9).

To enable a view beyond the horizon of the misinformation context, this dissertation provides insights into how user interventions can be applied to other contexts with sensitive information. Thus, I took a look at (hybrid) nudging in cybersecurity, focusing on user preferences and personalization potentials regarding transparency, attitudes, and influencing factors (see Chaps. 12 and 13). This complements findings on the previously discussed nudges to reduce misinformation sharing in Chap. 11. A representative survey in Germany ($N = 1,012$) revealed that over half of the participants assessed nudging in cybersecurity in a positive way, while 42% were undecided and 20% disagreed. Interested in the perceived risks and limitations, we found that some perceived it as patronizing or annoying or even feared interventions to spy on personal data. The commonly non-transparent nature of nudges was sometimes perceived as manipulative. This aligns with the impressions we gained from our misinformation research. Similarly, our study on the context of cybersecurity suggests focusing on transparent interventions where benefits are evident. Consistent with our findings on misinformation (especially Chap. 11) and related research, our findings suggest the necessity of personalized interventions instead of one-size-fits-all solutions to address user characteristics, needs, and preferences. Similarly to the personalization approach in Chap. 11, this has been partly investigated by applying two psychometric scales: the *General Decision Making Style* scale (GDMS) and the *Rational-Experiential Inventory* (REI), to identify potential influencing factors on user perceptions and attitudes. In this context, we found small significant differences in attitudes toward nudging (e.g., participants who tend to make important decisions consulting other people rated nudging interventions as slightly more necessary than others).

Our findings were complemented by an additional study on nudges in cybersecurity (see Chap. 13), where we compared multiple nudging interventions for stronger passwords with different levels of gamification in experimental settings ($N = 16$ and $N = 1,000$), again applying user characteristics as means of personalization. Interestingly, our findings reveal that gaming frequency as a user characteristic did not impact the efficacy or acceptance of gamified nudge interventions in our specific context. This underlines the necessity to find more successful means of personalization. While the outlook on the context of cybersecurity and password strength, in particular with its focus on nudges as a less transparent intervention, strongly differs from the main components of this dissertation focusing on transparent misinformation interventions, there are insights that align and complement each other. *First*, top-down interventions or less transparent user feedback tend to be perceived as patronizing and potentially manipulative—this can be true both for cybersecurity and misinformation research contexts. *Second*, personalized interventions according to user preferences, characteristics, and needs are promising to achieve better efficacy and acceptance. However, finding adequate segmentation measures (e.g., psychometric scales) that make a difference is complex.

Having thoroughly explored and answered the sub-questions with a systematic exploration of user-centered misinformation interventions, the in-depth assessment of user-centered indicators, the design and evaluation of indicator-based concepts and artifacts, and finally, the broader analysis of opportunities and limitations, these findings are now synthesized to address the main research question comprehensively.

Summary—Findings and Discussion

- This dissertation gained in-depth insights into the *design, implementation, and evaluation* of user-centered misinformation interventions for various modalities and user groups:

  I. A systematic literature review on user-centered misinformation interventions revealed the *heterogeneous landscape of interventions* and derived a *taxonomy* to cluster them regarding design, type of user interaction, and timing. The review demonstrates how research is mainly U.S.-based, focuses on textual content, and highlights transparent intervention (e.g., based on the display of indicators) as promising to address users' preferences and needs.
  II. Various empirical studies captured *autonomous assessment strategies* and evaluated the *perceptions of misinformation indicators* with users of diverse socio-demographic backgrounds.
  III. *Indicator-based interventions* (e.g., as a smartphone app and a web app) and contrasting nudging interventions were designed, implemented, and evaluated, revealing the potential to address users' needs and preferences.
  IV. The empirical findings highlight not only beneficial effects but also severe *limitations and risks* of indicator-based interventions.

Conclusion 5

Reflecting on the findings from the previous sections, this conclusion synthesizes these insights to provide a comprehensive answer to my main research question: **How can user-centered digital interventions assist users in navigating misinformation across modalities?**

Within a current digital landscape that struggles with an overabundance of credible information and misinformation, this dissertation contributes to the design, implementation, and evaluation of user-centered misinformation interventions as one piece of the puzzle, complementing other dimensions such as critical journalism and educational efforts for media literacy training at schools and afterward. The dissertation particularly focuses on indicator-based interventions that allow for a comprehensible understanding and potential learning effects, aiming to address users' needs and preferences while preventing unintended backfire effects of top-down interventions.

The key findings of this dissertation to answer the main research question can be divided into four dimensions that have been discussed in detail in Chapter 4. In summary, this encompasses the following insights.

5.1 Key Findings

First, the in-depth literature review of state-of-the-art misinformation interventions reveals a diverse landscape of intervention designs, types of user interaction, and timings where the intervention takes place. A taxonomy was derived to cluster existing interventions systematically. The various studies on user-centered misinformation interventions are shaped by their interdisciplinary

K. Hartwig, *Navigating Misinformation*, Technology, Peace and Security | Technologie, Frieden und Sicherheit,
https://doi.org/10.1007/978-3-658-47885-8_5

nature, encompassing diverse design and evaluation methods. The majority, however, is U.S.-based and focuses on mainly text-based social media posts on Twitter or Facebook, rarely covering other highly relevant modalities such as audio or video content. The results demonstrate that research is still far from a decisive solution, as the effects of the evaluated misinformation interventions (e.g., to impact sharing behavior) are often marginal. Transparent intervention types that provide users with additional comprehensible information (e.g., via alternative presentation of information or indicators for misinformation) have been highlighted as potentially promising to address users' preferences and prevent feelings of reactance, paternalism, or lack of control [299, 391]. However, these interventions are also controversially discussed, as they rely on rationality [62] and might create a false sense of confidence [75]. Within the subsequent empirical studies, this dissertation contributes insights into the opportunities and limitations of transparent interventions, answering questions raised within the literature review for future research on misinformation interventions.

Second, picking up on the positive perspective on transparency, empirical studies were conducted to investigate autonomous credibility assessment strategies and evaluate the user perception of misinformation indicators of various modalities for potential integration within digital interventions. Social media platforms each come with unique features and modalities, leading to unique misleading potentials. For example, a video on TikTok can mislead on diverse levels, including the content of the video itself, the spoken words, the background sound or music, or the comment section and interactions. The dissertation encompasses empirical studies for mainly textual content or social media posts, including text and images, voice messages, and short-videos on VSPs such as TikTok. It further addresses adult and adolescent participants of diverse socio-demographic backgrounds, crises, and everyday life contexts. For all these diverse conditions, the findings reveal an autonomous focus on easily visible and simple indicators (e.g., buzzwords or absence of references) and a general gut feeling when not assisted with any intervention. On the other hand, the findings revealed sophisticated autonomous skills, such as identifying artificially generated or manipulated videos for adolescents. Other misinformation indicators were not (or less) applied without interventions. This was particularly true for rhetorical indicators in tweets on Twitter/X, the tonality of voice messages, and emotion-evoking characteristics of TikTok videos (e.g., emotionalizing music). When presenting the derived misinformation indicators based on literature and our empirical studies, we found that participants indeed assessed this additional knowledge (e.g., on rhetorical indicators) as beneficial and rated many indicators as useful and comprehensible. This supports the potential of an indicator-based approach.

Third, building on the previous findings, this dissertation designed, implemented, and evaluated indicator-based interventions and contrasted nudging interventions with a less transparent approach. This includes artifacts on a conceptual level and low-fidelity prototypes, as well as high-fidelity prototypes and fully functional artifacts that have been designed, for example, as a smartphone app and a web app. Our participants intuitively interacted with the artifacts, positively rated their usability, and gave insightful feedback on potential improvements. While I gained rich qualitative insights and complemented quantitative evaluations of nudging interventions, the dissertation does not include a large-scale representative experiment to make conclusions about the statistically significant effects of indicator-based interventions, suggesting future research on the statistical efficacy compared to other state-of-the-art intervention types. The key findings lead to implications that, among others, focus on rather general and simple indicators by default and immediately visible indicators. This is a particular HCI challenge for voice messages, where visual anchors are naturally rare and cover solely information about the author's name or additional text that was sent together with a voice message. An implication that was repeatedly derived from the empirical findings concerns the personalization of interventions instead of one-size-fits-all approaches, as users significantly differ in terms of needs and preferences for transparency, and balancing between detailed, comprehensible information and information overload is necessary. The dissertation provides insights into how this could be applied using psychometric dimensions, both for the misinformation context and the additionally sensitive context of cybersecurity.

Fourth, for the application of user-centered misinformation interventions, it is crucial to consider their realistic opportunities and potential risks or limitations thoroughly. Overall, the findings demonstrate positive perceptions and impacts of indicator-based interventions. For example, the findings suggest that users learn to apply gained knowledge on novel content, and credibility assessments improve. Less transparent interventions have also revealed positive effects, such as various nudges when applied in a personalized form. However, presenting indicators was crucial for building trust and giving a sense of autonomy among the participants of the indicator-based studies. Despite the revealed potentials, this dissertation further identified several risks and limitations that come with transparent interventions. For example, we observed that participants did not only learn indicators for misleading content but tried to apply them to credible content as well, potentially increasing distrust in authentic social media posts. This needs to be further investigated in future research. The findings further reveal, especially among adolescent participants, a tendency to blindly trust in the (simulated) algorithm and a lack of realistic concerns or capabilities to imagine the consequences of the intervention making mistakes.

5.2 Limitations and Future Work

Considering the indicator-based approach and the qualitative and quantitative methods taken in this dissertation, inherent limitations must be acknowledged. The individual limitations of the studies included in this work are emphasized in the corresponding papers. However, this dissertation is also subject to limitations on a broader level that further offers potential for future work.

First, most evaluations within this dissertation focused on the user-centeredness of misinformation indicators and resulting interventions (e.g., a smartphone app). For instance, the interventions' perceived usefulness, comprehensibility, and overall perception were assessed through think-aloud studies and surveys. Potential learning effects were observed in short-term studies. Additional standardized usability scales complemented these insights with a quantification. However, the statistically significant effectiveness of these indicator-based interventions, such as their impact on reduced misinformation-sharing intention, was not quantitatively measured. These measurements were included for the nudging interventions in this dissertation and should be addressed in future work for indicator-based interventions as well, preferably observing real-world behavior on social media in field studies.

Second, while this dissertation included adults and adolescents from diverse socio-demographic backgrounds, including non-native German speakers, it did not include all relevant groups that might be particularly vulnerable to misinformation. For individual studies, the dissertation covers adults aged up to 72 years when conducting representative online experiments and surveys. However, it would be interesting to focus particularly on the age group of older adults in future work, who are potentially vulnerable to misinformation. This was emphasized within our studies with adolescents who reported being regularly asked for advice on credibility assessments (especially of video content) by older relatives.

Third, the dissertation focuses on the German context. This encompasses conducting user studies with German citizens and using stimuli representative of the German social media landscape. The dissertation offers a basis for transnational discourse and complements predominantly U.S.-based research by providing insights from a different national context. While we observed commonalities, for example, regarding the buzzwords and other characteristics of misinformation on Twitter/X between German tweets and related U.S.-based studies, the results may not be fully applicable to other contexts, as social and country-specific factors influence social media usage and behavior.

Fourth, indicator-based interventions themselves come with risks and limitations, as discussed in Sect. 4.4. Indeed, like with other types of interventions, they cannot reach everyone, particularly users already entrenched in conspiracy theories, who are very unlikely to download and use such artifacts. This is the case even if indicator-based interventions potentially create less reactance than top-down approaches. Instead, these interventions are more suitable for users who are generally open to learning more about credibility assessment and prefer keeping the autonomy within informed navigation of information in contrast to censorship. Future research might explore how these interventions can be made more accessible and appealing to a broader audience and find ways to address users resistant to lightweight digital measures.

Summary—Conclusion

Our contemporary world, characterized by complex global crises, is interrelated with the digital information landscape regarding its challenges. An overabundance of misinformation and credible content highlights the need to equip users with skills for informed information navigation. This dissertation underscores the significant potential of *indicator-based interventions* in the complex digital landscape, particularly for addressing the informed navigation of misinformation. While these interventions *can promote comprehension and learning*, they also have *limitations and risks*, including over-reliance on user rationality and the possibility of creating a false sense of trust.

Given the *German context* of the empirical studies within this dissertation, the findings may not be universally applicable. Nevertheless, this work provides an important piece of the puzzle that complements critical journalism and media literacy education efforts and lays the groundwork for transnational discourses and further research in diverse contexts.

Part II
Publications

The Landscape of User-centered Misinformation Interventions—A Systematic Literature Review

6

Abstract Misinformation is one of the key challenges facing society today. User-centered misinformation interventions as digital countermeasures that exert a direct influence on users represent a promising means to deal with the large amounts of information available. While an extensive body of research on this topic exists, researchers are confronted with a diverse research landscape spanning multiple disciplines. This review systematizes the landscape of user-centered misinformation interventions to facilitate knowledge transfer, identify trends, and enable informed decision-making. Over 6,000 scholarly publications were screened, and a systematic literature review ($N = 172$) was conducted. A taxonomy was derived regarding intervention design (e.g., labels, showing indicators of misinformation, corrections, removal, or visibility reduction of content), user interaction (active or passive), and timing (e.g., pre or post exposure to misinformation or on request of the user). We provide a structured overview of approaches across multiple disciplines and derive six overarching challenges for future research regarding transferability of approaches to (1) novel platforms and (2) emerging video- and image-based misinformation, the sensible combination of automated mechanisms with (3) human experts and (4) user-centered feedback

---

Original Publication *Katrin Hartwig, Frederic Doell, and Christian Reuter. "The Landscape of User-centered Misinformation Interventions—A Systematic Literature Review". In: ACM Computing Surveys (CSUR) 56.11 (2024), pp. 1–36. DOI:* 10.1145/3674724

Supplementary Information The online version contains supplementary material available at https://doi.org/10.1007/978-3-658-47885-8_6.

K. Hartwig, *Navigating Misinformation*, Technology, Peace and Security | Technologie, Frieden und Sicherheit,
https://doi.org/10.1007/978-3-658-47885-8_6

to facilitate comprehensibility, (5) encouraging media literacy without misinformation exposure, and (6) adequately addressing particularly vulnerable users such as older people or adolescents.

6.1 Introduction

The fast spread of misinformation is an enormous challenge both for society and individuals, with a great impact on democracy. Severe and fatal consequences can be observed about misinformation shared on social media related to COVID-19, with mistrust sowed in health measures required for combating a pandemic. With this in mind, [422] even go so far as calling it a 'matter of life and death'. In light of the grave consequences, the need for digital misinformation interventions to slow down its propagation is evident. Those technical approaches can be divided roughly into two main steps [433]: The (automatic) detection of misinformation and, second, the implementation of countermeasures as a concrete decision on what to do after successful detection. A great deal of research exists on detecting misinformation, which is often based on machine learning algorithms [e.g., 65, 230, 518]. Such algorithms are typically so-called 'black box' algorithms, which—while producing promising results with regard to detection accuracy—are not transparent in their reasoning. In order to make an algorithm's decisions transparent to users, interventions may profit from using 'white box' algorithms/explainable AI, which give greater insights into how the algorithm behaves and what variables influence the model [97]. The reliance on these automatic detection measures is increasing. For example, because of the COVID-related increase in traffic, Twitter (now X) has increased its use of machine learning and automation against misinformation [172]. Complementary research has been done on the implementation of concrete interventions after successful automatic detection [e.g., 53, 468, 479]. Those interventions are available in a wide range: Some aim at efficiently and automatically deleting content before exposure, while others try to educate users by showing corrections or flagging problematic content. While there is a large and heterogeneous field of interventions, they have a direct impact on end users of social media, as they focus on whether and how to communicate their output and findings, for instance, via information visualization. Although promising approaches have been established, the ongoing challenge of users being confronted and influenced by misinformation on diverse social media platforms such as TikTok, Twitter/X, Instagram, Facebook, and Co. suggests a need for further systematic design, implementation, and evaluation of effective digital interventions.

This review study aims to systematize knowledge on digital user-centered misinformation interventions. The term 'misinformation' is often used as an umbrella term for better readability, encompassing misleading information that has been created deliberately (frequently referred to as 'disinformation' or 'fake news') as well as unintentionally (frequently referred to as 'misinformation') [15, 97, 597]. For example, Li et al. [330] justify the use of misinformation as an umbrella term by stating that the majority of studies use this term to encompass different types of misleading information in general without denying the proper distinction between disinformation and misinformation. Indeed, misinformation and related phenomena such as rumors and conspiracy theories can all lead to severe consequences, even if those were not intended. Thus, in accordance with other research and systematic reviews [97, 330], in this paper, we will use 'misinformation' for better readability and to allow for a broader perspective on different kinds of misleading information while not denying the significant differences of phenomena. While the term 'misinformation intervention' has already been established by other researchers [35, 479, 480], we define *user-centered misinformation interventions* as digital countermeasures that go beyond a purely algorithmic back-end solution and exert a direct influence on the user in the form of information presentation or information withholding. Accordingly, we do not include approaches that deal exclusively with the automatic detection of misinformation without describing the subsequent communication to the user.

We provide a taxonomy that classifies and aggregates interventions regarding multiple relevant dimensions, such as time of intervention, addressed platform, and the thorough differentiation between intervention categories (e.g., correction, (binary) labeling, transparent indicators) to help identify promising research directions and encourage cross-disciplinary transferability. Researchers are faced with a very diverse research landscape on user-centered countermeasures, which is spread across multiple disciplines, such as computer science, Human-Computer Interaction (HCI), information systems, psychology, communication sciences, journalism, and even medical research. Hence, we address the challenge of gaining an overview and help build on existing research while considering and learning from current insights of different relevant disciplines as well as research on different social media platforms. Thereby, we seek to facilitate informed decision-making of researchers and practitioners when analyzing, designing, and evaluating (novel) digital countermeasures to combat misinformation. We are especially interested in approaches communicating to users how an algorithm arrives at its results (e.g., white box algorithms) instead of giving a top-down answer (e.g., misleading, not misleading) without explanation. In our paper, we understand a 'transparent' intervention as an intervention that allows for informed decisions and the ability to comprehend why the content potentially contains misinformation, for example, via explanations of varying degrees, and can, thus, be considered as more user-centered than top-down interventions. There is evidence that transparently assisting users in their own assessment of misinformation is more promising than a top-down approach that provides social media posts solely with a label stating 'This is/isn't misinformation' without cues to help comprehend the decision [299] or simply removes misinformation [27]. Research indicated that giving explanations or comprehensible cues can be significant to establish trust in the intervention [299], and counteract feelings of reactance or related backfire effects [391] that are controversially discussed in research [610].

While the topic of misinformation has been studied in systematic reviews, e.g., regarding specific contexts such as health [330] or political misinformation [269], existing literature reviews on interventions against misinformation and similar phenomena focus on more general overviews. For example, a related literature map by Almaliki [15] focuses on the research field of misinformation. It provides a general overview rather than analyzing the characteristics of concrete interventions and comparing the different approaches. They state that "less than 2% of the selected papers proposed digital intervention techniques", while our focus lies on those studies that fall into the 2% as a promising subgroup of interventions with growing research interest. Furthermore, when literature reviews, or meta-analyses deal with concrete interventions, they often focus on the detection step and machine learning interventions [e.g., 82, 198, 370, 447, 616, 631, 636] instead of user-centered interventions or focus on a specific subgroup like corrections [91, 439], warnings [358], accuracy prompts [418], or contexts like COVID-19 [265]. A first systematic overview of strategies against misinformation, including countermeasures with a direct influence on end users, was given by Chen, Xiao, and Kumar [97], who differentiate between five broad categories of solutions according to communication elements: message-based, source-based, network-based, policy-based, and education-based approaches. This contrasts our approach, which is not based on communication elements (e.g., message versus source) but considers interventions more in terms of their in-depth design. This design can be applied to the content itself, within a network-based approach, or to sources (e.g., by highlighting components in color as a passive intervention during exposure to misinformation). The authors give an overview of exemplary implementations within the four clusters. We build on that by providing an in-depth analysis of the design, interaction type, and timing of interventions as central aspects for user-centered implementations. In addition, we provide an overview of the methodological characteristics of intervention studies. Furthermore, Aghajari, Baumer, and Difranzo [7] reviewed misinformation interventions with a focus on underlying driving factors of misinformation like social contexts and beliefs. Thus, in a constrained search process around the term 'misinformation', they categorize interventions according to content-based, source-based, individual user-based, and community-based strategies. Our study complements the analysis of strategies in terms of their driving factors (e.g., content-based strategies like corrections) with an HCI perspective detached from individualistic or community-based emphasis. Differentiating misinformation interventions on an individual trial system level, Roozenbeek, Culloty, and Suiter [467] review boosting interventions, nudging, debunking, and content labeling in comparison to interventions that rely on algorithms, business models, legislation, and politics. We complement the findings of related reviews by (a) shedding light on user-centered aspects of concrete misinformation interventions by performing an in-depth analysis of their design, implementation, and methodological evaluation for a broad perspective that offers a more comprehensive understanding of misinformation interventions. Thereby, we (b) specifically discuss and categorize characteristics impacting end users, such as the intervention design, user interaction, and timing of the intervention. We further complement the existing research landscape by (c) performing a review on publications of diverse disciplines and not limited to a specific period until 2024, searching three major databases. In doing so, we address calls for future work on a review including multiple disciplines and phenomena [7], and, when combined with findings of existing reviews, we provide a different perspective and more nuanced understanding of the research landscape. To our knowledge, a systematization of knowledge on specific user-centered misinformation interventions has yet not been conducted to this extent and with this perspective.

Our overarching goal is to deeply examine and classify misinformation intervention studies in terms of methodological characteristics in study design and evaluation, content characteristics of user interventions, and derived trends and challenges for future research. Addressing that goal, all our considerations lead us to the following research questions:

RQ1: *What are the typical methodological characteristics of existing studies on misinformation interventions?*
RQ2: *How do existing forms of user-centered misinformation interventions assist users in dealing with misinformation online?*
RQ3: *Which trends and opportunities for future research can be derived from the existing literature?*

The paper is structured as follows: First, we present our methodology of a systematic literature review and the procedure of deriving a taxonomy of user-centered misinformation interventions (see Sect. 6.2). Then, we present our results, including methodological aspects of analyzed publications such as addressed formats and platforms, applied methods of user studies, sample size, and participant details (see Sect. 6.3.1). Then we present our taxonomy (see Sect. 6.3.2), distinguishing nine intervention designs, active versus passive user interaction, and five points in time at which an intervention can be applied. We further discuss transparency as a specific measure to facilitate users in dealing autonomously with misinformation (see Sect. 6.3.3). Lastly, we present 'nudging' as a concept applied in many extracted publications (see Sect. 6.3.4). In Sect. 6.4, we answer our research questions regarding the design of user-centered misinformation interventions, methodological characteristics of existing studies, and the derived trends, open questions, and challenges for future research.

6.2 Methodology

In this section, we present our methodological approach of performing a systematic literature review, comprising of the identification and screening of relevant literature (Sect. 6.2.1) and the thorough analysis and structuring of publications and interventions included therein (Sect. 6.2.2).

6.2.1 Identification of Literature

To identify and categorize relevant literature on misinformation interventions, we performed a systematic literature review, following the PRISMA guidelines [403] (see Fig. 6.1). Schryen et al. [497] state that literature reviews are important for "developing domain knowledge" and to identify knowledge-building activities, such as synthesizing, aggregating evidence, criticizing, theory building, identifying research gaps, and developing a research agenda. In accordance with these principles, we set up our literature search as follows: The initial search spans the ACM Digital Library, Web of Science, as well as the IEEE Xplore database. With this set of databases, we encompass a broad corpus of diverse literature as well as the ten conferences and journals listed by Google Scholar as the best regarding HCI: ACM Conference on Human Factors in Computing Systems (CHI), IEEE Transactions on Affective Computing, Proceedings of the ACM on Interactive Mobile, Wearable and Ubiquitous Technologies (IMWUT), Proceedings of the ACM on Human-Computer Interaction (PACM), International Journal of Human-Computer Studies (IJHCS), ACM/IEEE International Conference on Human-Robot Interaction (HRI), ACM Conference on Computer-Supported Cooperative Work & Social Computing (CSCW), IEEE Transactions on Human-Machine Systems, Behaviour & Information Technology (BIT), and the ACM Symposium on User Interface Software and Technology (UIST). The search took place until June 2024, and only papers that were published until that date could be considered. All publication years up to this date have been included, however, there were no relevant publications for our final set concerning content-wise inclusion and exclusion criteria before 2011.

The search term consists of two parts: The first part are terms included in or related to our umbrella term of 'misleading information'. The second part includes synonyms for 'intervention' and related concepts addressing user-centered measures. Only papers containing at least one term of each part in their title or abstract were included in our search. We did not filter for a publication year. The complete search term is the following:

((rumour OR rumor* OR "misleading information" OR "fake news" OR "false news" OR misinformation OR disinformation OR "news credibility") AND (combat* OR correct* OR interven* OR countermeasur* OR counteract* OR treatment OR relief OR educat* OR warning OR nudg* OR user-centered OR "media literacy"))*

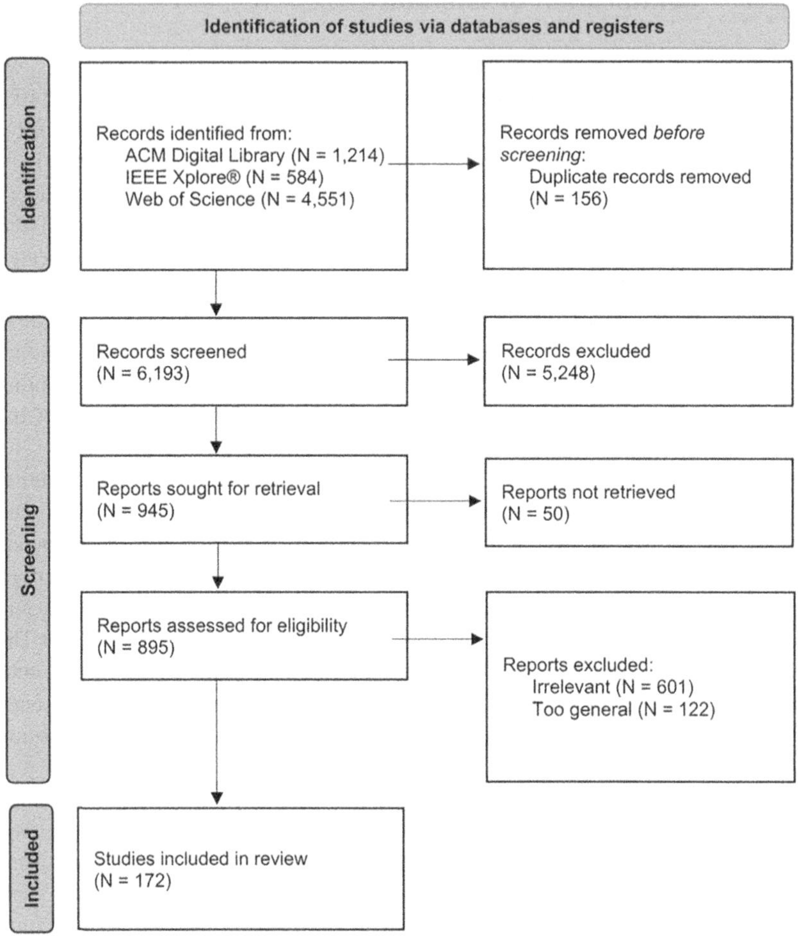

Fig. 6.1 PRISMA Diagram demonstrating our data flow within the systematic literature review (created with the template from [403])

The term was modified to adhere to database requirements and to run comparable searches. Furthermore, because Web of Science returned many results, the term was adjusted to exclude obviously irrelevant disciplines (e.g., chemistry). The broad interdisciplinary nature of Web of Science explains its large amount of 'false positives' during the initial search in comparison to the other two databases that already focus on disciplines relevant to digital misinformation interventions (e.g., computing and information technology). The search returned 1,214 results from the ACM Digital Library, 584 results from IEEE Xplore and 4,551 results from Web of Science, in total 6,349 results. After removing 156 duplicates, we screened 6,193 records, of which we excluded 5,248.

Records were removed for a multitude of reasons: First, some words of the search term had multiple meanings, which is why papers using a different interpretation were excluded. While we decided to take a broad perspective on diverse kinds of misleading information, including misinformation, disinformation, rumors, and related phenomena (e.g., conspiracy theories), as all types can have severe consequences, papers were excluded when the phenomenon investigated was not referring to our definition of the umbrella term 'misinformation' (see Sect. 6.1). This was particularly the case for misinformation referring to eyewitnesses remembering something inaccurately (e.g., due to suggestion) during a testimony in court. For phenomena included in our broad definition, there was a variety of terms included in our sample (see row 'Concept' in Table A.1, Electronic Supplementary Material (ESM)). Furthermore, some technical terms have different meanings in different fields, for example, network science uses the term rumor in the context of nodes spreading information [e.g., 73]. Second, when papers concentrated solely on the technical detection step with no involvement of the user at all, e.g., machine learning approaches focusing on increasing the detection rate, they were excluded. Furthermore, interventions that took a network-based approach, for example, by simulating which nodes to delete in order to reduce the spread of misinformation, were also excluded. In addition, we excluded psychological experiments without concrete reference to misinformation as well as surveys and questionnaires exploring background information (e.g., Which demographics are susceptible to misinformation?). Additionally, we decided to exclude reviews.

945 papers were sought for retrieval, of which 50 were removed because they could not be accessed, leaving us with a total of 895 publications that were assessed for eligibility. In the last step, a total of 723 papers were finally sorted out, 601 thereof because of the aforementioned criteria. Another 122 papers were excluded because they were too general, including papers that did not meet this review's focus because the intervention occurred long before the actual usage of social media, like educational school lectures, trainings, or serious games, but also implementations that focus exclusively on psychological phenomena (e.g., Do corrections of content generate reactance?). Particularly in the context of corrections or debunking of misinformation (e.g., in comment sections), there is a lot of in-depth research on factors impacting user reactions and interactions. Often,

these studies focus on psychological or social phenomena that are particularly valuable to consider when designing interventions tailored to a specific persona. To receive a reasonable number of publications and thus allow for a thorough focus on research regarding the design and evaluation of digital interventions, we decided to exclude studies that rather address (psychological or social) impact factors without a particular focus on intervention design and evaluation. The final set of papers contained 172 items which were included in our analysis and were categorized according to our taxonomy.

6.2.2 Development of a Taxonomy

For the development of the taxonomy, we first collected different relevant dimensions to compare and differentiate studies on user interventions within the context of misinformation. We developed and applied those categories in an iterative process of brainstorming sessions with two researchers with expertise in computer science, psychology, and HCI based on already familiar studies within the field of interest [e.g., 53, 299]. The coding process was initiated by a training phase where a common understanding of each category was obtained. When disagreeing on a categorization during the coding phase, the study was discussed to achieve a consensus. This approach of consensus coding is commonly applied in other research [606]. First, we defined our target group: researchers and practitioners interested in analyzing, designing, and evaluating digital countermeasures to combat misinformation that may potentially benefit from our taxonomy. To assist the target groups, several characteristics are particularly relevant as they provide information on (1) the intervention design, (2) the form of user interaction, and (3) the timing of the intervention. Categories were complemented and adjusted iteratively while identifying and reading new papers. For instance, when reading multiple papers that differed regarding the time of intervention, this category was included, and all relevant papers were categorized accordingly. Additionally, minor modifications to the categories were made during the process of reading and categorizing the articles when deemed necessary. Further, we looked at how user-centered interventions were categorized in other contexts with sensible information (e.g., cybersecurity) in systematic reviews [164]. The resulting final table can be found in the ESM (see Table A.1). A study can be sorted into several categories, and the subcategories are generally not mutually exclusive (e.g., some interventions may combine the intervention categories 'highlighting design' and '(binary) label' and others compare a 'correction' with 'showing indicators').

6.3 Results: The Landscape of User-centered Misinformation Interventions

In this section, a detailed analysis of the literature review is presented. First, we give an overview regarding *methodologies* used by studies on user interventions (see Sect. 6.3.1). We then provide a *taxonomy of interventions to assist users in dealing with misinformation* by categorizing and clustering the identified research sample in distinct dimensions (see Sect. 6.3.2). Furthermore, we highlight how *transparency* (see Sect. 6.3.3) is used to assist users in dealing autonomously with misinformation, present the concept of *digital nudging* (see Sect. 6.3.4) as a trending digital countermeasure, and finally discuss the impact and perceptions of reviewed misinformation interventions (see Sect. 6.3.5).

6.3.1 Methodological Characteristics

To provide an overview of research methods typically used in the field of user-centered interventions to assist in dealing with misinformation, details of the respective study designs were collected. All studies were published between 2011 and 2024. First, we were interested in the different concepts included under the umbrella term 'misleading information'. In total, 149 publications referred to either *misinformation, disinformation or misleading information.* Furthermore, 5 publications were specifically interested in *rumors*, and 10 publications that addressed the concept of *news credibility.* Other publications referred to myths, propaganda, or controversial topics. Out of a total of 172 included papers, 17 present exclusively *conceptual* ideas of interventions. In contrast, the remaining studies collected empirical data in the form of *laboratory experiments* (14 publications), *online experiments* (106 publications), *field studies* (9 publications), *surveys* (28 publications), and *interviews* (20 publications). In our study, we understand a field study as an evaluation type that specifically observes the natural behavior of participants in a real-world scenario, in contrast to experiments that encompass a controlled setting designed by the researchers. In the context of misinformation, research experiments rarely take place within an actual lab of the researcher (lab experiment) but typically remotely in an online setting (online experiment), for instance, as a link to the researcher's experimental website, as these experiments often do not require physical presence for additional hardware items. In some cases, there is a combination, e.g., of survey and interview or of laboratory experiment and online experiment within one publication. Regarding sample size, the empirical studies range from small groups of participants ([<20 e.g., 64, 90, 173, 306]) to large-scaled representative groups with far over 1,000 participants [e.g., 29, 271, 503]. You can find a visualization of sample sizes in Fig. 6.2. A closer look at the participants reveals a clear bias, with the majority of students reporting having U.S. adults and college students as participants. However, there are also isolated studies that either address a very specific (vulnerable) target group (e.g., teenagers [30, 224], marginalized communities [577], or blind/low vision social media users [507]) or make a comparison between several countries [e.g., 14].

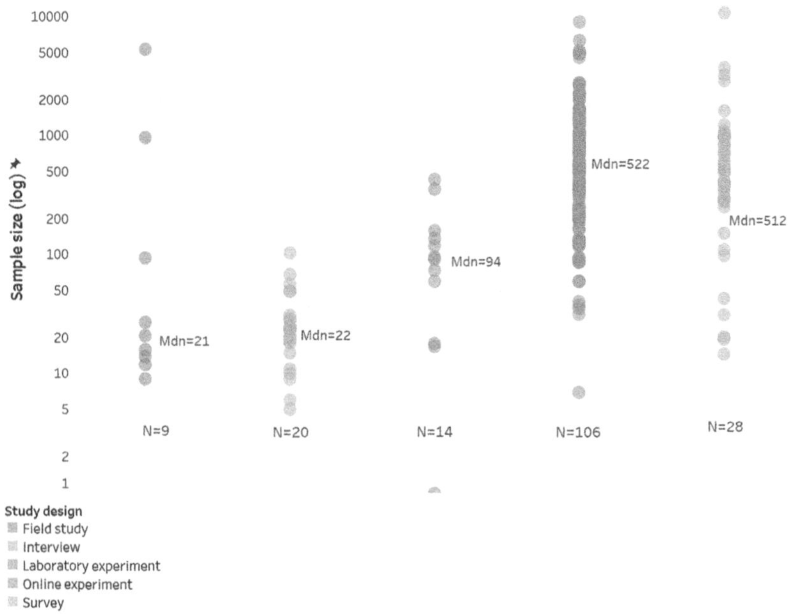

Fig. 6.2 Sample sizes (log) of the individual studies broken down by study type including the median (Mdn). Note that a publication often contains multiple user studies

While 71 publications generate their interventions or concepts generically for all online content and platforms (category *General*), others are developed and evaluated for specific platforms (see Fig. 6.3 for temporal distribution regarding platforms). Nevertheless, transferability to other platforms is often not excluded. 36 publications address interventions for *Facebook*, 33 publications for *Twitter/X*, and 3 publications for *Instagram*. Another 21 publications deal with platforms that do not fall into one of the categories already mentioned (e.g., Reddit [63, 592], TikTok [199, 224], messengers like Telegram [222], websites [322], arguments over an audio speaker [120], text documents [171], messenger forwards [412]) and therefore were categorized as *Other*. This corresponds to known research biases that show a focus on much-researched platforms such as Twitter/X. This is often justified by the already developed data situation and easier linkage to existing literature. Especially the great relevance of misinformation on newer social media platforms like TikTok in crises like the Russian-Ukrainian war shows that there is still a great need for research. When looking more closely at the addressed content format also see Figure A.1 in the ESM), we can see that most publications focus on *social media posts* (91 publications), 49 on *articles or text in general* and only a few on *images* (8 publications) and *videos* (9 publications) while we observe a growing relevance of misinformation of exactly these formats. Additionally, there are a few exceptions that address a very specific format, such as *audio* [e.g., 120] or misleading graphs [605].

A minority of the publications describe interventions that go beyond a low-fidelity prototype (e.g., in the form of screenshots) to include an actual implementation. Thus, *no implementations* can be read from 123 publications, while 14 publications deal with *browser plugins or browser extensions* [e.g., 53, 292]. Fourteen publications describe the implementation of a *custom platform* [e.g., 20], and 3 publications show a *game-based implementation*. Elaborate tools are an important part of mitigating the spread of misinformation and can be part of a holistic solution. An example is 'Verifi!' [286], which provides an interface for dealing with misinformation on Twitter (now X). The system consists of five display options, allowing for easy comparison between how real and questionable news sources report on a subject, for example, by comparing the words or images used. Another example would be 'Prta' [350], which provides the user with a tool that takes a text or URL as input and highlights propaganda techniques.

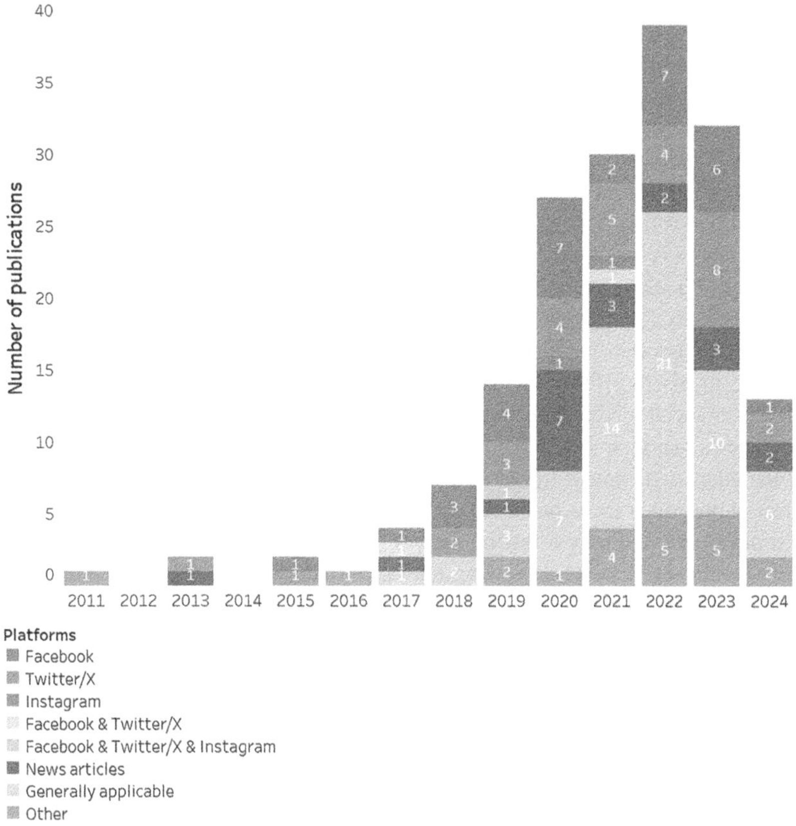

Fig. 6.3 Number of papers published according to their addressed platform

6.3.2 A Taxonomy of User-centered Misinformation Interventions

The wide range of addressed concepts, platforms, and research areas shows that, on the one hand, a large number of conceptual ideas and empirical findings already exist for digitally supporting users in dealing with misinformation; on the other hand, these often differ fundamentally. In order to distinguish existing approaches from each other and to cluster commonalities, we have derived a taxonomy based on the identified literature. Therefore, we performed an in-depth analysis of interventions. For the interpretation of the following results, it is important to notice that publications can contain multiple interventions – in total, there were 237 interventions within the 172 publications. Those interventions were analyzed individually regarding the taxonomy characteristics, while previously reported methodological findings are valid for the entire publication and, therefore, did not distinguish between individual interventions. In the following, the literature-based categories of the taxonomy are explained in detail (see also Table 6.1 and Fig. 6.4):

Table 6.1 A taxonomy of user-centered misinformation interventions

Category	Definition	Intervention examples	Publications
Intervention design	The intervention design distinguishes different countermeasures after the successful detection of misinformation. This includes general actions such as deleting content as well as concrete design decisions to encourage a learning effect.		
Warning	Interventions that give an explicit warning that the content is (potentially) misleading	Warning label; stop sign; "This post was disputed"	[22, 24, 38, 52, 53, 74, 104, 133, 159, 162, 166, 173, 187, 199, 205, 206, 266, 299, 301, 310, 320, 334, 335, 357, 374, 382, 400, 407, 410, 424, 432, 437, 479, 489, 503, 507, 508, 513, 572, 581, 594, 605, 627]
Correction/ debunking	Interventions that correct/debunk misinformation	Naturally occurring or artificially generated user comments or comments from officials that correct misinformation; links to fact-checking websites; expert sources; corrected headlines by users	[11, 29, 33, 38, 59, 60, 63, 86, 90, 112, 114, 115, 125, 126, 136, 141]–[144, 179, 199, 205, 207, 245, 251, 263, 263, 264, 291, 296, 299, 310, 318]–[320, 331, 333, 340, 341, 349, 373, 400, 412, 429, 431, 432, 464, 477, 483, 488, 490, 511, 525, 528, 529, 544, 550]–[552, 557, 560, 562, 563, 572, 573, 577, 582]–[589, 594, 596, 601, 605, 609, 627, 628]

(continued)

Table 6.1 (continued)

Category	Definition	Intervention examples	Publications
Showing indicators	Interventions that display indicators for misinformation to achieve transparency	Showing how old a video actually is; color relevant words for misinformation classification; generic tips to detect misinformation; infographic	[31, 33, 51, 63, 86, 99, 135, 156, 171, 188, 194, 205, 206, 222, 224, 225, 234, 276, 286, 297, 301, 322, 350, 437, 438, 472, 477, 491, 511, 513, 581, 592, 605, 621, 623]
(Binary) labels	Interventions that label content as misinformation or true information; often binary	Tagging post as true or false; thumbs up/thumbs down; "Prediction: It is Fake News!"; traffic light symbols	[25, 40, 63, 64, 133, 142, 159, 187, 227, 260, 262, 271, 292, 310, 318, 320, 322, 334, 335, 344, 374, 407, 409, 412, 424, 432, 438, 477, 479, 489, 491, 507, 511, 513, 529, 557, 562, 576, 592, 627]
Highlighting design	Interventions that visually highlight relevant parts of a post for misinformation classification	Highlight relevant words by color or size; color code tweets according to accuracy; highlight propaganda techniques using colors	[20, 25, 51]–[53, 171, 222, 224, 225, 234, 260, 262, 276, 297, 322, 331, 350, 472, 511, 513, 557, 605, 621, 629]
Visibility reduction	Interventions that reduce the visibility of misinformation visually	Reducing opacity or size	[24, 52, 53, 199, 297, 299, 310, 432, 479, 513]
Removal	Interventions that hide or remove misinformation	Deleting misinformation	[479]
Complicate sharing	Interventions that include additional user effort before allowing to share misinformation	Additional confirmation before sharing; require users to assess accuracy before sharing	[20, 261, 297, 299, 581]
Specific visualization	Interventions that use creative visualizations of relevant information	Visualizing sentiment and controversy score of news articles; visualizing fact-checker decisions; platform based on social network analysis visualization; aggregate authentication measures; visualization of number of unvaccinated children with measles as fear correction; visualizing fact-checker decisions; infographic	[93, 135, 259, 260, 263, 263, 267, 286, 291, 295, 322, 410, 429, 435, 438, 444, 476, 491, 513, 595, 623, 627, 629]

(continued)

Table 6.1 (continued)

Category	Definition	Intervention examples	Publications
User interaction	Interventions require varying degrees of interaction with the countermeasure		
Active	Active interventions require users to actively interact with a countermeasure	Click to confirm before sharing; overlay on Facebook/Twitter (now X) post; pop-up	[8, 20, 22, 24, 74, 86, 96, 133, 162, 187, 199, 206, 260, 261, 292, 297, 299, 310, 320, 422, 432, 444, 479, 490, 506, 561, 581, 588]
Passive	Passive interventions can be potentially ignored while using social media	A label below a social media post; tooltip; correction in comment section; warning next to a post	[11, 25, 29, 31, 33, 38, 51–53, 59, 60, 64, 74, 104, 112, 115, 125, 126, 133, 135, 136, 138, 141–144, 156, 159, 166, 171, 173, 179, 187, 194, 199, 205, 207, 222, 224, 225, 234, 245, 251, 259, 263, 263, 264, 266, 267, 271, 276, 291, 294, 296, 297, 299, 301, 318–320, 331, 333–335, 340, 344, 349, 357, 373, 374, 382, 400, 407, 409, 410, 412, 423, 424, 427, 429, 431, 435, 437, 438, 444, 464, 472, 476, 479, 483, 488–490, 503, 506–508, 511, 513, 525, 528, 544, 550–552, 557, 560, 562–564, 572, 573, 577, 582–586, 586, 588, 589, 592, 594, 596, 601, 605, 609, 621, 623]
Timing	Digital misinformation interventions can address varying points in time within the social media usage		
Pre exposure	Interventions that take place immediately before the exposure to (mis)information	Accuracy nudge before a news-sharing task; Pro-Truth pledge to engage in more pro-social behavior; narrative fear appeal message to encourage health experts to correct health misinformation online; general warning message about misleading articles; generic infographic; protective message: "Warning! Note: fake news can occur on Facebook. [...]"	[8, 64, 74, 86, 96, 104, 112, 115, 133, 135, 162, 187, 194, 199, 205, 206, 286, 297, 299, 320, 374, 422, 423, 444, 489, 490, 551, 561, 564, 587, 588, 609]

(continued)

Table 6.1 (continued)

Category	Definition	Intervention examples	Publications
During exposure	Interventions that take place during the usage of social media (and during the exposure to misinformation)	Algorithmic corrections next to a post; user comments underneath a post; warnings; adding "Rated False" tag to article headline; credibility labels; highlighting indicators for propaganda; wearable reasoner giving AI-based feedback on claims	[11, 20, 22, 24, 25, 31, 33, 38, 51–53, 59, 60, 63, 64, 104, 114, 120, 133, 136, 138, 142, 156, 159, 173, 187, 199, 205, 207, 222, 225, 234, 241, 251, 259, 260, 262, 263, 263, 264, 266, 271, 276, 294, 296, 299, 301, 310, 319, 320, 331, 334, 335, 341, 344, 357, 373, 374, 382, 400, 407, 409, 410, 424, 429, 431, 432, 437, 438, 472, 489, 503, 506–508, 513, 525, 528, 544, 557, 562, 563, 582–585, 587, 592, 594, 605, 621, 623, 627, 629]
Post exposure	Interventions occurring after seeing misinformation	Warnings after exposure to misinformation; responses by official health authority; debunking text based on debunking handbook including plausible scientific explanations to close gaps in mental models	[29, 64, 74, 112, 115, 125, 126, 141, 143, 144, 166, 179, 187, 245, 291, 318, 319, 340, 344, 349, 412, 423, 427, 464, 483, 488, 490, 550–552, 560, 572, 573, 577, 586, 588, 589, 596, 601, 609]
At the moment of sharing	Interventions taking place directly at the moment of sharing misinformation	Encouraging to reflect on content before sharing; endorsing accuracy prompt: "I think this news is accruate" placed into sharing button; behavioral nudges using checkboxes to indicate whether a heading is accurate and to tag reasons via checklist at posting time; report of linguistic analysis as immediate feedback when sharing and possibility to cancel a tweet within 30 seconds	[20, 84, 260, 261, 299, 581]

(continued)

Table 6.1 (continued)

Category	Definition	Intervention examples	Publications
On request of the user	Interventions that take place detached from social media platforms and have to be reached out to actively	Web app based on social network analysis for user exploration; Android application where user can enter URL or text for credibility assessment; system to analyze articles or URLs via interface or API	[40, 93, 99, 224, 227, 267, 286, 292, 295, 297, 322, 350, 435, 476, 477, 491, 511, 529, 576, 595]

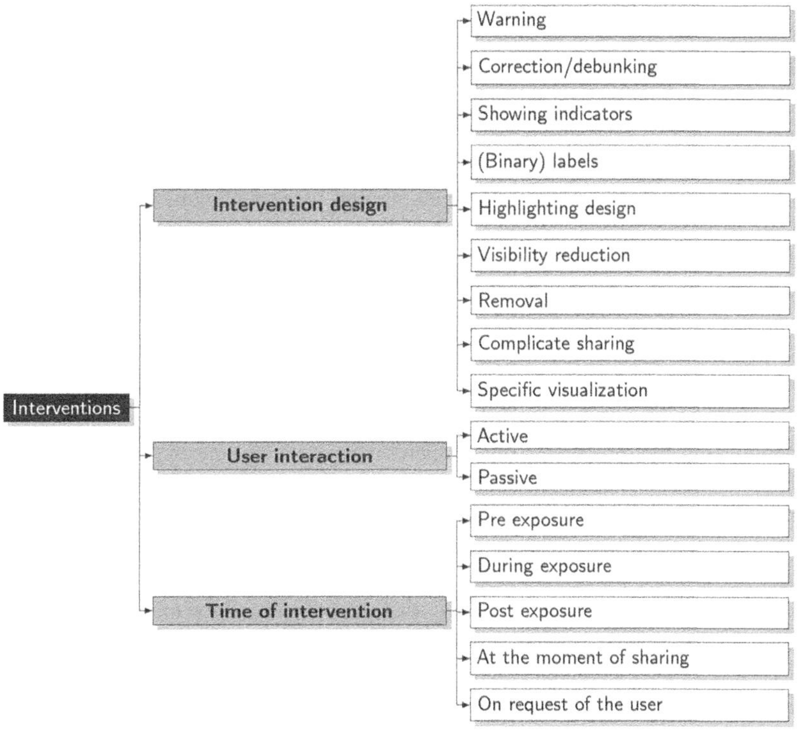

Fig. 6.4 A taxonomy for user-centered misinformation interventions

6.3.2.1 Intervention Design

The identified interventions on user-centered misinformation interventions vary greatly in their starting point. The digital support approaches and concepts are as diverse as the possibilities for protecting users from the effects of misinformation (e.g., deleting problematic content, warning, or strengthening media literacy). In an iterative process, nine intervention designs were identified based on the literature. Interventions could be assigned to multiple intervention designs, as they often used combinations. The majority of interventions propose or evaluate *correction/debunking* of misleading contents (66 stand-alone interventions and 30 interventions in combination with other intervention designs within 80 publications) and often represents a quite natural behavior of social media usage rather than an artificially generated technical countermeasure. For instance, many publications in that context evaluate whether corrections by users in the comment section of a post are effective in reducing belief in misleading content [e.g., 349]. Some of those interventions include a link to fact-checking websites, where the misleading content is debunked. This can be implemented both naturally by users in the comment section posting debunking links or digital interventions automatically exposing users to debunking (e.g., link to correcting source or user rebuttal within a comment/reply to a social media post or exposure to automatically generated counterfactual explanations [114]). Thus, interventions vary in terms of who is the arbiter of credibility assessment. While some are expert-based or rely on algorithm decisions, others rely on crowdsourcing of the community [136]. For instance, a browser extension allows users to suggest alternative headlines as a crowdsourced odd case for corrections, which are then presented to other users, empowering them to more actively participate in news consumption [263]. The dimension of who decides what is wrong or right within corrections and other intervention types was not systematically covered by our taxonomy but constitutes a relevant research area that has already been addressed by several studies [210, 263, 583]. Many interventions do not only expose users to content debunking or rebuttals but give an explicit *warning* that the content is (potentially) misleading (19 stand-alone interventions and 39 interventions in combination within 42 publications). Those warnings reach from warning labels like stop signs to textual warnings, e.g., "This post was disputed!".

Misinformation interventions can have different objectives. One of these objectives is to strengthen media literacy. In these types of interventions, concrete assistance in the form of indicators, for example, is typical. By *showing indicators* that support users in evaluating the credibility of content, the aim is to achieve a learning effect. Thirty-seven interventions correspond to this intervention design (including 7 as stand-alone interventions; 35 publications), for example by showing how old a video actually is [513] or by deriving words in the text that were particularly relevant for automatic detection as misleading [31]. Other intervention designs of this type compile more generic tips that users can apply to detect misleading content [135, 194, 205]. Research on indicators for misinformation, for instance, from the perspective of journalists as annotators [626], can be considered a significant foundation to inform indicator-based interventions. This intervention design is especially relevant, as studies have shown that users prefer transparent approaches where there is a potential learning effect [299]. However, in contrast to showing indicators for misleading content in a nuanced way, 48 interventions take the approach of assigning *(binary) labels* to contents (including 12 stand-alone interventions; 41 publications). This can be implemented, for example, by tagging content with a true or false tag, or thumbs up, thumbs down [40]. Other interventions give a probability in percent that content is misleading [292], and thus extend the framework of binary labels with richer information or provide more nuanced labels e.g., using traffic light colors. Often, these labels have in common that they do not provide a transparent explanation. However, binary and more nuanced labels do not, per definition, rule out a user-centered approach as they may be sensibly combined with explanations and can be applied for simplification of a complex underlying rating system.

Similarly, other ideas are concerned with increasing transparency. Some approaches use *highlighting design* as an intervention design which aims at facilitating potential learning effects. Twenty-four interventions (24 publications), for example, visually highlight relevant words for automatic classification within a social media post by color or size. For instance, Bhuiyan et al. [53] color code tweets on Twitter (now X) according to their computed accuracy. In contrast, Martino et al. [350] highlight propaganda techniques (e.g., exaggeration, loaded language, or oversimplification) detected within a text using different colors. Often, interventions that show indicators use some kind of highlighting design to do that, resulting in a common combination of both intervention designs. Indeed, despite highlighting components of content with colors, more *specific visualizations* (23 interventions; 22 publications) can be considered a distinct intervention design. Visualization is a very effective way to provide information as it provides *"the highest bandwidth channel from computer to the human"* [598, p. 2] and is used for interventions in different contexts [218]. Within the literature sample, there is a diverse set of creative visualizations of information. For example, Kim et al. [295] visualize the sentiment and controversy score of news articles within a conceptual study. In contrast, Park et al. [410] visualize fact-checker decisions regarding textual rumors. Moreover, Schmid et al. [491] developed and evaluated a platform based on social network analysis of contents on Twitter/X for users to proactively assess misinformation through visualization, and Chen, Ciuccarelli, and Colombo [93] designed visualizations of filter bubbles, exposing users visually with topics, sources, and opinions outside of their own bubble.

While previous intervention designs tended to provide verified feedback together with the problematic content, one study evaluates the effect of the *removal of misinformation* by hiding or removing a questionable post altogether [479]. Similarly but less rigorous is the attempt of *visibility reduction* (14 interventions in 10 publications), e.g., by reducing opacity or size. While we focus on visibility reduction that takes place visually, there are other (often network-based) approaches not covered by our more narrow understanding of user-centered misinformation interventions, as long as a user study demonstrating a direct impact on users is not included. For instance, studies reduce the visibility of misinformation in an algorithmic approach by reducing its flow [257]. Similarly, Epstein, Pennycook, and Rand [149] examine how layperson crowdsourcing of source credibility may be applied as input to social media ranking algorithms with promising results, leaving the potential for future research to investigate how this approach may be implemented regarding user feedback. Many intervention designs aim at preventing negative effects on people when confronted with misinformation or educating them to detect those contents themselves, as presented in the previous intervention designs. However, this can be extended to specifically preventing the spread of problematic content altogether, for instance, via *complicating sharing* (5 interventions in 5 publications). This may be implemented, for example, by including an additional confirmation before sharing or by nudging users to assess the accuracy of the content as they share it [261]. These user-centered approaches stand in contrast to network-based approaches to prevent the spread of misinformation through computational techniques as described by Chen, Xiao, and Kumar [97].

Many publications used multiple intervention designs, for instance as combinations or comparing interventions of different types against each other [205]. For example, showing indicators for misinformation often comes with some sort of highlighting design or a specific visualization (of the indicators).

While most interventions could be assigned to one or more of the intervention designs listed above, 50 interventions additionally used an intervention design that did not clearly fit the scheme (including 30 stand-alone interventions; 39 publications) while not appearing often enough as a distinct intervention design to represent an own type within the taxonomy. There are particularly unusual approaches, such as the development of wearable glasses that provide audio feedback on the truth of content [120] or a study that evaluates the effect of priming participants by letting them rate the accuracy of a headline before exposure to more potentially misleading content as a nudge to think more sufficiently [422] or on a similar basis, an explanation prompt that lets users explain why headlines were true or false [427]. Among the *other* category, there are nine interventions giving diverse kinds of information about misinformation and its detection immediately before exposure, e.g., in the form of an infographic [8], a video tutorial [30], a text about negative consequences of misinformation [104], a debiasing message [115], an awareness training [374], or a Pro-Truth Pledge [561]. Three interventions display a star rating or score, for instance, regarding credibility or sentiment [138, 294, 295]. Two interventions explicitly state to have integrated gamification elements [14, 529]. You can find a visualization of intervention designs in Fig. 6.5.

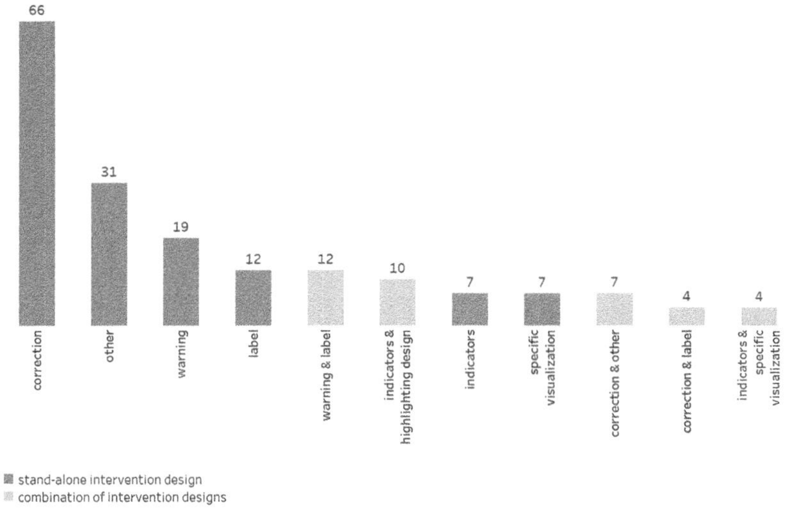

Fig. 6.5 Number of interventions addressing common intervention designs while differentiating between approaches that use a single intervention design versus multiple types in combination. Instances < 4 were excluded. A more detailed breakdown of the category 'other' can be found in Sect. 6.3.2.1

6.3.2.2 User Interaction

We compared whether an intervention required users to *actively* interact with the countermeasure (e.g., having to click to confirm sharing) or whether they could *passively* ignore the countermeasure (e.g., a label below a post). Furthermore, some interventions cannot clearly be labeled as active or passive as the actual form of implementation is not (yet) defined (30 interventions in 27 publications). For instance, some approaches make a more generic proposal of a warning without stating if it can be ignored or not. Other approaches present an intervention that takes place only at the request of the user, for example, as a separate smartphone app [e.g., 40] and, thus, is neither active nor passive during the actual usage of social media. In cases where the intervention is designed as a mandatory one-time experience [e.g., 8, 422, 561], users do not have to interact with the intervention during the following usage of social media but definitely once before the usage. Thus, those interventions are classified as active. We found that the majority of interventions are passive (175 interventions in 129 publications) while only 32 interventions (in 28 publications) deal with active interventions. A representation of the number of interventions regarding different interaction types can be found in Fig. 6.6.

Fig. 6.6 Number of interventions regarding each type of user interaction. Publications can contain multiple interventions with different user interaction types

6.3.2.3 Time of Intervention

Misinformation interventions can address different points in time in the context of social media use. While countermeasures that take place long before the actual usage of social media (e.g., trainings, educational games for school lessons, inoculation) were excluded from analysis, we included countermeasures that take place immediately before the exposure (e.g., short messages when logging in to a social media platform). We found that 37 interventions take place *pre-exposure* to misinformation (including 5 interventions coming with a combination of timings; 32 publications). For example, Pennycook et al. [422] nudged participants of a large-scale online study to think about accuracy before getting a news-sharing task. They asked participants to rate a headline's accuracy before exposing them to multiple other headlines and measuring their sharing intentions. Indeed, they found that giving a simple accuracy induction resulted in increased sharing discernment [422]. The majority of interventions (120 interventions in 92 publications) however, is designed to take place immediately *during* the exposure to misinformation while users are engaging with a social media platform or other content and encounter misinformation. This includes most of the corrections, warnings, labels, and highlighting approaches. For example, Bode and Vraga [60] compare algorithmic corrections via related articles from Snopes with social corrections via user comments underneath a Facebook post referring to the same debunking link on Snopes.

Forty-nine interventions in 42 publications deal with countermeasures *post-exposure*, e.g., Grady, Ditto, and Loftus [187] compared warnings after exposure to an article with warnings during other points in time. To specifically combat the spread of misinformation, a few interventions intervene directly *at the moment of sharing* (6 interventions in 6 publications [e.g., 581]). These presuppose that the user is about to share misinformation and he or she has already slightly passed the timing dimension 'during exposure'. Detached from the actual social media platforms, some approaches offer their own platforms. Accordingly, the intervention here takes place *at the request of the user* at any time (21 interventions in 19 publications; [e.g., 40]), when users have to actively reach out to the intervention on a separate platform. A few interventions could not be assigned to one of those points in time (10 interventions in 8 publications). For example, Furuta and Suzuki [171] present a countermeasure to take place during article creation. While the majority of interventions specifically take place at one exact moment in time, eight interventions are designed to take place at multiple points in time, combining, for example, a pre-bunking message with a warning during exposure to misinformation. See Fig. 6.7 for a visualization of the number of publications for each timing of the interventions.

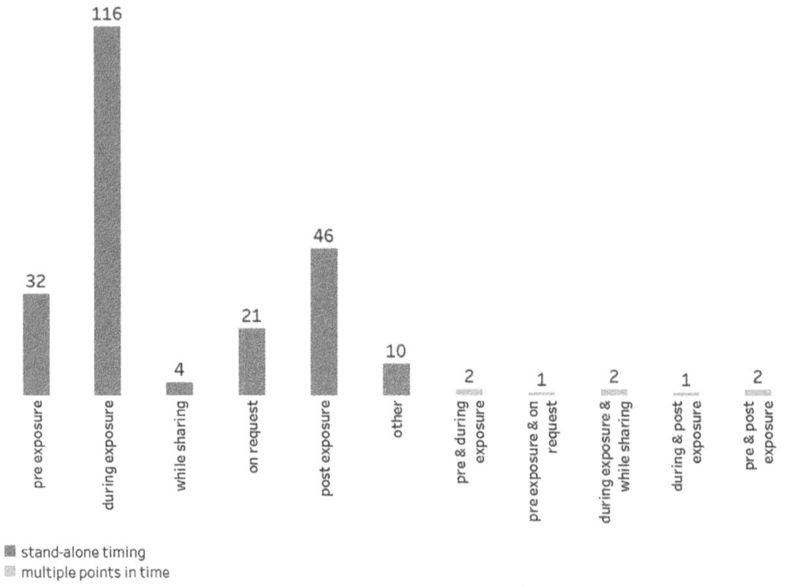

Fig. 6.7 Number of interventions addressing different points in time

6.3.3 How is Transparency used to Facilitate users Autonomously Dealing with Misinformation?

As highlighted in Sect. 6.3.2.1, digital countermeasures in the context of misinformation can have different objectives. While some interventions aim to reduce the spread of such content itself, others aim to communicate the findings of the digital countermeasures to the end users, sometimes resulting in an environment that facilitates the strengthening of media literacy skills. Within our systematic literature review, our special focus of interest is on transparent approaches that offer some form of explanation, as opposed to (binary) labels without explanations or deletion of problematic content. Transparency can be achieved by different distinct intervention designs. One very common way of explanation is exposure to corrections or debunking. Indeed, the majority of publications within our scope deal with some sort of correction or debunking. While some approaches investigate the effect of user corrections within comment sections of social media posts, others focus on corrections of authorities. Interestingly, there is a very specific scientific discourse on the effectiveness of corrections, which is conducted in different disciplines. Corrections and debunking can be considered a central part of combating misinformation online. This type of intervention provides an opportunity to thoroughly confront officially refuted content with facts. Often, this takes the form of a more detailed article, which backs up its corrections with official sources. Official fact-checking websites, which are linked by the intervention, are usually used for this purpose. On the other hand, there are approaches that aim for transparency through media literacy training, for example, in the form of showing indicators and using a highlighting design of those (see examples in Fig. 6.8). Here it is examined which components of a social media post comprehensibly indicate that it is misleading content. We discussed this type of intervention in more detail in Sect. 6.3.2.1. In addition, while labeling content as false or true without explanations typically comes as a top-down approach not addressing users' needs for transparency, labeling interventions can indeed provide comprehensibility and transparency when applied as a simplification of an otherwise too complex rating system as a combination with additional explanations.

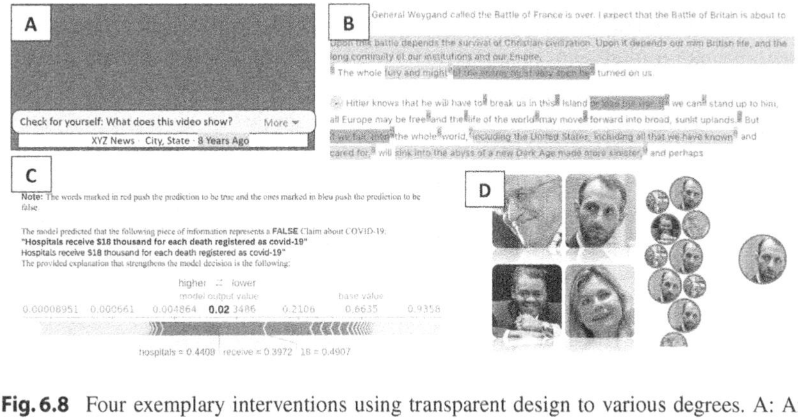

Fig. 6.8 Four exemplary interventions using transparent design to various degrees. A: An intervention showing the publish date of a video as indicator by Sherman, Stokes, and Redmiles [513]. B: An intervention highlighting parts of a text containing propaganda techniques by Martino et al. [350]. C: An intervention used to affirm and refute claims using explainable machine learning by Ayoub, Yang, and Zhou [31]. D: The image comparison view as part of a larger system designed by Karduni et al. [286]. The screenshots were taken from the respective papers

6.3.4 Nudging as an Ambivalent Trending Countermeasure

During the content analysis of the identified relevant literature, one specific form of user intervention particularly caught the eye: 22 publications refer to their form of intervention as a digital nudge. A nudge is defined as an intervention that *"alters people's behavior in a predictable way without significantly changing their economic incentives"* [554, p. 6]. It is a concept that has already been applied to many contexts, such as cybersecurity and health. The concept of nudging is controversially discussed in research. Thus, under certain circumstances, it represents a potential for subconscious manipulation also in harmful directions. Like many digital countermeasures, digital nudges, while promising, often offer little transparency and may run the risk of steering users in the wrong direction in ambiguous situations if they are not critically engaged with but rather trusted blindly. For instance, Lu et al. [344] show that AI-based credibility indicators can be used to steer participants in a certain direction, even if the AI is wrong. Since we did not specifically review which of the 237 interventions within 172 publications actually fit the definition of a nudge, we would nevertheless like to provide an overview of the publications that refer to their interventions as nudges themselves. Nudging has been applied by other reviews as a category of misinformation interventions on an individual level itself, complementing countermeasures like boosting, debunking, and content labeling [467]. In our review, we understand nudging as a concept that can be applied in diverse intervention designs. Some publications present intervention as "accuracy nudges" and introduce concepts in which users are specifically nudged to reflect on the accuracy of the content and to act more thoughtfully accordingly [e.g., 25, 84, 271, 382, 422]. For example, Capraro and Celadin [84] report promising results that indicate positive effects on sharing behavior when using an accuracy prompt. Similarly, von der Weth, Vachery, and Kankanhalli [581] developed 'ShareAware' as a nudge for more conscious posting and sharing behavior. In a different approach, Andi and Akesson [22] developed a social norm-based nudge to effect sharing behavior by exposing participants to the message: *"[...] Most responsible people think twice before sharing content with their friends and followers"*. In contrast, there are attempts to nudge users not only away from misinformation [138] but towards the consumption of credible news [e.g., 173, 241] or a habit of assessing accuracy of information [52, 53, 261]. In that context, Thornhill et al. [557] developed a nudge to steer users into fact-checking news online.

6.3.5 Impacts and Perceptions of Digital Misinformation Interventions

Our systematic review revealed a wide range of intervention designs addressing various types of user interaction and timings. For future research, it is important to determine which interventions are most promising and should be given more consideration. This paper aims to provide a comprehensive overview of misinformation interventions from diverse disciplines. These interventions strongly differ in their target of behavior change, such as improving credibility assessment, reducing the sharing of misinformation, helping users distinguish between misinformation and credible content, or decreasing the overall flow of misinformation on social media. The focus is on the design characteristics of the intervention and the user-centered evaluation method, including qualitative and quantitative evaluations. Many studies evaluate the efficacy of an intervention within a specific context (social media platforms, user groups, format of content etc.) and in comparison to a specific condition (control group without intervention, state-of-art intervention of the specific platform etc.). Others derive rich qualitative insights, e.g., into how users perceive an intervention in terms of concepts like trust, reactance, or comprehensibility.

While our review does not include a meta-analysis to derive statistical evidence of efficacy, we provide some initial insights into what appeared to be promising based on a qualitative overview, complemented by statistical effect size information that was explicitly stated in the corresponding publications. However, it is important to note that due to the nature of our review it does not allow for direct comparisons or objective evaluations of which interventions worked best or worst. Instead, it provides initial insights from a broad interdisciplinary perspective. We present our overview of central findings regarding (positive and negative) impacts and perceptions of interventions in each publication in Table A.2 (ESM). There, we summarize the beneficial effects of interventions that were mostly collected quantitatively, beneficial perceptions of interventions that were derived from qualitative studies, and insights on measures that were not effective or even resulted in counterproductive or unintended effects. Taking a closer look at the studies, they identify measures and characteristics that do or do not impact efficacy - sometimes with contradictory findings that demonstrate a necessity for further investigations. For instance, the timing of the correction does sometimes but not always seem to matter [115, 464], and there are indications that efficacy is sometimes but not always impacted by whether the correction is narrative or non-narrative [144, 318, 483]. There are further controversial findings on whether transparent information and explanations have a significant impact on efficacy (e.g., rather yes: [179, 299]; rather not: [349]). However, when considering findings on the role of transparency and explanations over all publications, the general tendency (including qualitative insights) indicates its impact on efficacy, user perception, and acceptance as promising.

Further controversial findings discuss which intervention mechanism/source type (social versus algorithmic correction or warning by citizens versus news agency or (e.g., health) experts) matters in terms of efficacy [e.g., 60, 207, 245, 340, 344], for example, by emphasizing potential unintended over-reliance on AI predictions even if they are not correct [344]. Other studies evaluate the impact of modality (e.g., images, videos, voice messages) of interventions on efficacy [412, 528, 560, 605, 627]. For instance, Pasquetto et al. [412] found that audio files were more effective in correcting beliefs than text or videos and Karduni et al. [286] revealed that corrections using images are more effective in correcting myths than corrections without images, independent of the image type (machine-technical image, expert image, diagram). When looking into the effect sizes (e.g., Cohen's d, (partial) η^2, Spearman's r and ρ) explicitly stated in the publications, they are small (e.g., (partial) $\eta^2 < 0.06$ or $d < 0.5$) for the majority of publications [e.g., 30, 585], and medium to large ((partial) $\eta^2 > 0.06$) in fewer cases [e.g., 587]. As the interpretation is highly dependent on the research design, a future meta-analysis may complement our findings with statistical comparisons of efficacy that aim at controlling influencing factors revolving around the context of data in more narrowly defined domains. Comparability is often not possible due to the very diverse settings in which studies take place, addressing different social media platforms, formats of content, and participants (e.g., students with potentially higher levels of media literacy, elderly, adolescents, representative studies in different countries), and applying a variety of research designs to measure efficacy, e.g., asking participants to state whether they would share specific content versus asking them to rate the credibility [193]. Indeed, finding a consensus in research to measure the efficacy of misinformation interventions has been emphasized as an important step towards more successful interventions, and possible frameworks have been proposed [193]. Some meta-analyses have already reported on the efficacy of specific misinformation interventions like corrections, where deriving a subgroup of studies with a similar research design and conditions is sometimes achievable, allowing for comparisons or comparable interventions in different contexts. For instance, Chan and Albarracin [91] conducted a meta-analysis on the efficacy of corrections/debunking in the context of scientific misinformation, examining over 200 effect sizes and revealing that corrections are more successful when detailed. Still, in general, the debunking effect was not significant. Given the overall estimated lower impact of corrections/debunking and the strong research focus of the majority of studies on this type of intervention revealed in a related meta-analysis [58], which was confirmed in our review, this suggests scholars should not disregard other intervention types that might be less studied but more promising.

Due to the publication bias, most studies report statistically significant or qualitatively promising results. Only a few exceptions exclusively report what did not work in general [25, 86, 297, 506] or for specific user groups [320, 507]. For instance, Aslett et al. [25] report that dynamic source reliability labels placed in-feed did not reduce misperceptions, and Caramancion [86] demonstrates how preventive infographics had trivial to no effect. Despite non-efficacy of interventions, in some cases, studies reveal other unintended or counterproductive effects such as over-correction and other spill-over effects on accurate content [156, 206], over-reliance on interventions [344], increased belief in misinformation under certain circumstances like subjective messages or repeated exposure to content [29, 429, 550], priming of general mistrust in authentic content due to warnings [572], or lowered perception of extremeness due to stance labels on political ideology [173].

6.4 Discussion

In Sect. 6.3 we have systematically categorized a variety of existing misinformation interventions to assist in dealing with misinformation online, providing concrete examples of identified dimensions. The analysis of our systematic literature review underscores the impression that a variety of diverse approaches have emerged in recent years and continue to emerge. It can be observed that these often differ significantly in their characteristics. In this section, we discuss and summarize our findings regarding our research questions.

6.4.1 RQ1: What are the Typical Methodological Characteristics of Existing Studies on Misinformation Interventions?

Methodologically, studies of misinformation interventions differ in various dimensions, although several emphases and typical patterns are also apparent. Due to the inclusion and exclusion criteria of our review, the sample contains mainly studies that collect empirical data and only a few publications with an exclusively conceptual approach. Typically, publications on misinformation interventions evaluate novel or already established interventions in user studies, often in comparison to other existing approaches. A particular focus is on online experiments with a collection of quantitative and qualitative data, as this method is suitable for large-scale samples and a controlled environment. In order to examine the interventions in a realistic environment and to minimize biases, more evaluation in the form of field studies would be desirable for future studies. It is striking that mainly U.S. adults and college students are surveyed as study participants, while specific (vulnerable) target groups such as teenagers, persons of older age, or non-native speakers are largely neglected. This can be explained by the better accessibility of different user groups and represents a common problem known from other user studies in contexts of HCI and similar disciplines.

Not surprisingly, most publications deal with Facebook and Twitter/X as social media platforms or address news articles in general (see Sect. 6.3.1). Considering current and emergent social media platforms like TikTok and Instagram as image- and video-based platforms is still largely missing within the research landscape. However, the impact of those platforms and content types has shown to be highly relevant. Looking closer at our publication sample, we can note that there are already isolated publications for addressing exceptional formats such as image, video and audio. We can see that there is a positive correlation between the addressed formats 'video' and 'image'. Indeed, three out of five interventions for video formats are specifically addressing images as well [e.g., 30].

6.4.2 RQ2: How do Existing Forms of User-Centered Misinformation Interventions Assist Users in Dealing with Misinformation Online?

Looking closely at misinformation interventions, one notices a publication emphasis on corrections and debunking. This form of intervention can be artificially controlled or occur naturally in the form of user comments. It is striking that corrections/debunking are examined in great detail in the literature from a wide variety of perspectives and with a focus on the smallest details, e.g., regarding timing or repetition [112, 125]. This finding is also supported by the review study by Chen, Xiao, and Kumar [97], who identified fact-based corrections as "the most common type of corrective communication strategy", classifying it as a part of message-based approaches. Often, corrections from the official side are based on thorough and elaborate journalistic work. For example, social media articles are linked to an official correction once the content has been thoroughly checked manually by experts. In the fast pace of social media and especially in emerging crisis situations, there is an overflow of accurate and misinformation that needs to be reacted to quickly. This is where expert-based corrections as digital countermeasures sometimes reach their limits as stand-alone interventions. Other approaches pursue corrections based on the assessment of users themselves. While there is no expert review here, active user participation in news consumption is facilitated and studies reveal promising findings. For instance, participants preferred suggested headlines by laypersons that corrected the original ones [263]. Indeed, the effects of corrections by laypersons versus experts have been studied in prior work [210, 583] and constitute a relevant dimension of interventions beyond this work's scope. While many correction interventions provide users with additional (fact-based) knowledge and can thus create transparency, other types of interventions aim to increase transparency and thus media competence, for example, through linguistic or content-related indicators. An advantage of those interventions is the scalability of using automatic detection algorithms in real-time during emerging crisis situations and on large data sets, often based on machine learning approaches. However, when automatically showing indicators such as a missing verification seal or semantic propaganda techniques, the final decision on whether content is misinformation or not either lies with the user or is taken over by the algorithm based on (potentially biased) training data, missing the expert knowledge of professional fact-checkers. Transparent misinformation interventions, independent of their implementation as correction or display of automatically detected indicators or other types, may offer the opportunity to counteract reactance of end users in contrast to approaches that lack an explanation and, in some cases, facilitate a feeling of censorship, paternalism, and loss of control.

In contrast to transparent approaches, there are also fewer educational interventions with the goal of reducing the consumption of misinformation through removal or visibility reduction. Both intervention designs can be considered helpful when considering the bias of people remembering content itself without a potentially shown correction or warning when exposed to misleading content [187]. On the other hand, this intervention design may lead to (a feeling of) censorship and a resulting migration to other platforms or tools that take less rigorous action against problematic content. While deleting/censoring dangerous or explicit content is a legitimate and important responsibility of social media platform operators, applying this solution of deletion to all problematic content, such as disinformation and misinformation, would not only lead to a migration of users to less restrictive platforms. In particular, it would represent a missed opportunity for media literacy education, some of which can be achieved through transparent digital countermeasures as a complement to school lessons.

In order to develop misinformation interventions in a user-centered way and to be able to achieve an actual effect, the early inclusion of the needs and requirements of different target groups is indispensable. A particular challenge is the accessibility of people who have no trust in official bodies. In this context, limits certainly emerge as to who can be reached at all by the corresponding technical tools. In order to avoid reactance, the timing of the intervention certainly plays a role in addition to the transparency of approaches. In our systematic literature review, we identified interventions that can be used at the user's request and others that are permanently present during the normal use of social media. It is an interesting research question: which point in time or which regularity of an intervention is suitable for which target groups? Considering the broad variety of misinformation interventions, we hope to provide a helpful overview of existing forms. We propose our taxonomy as a starting point to systematically capture intervention categories and identify relevant dimensions. It is intended to provide researchers with a framework to develop new interventions, to pool knowledge from different disciplines for the promotion of cross-disciplinary research, and to reveal promising research directions.

6.4.3 RQ3: Which Trends and Opportunities for Future Research can be Derived from the Existing Literature?

In this paper, we have systematically analyzed 172 publications with 237 user interventions to assist in dealing with misinformation online. Our findings reveal current trends and movements in HCI, psychology, information systems, and communication sciences. As potential avenues for future research, we propose the following questions and interests:

(1) Are approaches for specific platforms transferable to other new platforms? With regard to particularly relevant contexts of use, it is also necessary to consider current and emergent social media platforms. Social media platforms are constantly changing. For some time now, there has been a noticeable trend toward TikTok, and Facebook, in particular, is losing a great deal of its importance, especially among younger people. In order not to have to reinvent the wheel again and again, studies on the transferability of findings to new types of platforms are important. While the majority of studies surveyed much-researched platforms such as Twitter (now X) and Facebook [e.g. , 52, 296, 357, 400] (see Fig. 6.3 and

Sect. 6.3.1), there has been little research on misinformation on image- and video-based platforms such as Instagram and TikTok. Given the usage rates of these media, particularly among youth, and the increasing relevance of the platforms for misinformation (e.g., concerning the Russian-Ukrainian war), addressing this research gap is considered particularly relevant. At the same time, there are major obstacles to overcome here, especially with regard to the availability of labeled datasets, as they typically already exist for Twitter/X but are very time-consuming and complex to establish for video data.

(2) How can collected findings and technical approaches for text-based interventions be applied to emerging video- and image-based misinformation? With regard to transferability to new platforms, transferability to other information channels is also particularly central. Can text-based user interventions ([e.g., 171] be adapted for video- and image-based channels? How can new indicators and measures for detecting misinformation (e.g., image reverse search) be integrated into misinformation interventions? Challenges researchers are confronted with include the fast-evolving trends in social media. For instance, emotion-evoking content features on TikTok might solely constitute a characteristic of the platform's content while it might be considered a more valuable indicator for misinformation in other modalities.

(3) How can opportunities of digital misinformation interventions be effectively combined with the advantages of human experts? Fully automated mechanisms, e.g., for machine learning-based detection, can handle large amounts of data better than humans. In contrast, trained humans as experts (e.g., journalists [529]) can handle specific case decisions better than algorithms when the boundary between true and false is blurred and information is missing. How can human expert knowledge be used within digital countermeasures without losing the performance of the automatic tool? In which steps of the countermeasure can human intervention be integrated?

(4) How can automatic detection be combined with user-centered feedback? Automatic detection is often a black-box procedure and, therefore, cannot explain its decision-making. While efforts in explainable artificial intelligence already reveal promising results to make detection approaches more transparent without the human directly in the loop, they are still challenged with how to present these explainable outputs in a way that is valuable for a layperson, especially for people with low media literacy. How can the advantages of accurate automatic detection be combined with transparent and comprehensible explanations as user-centered feedback (see Sect. 6.3.3; initial attempts e.g., by Schmid et al. [491])?

(5) How can media literacy be encouraged without exposing users to misinformation? Approaches to increasing media literacy are often based on a display of misinformation with additional reference to comprehensible indicators or debunking. Nevertheless, the user is still exposed to the misinformation in this case. Studies suggest that even with a simultaneous warning, the misinformation content may be remembered at a later point in time, which speaks for less exposure to misinformation [187] and constitutes an effect that is controversially discussed in literature [482]. However, users tend to feel reactance and paternalism when content is hidden or deleted [299]. How can media literacy be trained within misinformation interventions without continuing to expose users to misinformation?

(6) Can vulnerable people profit from the general findings of participants with high media literacy? How can we reach vulnerable people with official tools? The bias toward U.S. adults and college students as study participants continues to be striking. Since not all individuals are equally affected by misinformation [515], but rather particularly vulnerable groups exist, the inclusion of individuals with lower levels of media literacy in the iterative design and evaluation process of the interventions is essential. Initial approaches are already moving in this direction [30]. Still, researchers are often challenged with conducting user studies outside of the university bubble with convenience samples of participants, as particularly studies with children and teenagers in the context of misinformation come with additional ethical questions and recruiting challenges. For instance, confronting adolescents with misleading information during a user study is a sensible task that needs thorough consideration. How can the findings be applied to vulnerable people (see Sect. 6.3.1; e.g., older people [477], children, teenagers [30, 524], non-native speakers)? How can these target groups be meaningfully integrated into the iterative design and evaluation process?

With our systematic literature review we hope to provide a starting point for cross-disciplinary debates and knowledge exchange, as well as an inspiration for future research.

6.5 Limitations and Conclusion

First, the large amount of publications in the area of interest, including a variety of different disciplines, was a challenge to deal with. Therefore, we focused specifically on approaches that take place within the real-time usage of social media and excluded approaches (especially educational trainings, games, and presentations) that take place long before the actual usage of social media [e.g., 524]. However, those approaches may provide additional insights into effective and user-centered interventions and are, therefore, suggested for future research. In addition, we excluded studies on psychological or social phenomena (e.g., norms) to receive a reasonable number of publications that allows for a thorough focus on research regarding the design and evaluation of digital interventions. These studies are valuable to consider when designing interventions tailored to specific persona and are suggested for future reviews.

Second, our approach takes a broad perspective on types of misleading information referred to as 'misinformation' as an umbrella term and encompassing unintentionally and intentionally misleading information as well as related phenomena (e.g., rumors, conspiracy theories). Table A.1 (ESM) roughly demonstrates in clusters which concept was used in each paper. While we excluded papers that contained a related term but understood it as a phenomenon not fitting within our broad definition (e.g., eyewitnesses remembering something inaccurately as 'misinformation'), we did not perform an in-depth analysis of how each term was defined and utilized in each paper. This is a limitation that may have impacted our screening phase. In addition, there might be odd cases of papers not identified within our systematic screening phase as they use different terms to address the topic of misinformation not included in our search term. For instance, a study by Ennals, Trushkowsky, and Agosta [148] was brought to our attention during the review phase that refers to 'disputed claims' and was, thus, not detected.

Third, within our work, we thoroughly categorized the publications regarding multiple characteristics, involving two researchers with expert knowledge in that field of research. As the publications provide information on our categories in varying detail, we cannot exclude the possibility that some interventions were classified differently than they were intended by the authors themselves.

Fourth, while our work examines misinformation interventions from multiple perspectives, there are additional significant dimensions that have not yet been covered in this systematic study and are suggested for future work. In particular, in the context of user-centered interventions, looking at who is the arbiter of content credibility (e.g., decentralized decisions by the crowd versus experts or algorithmic decisions) is an important dimension that has significant effects on intervention perception and impact.

Misinformation remains a threat to the democratic order and the cohesion of society, and the fight against it remains important. It is a central goal to empower users in dealing with the overabundance of information online, especially during emerging crises. Digital misinformation interventions are one of several starting points to address that challenge, complementing professional journalistic work and media literacy training at schools. In this work, we have given an overview of existing countermeasures and have developed a taxonomy in order to systematize misinformation intervention research. Finally, we hope that this work—being a first step towards the systematization of misinformation intervention research—serves as an inspiration for future research and facilitates cross-disciplinary exchange of knowledge.

7 Misleading Information in Crises: Exploring Content-specific Indicators on Twitter from a User Perspective

Abstract Recent crises like the COVID-19 pandemic provoked an increasing appearance of misleading information, emphasizing the need for effective user-centered countermeasures as an important field in HCI research. This work investigates how content-specific user-centered indicators can contribute to an informed approach to misleading information. In a threefold study, we conducted an in-depth content analysis of 2,382 German tweets on Twitter (now X) to identify topical (e.g., 5G), formal (e.g., links), and rhetorical (e.g., sarcasm) characteristics through manual coding, followed by a qualitative online survey to evaluate which indicators users already use autonomously to assess a tweet's credibility. Subsequently, in a think-aloud study participants qualitatively evaluated the identified indicators in terms of perceived comprehensibility and usefulness. While a number of indicators were found to be particularly comprehensible and useful (e.g., claim for absolute truth and rhetorical questions), our findings reveal limitations of indicator-based interventions, particularly for people with entrenched conspiracy theory views. We derive four implications for digitally supporting users in dealing with misleading information, especially during crises.

Original Publication *Katrin Hartwig, Stefka Schmid, Tom Biselli, Helene Pleil, and Christian Reuter. "Misleading Information in Crises: Exploring Content-Specific Indicators on Twitter from a User Perspective". In:* Behaviour & Information Technology *(2024). DOI:* 10.1080/0144929X.2024.2373166

Supplementary Information The online version contains supplementary material available at https://doi.org/10.1007/978-3-658-47885-8_7.

K. Hartwig, *Navigating Misinformation*, Technology, Peace and Security | Technologie, Frieden und Sicherheit,
https://doi.org/10.1007/978-3-658-47885-8_7

7.1 Introduction

In an era of complex global crises, including the Russian war of aggression against Ukraine and the COVID-19 pandemic, the world is struggling with an overabundance of accurate and false information, leading to what the World Health Organization has coined an 'infodemic' [611]. Social media channels such as Twitter (now X), TikTok, and Facebook have emerged as essential and compelling platforms for individuals to share information [460]. However, the dissemination of misleading information is increasingly facilitated by social media because of characteristics such as low costs and anonymity [539]. Experts have studied misleading information during crises from various perspectives and disciplines. Our work is interested in misleading information that was deliberately created as well as unintentionally, following [178] and their understanding of misinformation to collect false information "agnostic to the intention of the original creator or poster" for our sample. In line with previous scholarship [178, 181], we define misleading information as an umbrella term encompassing both intentionally misleading information (i.e., "disinformation" or "fake news") and unintentionally misleading information (i.e., "misinformation"). In our understanding, misleading information comprises false information and misleading satire, whereas non-misleading information encompasses true information, opinions that are not absolute, or satire that is not misleading in the specific context.

Research in Human-Computer Interaction (HCI), psychology, and information systems has started to investigate how to tackle the pervasive issue of misleading information using technology. This multifaceted exploration encompasses the development of very technical, often machine-learning-based, automatic detection approaches [518, 613] alongside approaches with stronger user-centered HCI focus on post-detection decisions or aspects detached from detection (e.g., providing corrections).

Digital misinformation interventions are employed to assist users in processing misleading information online and are typically applied after automatic (and sometimes manual) pre-filtering problematic content. They build upon and expand educational and journalistic initiatives. The term 'digital misinformation intervention' has gained recognition within the research community [35, 479, 480], where user-centered misinformation interventions "go beyond a purely algorithmic back-end solution and exert a direct influence on the user in the form of information presentation or withholding" [214, p.2]. For instance, some approaches correct misinformation by displaying a link to a fact-checking website, debunking videos, or corrections within the comment Sect. [24, 49, 349], and others provide binary labels to flag false content [40]. Even after extensive research on interventions to combat misleading information and the contentious debates surrounding their effectiveness, the field has yet to come to a definitive solution. Therefore, it is vital to persist in examining users' particular requirements, abilities, and viewpoints in a user-focused manner.

Previous research suggests that transparency plays a critical role in establishing trust among users in digital interventions [299] and minimizing the likelihood of reactance or other backfire effects [377, 391]. In response to this need, some studies have started to develop and evaluate indicator-based interventions that present cues for users to evaluate information. These interventions have demonstrated promising outcomes and continue to be a topic of controversy and investigation. *Indicator-based misinformation interventions* are defined as digital countermeasures using identifiable misinformation indicators to assess if a piece of information is credible [224]. Typically, these indicators (e.g., emotional wording, hashtags including conspiracy theories) are displayed to users for immediate feedback and educational purposes, facilitating the development of skills to assess the credibility of content autonomously [224]. Often, indicator-based interventions involve highlighting indicators within the content using color-code [350].

The objective of our article is to contribute to the design of user-centered interventions based on indicators by determining which indicators are suitable in terms of perceived usefulness and comprehensibility. Addressing this research gap in HCI research, our paper offers a comprehensive exploration of qualitative and user-centered insights into an indicator-based approach. We apply this approach specifically to short-text content on Twitter, aiming to understand its applicability in real-world scenarios. As previous studies have suggested that misinformation during crises varies from everyday content, we apply our analyses and user studies to the particular long-term crisis of the COVID-19 pandemic. Additionally, we demonstrate how our approach can be applied to the Russian war of aggression against Ukraine as a distinct crisis. Our perspective on German content contributes to the predominantly U.S.-focused research on misinformation.

Our study consists of a threefold research design, comprising (1) a content analysis of Twitter data to get a sense of what misleading information in our crisis context looks like, (2) an online survey to investigate which indicators users have already been applying autonomously to assess tweets, and (3) a think-aloud study to evaluate the comprehensibility and perceived usefulness of previously identified indicators. The sequential studies offer a chance to triangulate data and obtain comprehensive findings.

We advance misinformation research by applying existing knowledge about indicator-based interventions to specific crises and preserving a user perspective of a diverse group of participants. Our core contributions (C) and findings (F) are first (C1) thoroughly assessing 2,382 German tweets within a realistic crisis context. In doing so, we identified (F1) topical indicators including context-specific themes like "Restriction of Liberty", formal indicators such as links and hashtags that occur in similar measure in credible tweets but are a valuable cue for misleading information when being nudged to take a closer look at what they refer to, and rhetorical indicators for misleading information like claim to "Absolute Truth". We then (C2) conducted an online survey to investigate how users autonomously assess the credibility of tweets. Through this survey, we (F2) gained insights into several indicators that users apply autonomously (e.g., buzzwords like "mainstream media" as topical indicator), and found that they are in particular *not* aware of rhetorical indicators of misleading information, revealing the benefit to display them in an assisting tool. Finally, we (C3) assessed the comprehensibility and usefulness of our previously identified indicators and discovered (F3) a positive perception of the overall indicator-based approach. Evaluating in particular how the topical, formal, and rhetorical indicators enhance the perception and evaluation of users when presented as a digital intervention, we gained (F4) positive feedback regarding comprehensibility and perceived usefulness for instance for topical indicators when kept rather general, emotional emojis, and capitalization, but also criticism for others (e.g., "Punctuation" being too excluding for non-native speakers). Additionally, we identified (F5) the opportunities, challenges, and limitations associated with the indicator-based approach for short-text content, such as indicators not being valid if within a quote and drawing attention to misleading content even more when highlighting specific parts.

7.2 Related Work and Research Gap

Our work contributes to the design of user-centered digital interventions to counter misleading information on Twitter/X, which is widely used by people during crises to access as well as share information. We discuss related work regarding misleading information during crises and shed light on crisis informatics as a research area (see Sect. 7.2.1). We further present different ways to assist users of social media in dealing with misleading information from an HCI perspective, including digital interventions of various forms that strongly motivated the user-centered proceeding of our work (see Sect. 7.2.2). Subsequently, we illustrate research gaps and our research questions (see Sect. 7.2.3).

7.2.1 Misleading Information in Crises

As we investigate misleading information on social media during crises like the COVID-19 pandemic and the Russian war of aggression against Ukraine, it is crucial to shed light on previous works of *crisis informatics* which have focused on different user perspectives and their motivation to create misleading content in both long- and short-term crises. While we focus on the user perspective and resulting implications for the design of digital interventions, many existing studies investigate motivations for creating misleading information which are partly founded in antisemitism (e.g., believing in an international elite controlling government) [211, 532] or an anti-globalist view [372, 532], or propagating actors and communities [356]. Other studies focus on specific conspiracy theories (e.g., 5G theory) [10] or contents of misleading narratives in English, Hindi, Bangla, or Telugu tweets during the COVID-19 pandemic [12, 510]. These contributions are especially valuable for our research goals as topics and, therefore, reflected motivations may allow us to derive comprehensible indicators for users with different backgrounds and varying degrees of vulnerability towards certain discourses and deception strategies. We conducted our study based on these findings, identifying similarities and complementing insights.

There are valid assumptions that misleading information during crises differs from misleading information in everyday life [244, 359], for instance in regard to the speed at which it is transmitted [625]. In addition, the information needs during a public health crisis differ from those of other types of crises [195]. Moreover, information, especially in the early stages of the crisis, is often incomplete and contradictory [195]. This period is particularly characterized by great uncertainty [77]. Considering worldwide spread of COVID-19-related misinformation, studies have frequently focused on English-speaking or US-centric debates [492]. For example, [151] identified themes such as "New World Order", "Wuhan Lab / Bioweapon", "Bill Gates", "5G", "Antisemitic Conspiracies", and "Plandemic" to be prominent. While reflecting several of these themes in our study as well, regional differences can be identified as our German-language data set did not comprise tweets referring to "Democratic Party Hoax" or "Dr. Anthony Fauci" [151]. Another study, concerned with the perception of legitimacy of vaccine-related disinformation in the context of the COVID-19 pandemic, is also based on data from the U.S. [129]. Connecting to works on the German COVID-19 social media debate [492, 624], it is indicated that overarching topics such as science denial or conspiracies can be seen as phenomena across the globe, with context-specific signifiers (e.g., politicians) diverging according to the national context.

Dedicated to the cooperative and collaborative use of information and communication technology in disruptive situations [404], *crisis informatics* has focused on the spread of false information in the course or aftermath of crises [79, 183, 579]. Further, Starbird, Arif, and Wilson [533] argue that information operations on social media, often led by political actors, are an important part of HCI research: such information operations "function to undermine the integrity of the information space and reduce human agency by overwhelming our capacity to make sense of information". Indeed, finding adequate ways to assist users in dealing with misleading information (in crises) is a socio-technological challenge within the HCI research community. Characteristics of misleading information identified in the previously presented studies are a central part of developing effective countermeasures. In the following, we present how HCI works and related research designs have evaluated user-centered digital interventions.

7.2.2 Assisting Users in Dealing with Misleading Information

The dissemination of misleading information represents a worldwide issue with efforts from multiple perspectives to combat its impact. Potential solutions encompass professional fact-checking, media literacy campaigns, policy implementations, and digital interventions on social media. The strategies differ depending on various factors, including geographical location, culture, socioeconomic context, and political climate [212]. Additionally, the type of assistance required can vary among individuals, for instance, based on their political ideology and age [194]. The computer science research community, including HCI, addresses the challenge of misleading information by supporting media literacy efforts, policy implementations, and professional journalism by implementing technological solutions. These may comprise of automatic detection approaches [518], especially applying machine learning techniques, but also user-centered decisions and implementations on what to do with successfully (manually or automatically) detected misleading information. In particular, post-detection digital misinformation interventions [35, 480] have been investigated taking various forms [214]. This includes corrections, for instance underneath a social media post, in the comment section, or as a link to fact-checking websites [49], visibility reduction by reducing opacity or size [24], giving (binary) labels to mark content as right or wrong [40], or highlighting specific characteristics of misleading content as cues for users [51, 224, 491].

While digital interventions vary widely; research suggests that users prefer those that come with a certain degree of transparency, allowing for informed decisions and the ability to comprehend why content is potentially misleading [299]. Giving explanations or comprehensible cues can be significant to establish trust in the intervention [224, 299] and counter feelings of reactance or related backfire effects [391] that are controversially discussed in research [610]. Indeed, a higher level of transparency can be used to enhance the media literacy and critical thinking competence of users to establish their own assessment strategies instead of forcing a total dependency on the intervention's feedback. In line with other researchers, we define media literacy as the ability to decode, evaluate, analyze, and produce media and internalize "critical autonomy" [28]. While the concept of media literacy has received some criticism for oversimplifying the complex information space [226], relying too much on rationality [62], and creating a false sense of confidence [75], research has demonstrated its beneficial effects to alter behavior for the better [530, 600] – an insight we draw on in our study.

In the context of different digital interventions to enhance media literacy (e.g., giving generic tips on how to detect misleading contents before users are even confronted with social media posts or news articles [135, 194, 205]), indicator-based interventions have gained some attention in related HCI research [224, 350]. These indicator-based approaches may be applied in two ways:

(1) Integrated into a fully automatic tool that aims to detect misleading information (e.g., using machine learning) and, in addition, automatically detects previously defined user-centered indicators on successfully detected misleading content. This allows us to make the output comprehensible and useful for laypersons.

(2) Integrated in combination with manual professional fact-checking, where fact-checkers manually label user-centered indicators in addition to a binary decision or conduct the detection of previously defined user-centered indicators automatically (e.g., via natural language processing).

Several studies have started to investigate how characteristics of misleading or credible content can be derived and utilized either for statistic-based automatic detection approaches or as a more user-centered intervention. These characteristics or indicators can refer to diverse parts of a social media post, for instance, information about the author [129] (e.g., suspicious profile, name, followers, other posts), interaction with the posts [224] (e.g., likes, comments), or the content itself [131] (e.g., rhetorical strategies, linking to questionable websites). For example, Shu et al. [518] have pointed out different topics, lexical, and syntactical features of fake news and Horne and Adali [240] evaluated how fake news articles differ from reliable content and found that fake news articles tend to be shorter, use less punctuation and more lexical redundancy, and titles can be considered a "strong differentiating factor between fake and real news" [240]. Schoenmueller, Blanchard, and Johar [494] have investigated the relations of socio-demographic factors, social media activity, and personality traits of disinformation-sharing of users with textual and linguistic features based on their post-histories. They identified distinctive characteristics such as linguistic markers, high-arousal negative emotions (e.g., anger and anxiety), religion, and power-related vocabulary.

Recent attempts to capture specific rhetorical structures and tactics have been pursued by Diaz Ruiz and Nilsson [131] who have identified how identity-driven controversies and pre-existing culture wars are rhetorically exploited by the disseminators of disinformation. Their focus is on the particular example of flat Earth conspiracies, but more fundamentally, they unravel interconnections with broader narratives of faith and religion, societal agency, anti-elitism, and anti-intellectualism, leaving a much more diverse group of social media users susceptible to such tactics. Central to this is their perception that disinformation intertwines with general beliefs [131]. [129] additionally found through their study of social media users' behavior in the context of vaccination misinformation in the U.S. that sharing behavior depends on how legitimate users consider a piece of information to be. They explain how the perception of legitimacy can be described by five dimensions (*cognitive*, *pragmatic*, *moral*, *expert*, and *algorithmic legitimacy*), which must be addressed differently in an intervention. Furthermore, they identify specific "themes" and cues through which the legitimization can be realized (e.g., "Power Dynamics," "Media-State Agenda-Setting" and "Outsiders Thinking" for cognitive legitimacy) and emphasize particularly how "expert cues" (i.e., the representation of an author as an expert in the field) increase perceived legitimacy of misinformation. Our work builds on existing studies on misinformation indicators by systematically applying the derivation and user-centered evaluation of content-specific indicators (differentiating topical, formal, and rhetorical forms) referring to a specific crisis from an HCI perspective. We thereby were strongly inspired by related work (see Table 7.1). We distinguish topical content, i.e., the meaning communicated by a post, from more formal indicators, such as punctuation, which is usually also considered as part of news content [518]. Additionally, we search for rhetorical indicators as related structures have been proposed to complement detection based on lexicosemantic analyses [470].

In addition to the deriviation of diverse indicators from text, images or videos, research has started to implement them within digital interventions, often using color-code to highlight indicators as cues for end users for immediate feedback and education [53]. Martino et al. [350] developed a system that analyzes text and automatically highlights propaganda techniques such as exaggeration or oversimplification within the text in different colors. Similarly, Schmid et al. [491] present a web app that allows users to explore and comprehend patterns of potentially misleading and non-misleading tweets based on a social network analysis, revealing the potential to encourage the development of their own assessment skills. Evaluating derived indicators from the perspective of end users (e.g., regarding their comprehensibility) is necessary to develop user-centered and effective interventions. Some studies have started to investigate within this research direction (e.g., for video content [224], for voice messages [222], or for expert annotators instead of layperson [626]).

Indeed, research on indicator-based interventions has revealed promising findings regarding trust [51, 299], utility [188], and the development of autonomous assessment skills [224, 491], but also revealed challenges and limitations (e.g., not being transparent enough or having a biased design [188]). We build on those findings by complementing them with a crisis-related perspective and a specific focus on social media users as laypersons.

The majority of existing research on indicators has focused on social media content in everyday life instead of crises. However, specific conditions such as indicators or user requirements, may vary during a crisis [428]. These conditions are essential for technology-based user assistance in both automatic detection and subsequent countermeasures. van der Meer and Jin [573] investigate the efficacy of correcting misleading information as a countermeasure against its spread specifically in crisis contexts. The authors posit that such measures may prove even more effective during crises, as the dissemination of misleading information concerns new information only, and therefore would not require individuals to alter their attitudes, identity, or ideology, but only their beliefs regarding the current crisis situation. It is uncertain whether this assumption is applicable to a crisis such as the COVID-19 pandemic, as a long-term and disruptive event. Furthermore, Plotnick et al. [428] conducted an online survey to assess general user behavior and attitudes towards fake news. The analysis results indicate that users consider grammar and the sender's trustworthiness as key indicators of trustworthy content. Additionally, the majority of participants expressed a desire for a color-coded graphic to aid them in evaluating the trustworthiness of information on social media. Although the authors did not specifically concentrate on crises, they underscore the significance of evaluating the credibility of social media content during such situations. For example, they propose examining whether grammar can still effectively indicate untrustworthy content during a crisis when content creation time is limited. These findings drive our research and seeks to understand user perspectives on comprehensible and useful indicators as potential interventions during crises, which represent exceptional and extraordinary circumstances for misleading information.

7.2.3 Research Gaps and Research Questions

Our work contributes to existing literature on HCI regarding misleading information during crises by shedding light on the user perspective, filling the following two gaps:

1st gap: Prior studies have investigated misleading information with respect to the COVID-19 pandemic, yet clearly focused on the U.S.. We offer a vantage point for research on transnational discourses of misleading information, revealing regional specifics or patterns of German tweets that are similar to U.S. conspiracist narratives [12, 510, e.g.,] on COVID-19. We differentiate between topical, formal, and rhetorical indicators. This also allows us to reveal relationships between these three types of characteristics and accompany studies that have been interested in social media behavior with regards to different topics.

2nd gap: Studies have already investigated features of misleading contents in terms of applicability for automatic detection [625], the motivations for the creation of misleading information [531, 532], and a user perspective on (digital) misinformation interventions [25, 240, 350, 423]. However, the user perspective regarding countermeasures is still to be addressed even more extensively [195, 299] regarding indicator-based approaches to encourage media literacy. By identifying user-centered, i.e., comprehensible and useful content-specific indicators for misleading tweets, we contribute to existing research [280, 299, 365, 433] that focuses on the subsequent countermeasures after (automatic or manual) detection. Recently, researchers have stressed the importance of designing user-centered measures that counter reactance [377, 391] and follow the needs and preferences of end users for comprehensible and transparent information [299]. Building on other approaches for user assistance ([52, 168, 221]), we follow up on those findings and complement them with a focus on specific crises. As the time at the beginning of a crisis is particularly relevant for the investigation of misleading information due to high uncertainty [77], we focus on this period in our study when thoroughly analyzing tweets. Instead of working towards an explanation of AI models, we are specifically interested in comprehensible content-specific indicators, independently from the preceding detection method.

We are convinced that approaches governing users' online activities need to offer the necessary ground for reflection or learning and that research on assistance for users in dealing with misleading information online during a crisis may help to achieve this. Thus, our overarching goal is to answer the following research questions:

RQ1: *What are topical, formal, and rhetorical characteristics of misleading tweets in crises, with the COVID-19 pandemic serving as an example?*
RQ2: *Which indicators are used autonomously to assess the credibility of tweets?*
RQ3: *What comprehensible and useful content-specific indicators can be derived with regard to the design of digital artifacts assisting users on social media in dealing with misleading tweets during crises?*

We present our contribution within a threefold study design. First, we perform a content analysis of Twitter data to get a sense of what misleading information in the context of crises such as the COVID-19 pandemic looks like and to identify content-specific characteristics. Second, we conduct an online survey to qualitatively assess which indicators users already apply autonomously to estimate the credibility of tweets. Third, our think-aloud study gains qualitative insights into how comprehensible and useful people from different backgrounds perceive the previously identified indicators to be, extending our approach to the Russian war of aggression against Ukraine as an additional recent crisis event.

7.3 Methodology

In this paper, we employed a threefold study design to provide the foundation for future digital misinformation interventions after (manual or automatic) pre-filtering. Our primary goal was to derive user-centered indicators to help users evaluate the credibility of tweets, specifically regarding misinformation during crises like the COVID-19 pandemic or the Russian war of aggression against Ukraine. The study did not develop any intervention but aimed to identify and evaluate potential indicators for integration into subsequent interventions. The threefold methodology is motivated by related work [474] that emphasizes using social media data as a source of information to provide an extension to traditional methods such as the think-aloud method.

Step 1 of the study entailed a quantitative Twitter analysis. Here, we established a thorough knowledge base to "get a sense of what [misleading information] looks like in the context that [we] are interested in" [421]. Tweets concerning COVID-19 were gathered and examined by researchers, who distinguished between misleading and non-misleading content. A content analysis was carried out, identifying several characteristics of misleading tweets, which were sorted into topical, rhetorical, and formal characteristics. These characteristics could be potential indicators assisting users in assessing the credibility of tweets. However, to evaluate this, the subsequent user-centered steps are still necessary.

Moving on to *Step 2*, the researchers aimed to discover the characteristics that Twitter users utilize when assessing the credibility of tweets autonomously. They conducted a qualitative online survey, where participants rated the credibility of a set of tweets and were asked to provide explanations for the characteristics or indicators used for their credibility assessments. This phase provided researchers with insights into the indicators that individuals autonomously use when evaluating the credibility of tweets and might be particularly beneficial for users with lower media literacy (e.g., due to young age) to be displayed as cues.

In *Step 3*, the study integrated the findings from Steps 1 and 2. The identified characteristics of misleading tweets from Step 1, both on the topical, rhetorical, and formal levels, were combined with the indicators autonomously used by participants in Step 2. In this final phase, we examined in a think-aloud study how users perceived the usefulness and comprehensibility of these indicators when presented alongside actual tweets regarding multiple current crises. This encompasses indicators that can be considered rather definitive for credibility assessment and others that are more subtle but still provide a fuller picture of creators' intentions or beliefs, and overall encourage critical reflection. In the following, we present Twitter as our case selection and a detailed description of the methodologies in Steps 1 to 3 (for a graphical representation of our threefold study design see Fig. 7.1).

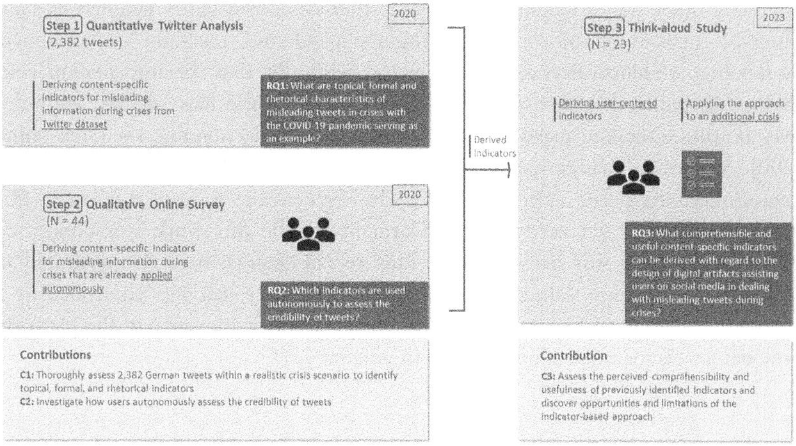

Fig. 7.1 Graphical representation of our threefold study design comprising of (1) a quantitative Twitter analysis, (2) a qualitative online survey, and (3) a think-aloud study

7.3.1 Case Selection

Twitter (now X) as an important social media platform [145] allows fast information dissemination across national borders. Assuming the network constitutes a sphere of public discourse, comprising users with different levels of educational and political backgrounds, we decided to select Twitter during crises as our research case. We addressed this case in three different case analyses that build on each other (see Sect. 7.3 and Fig. 7.1). We conducted an analysis of tweets regarding the COVID-19 crisis in 2020. Focusing on German tweets may reveal important insights as both conspiracy theories and vivid debates by civil society have a long history. It further allows for the derivation of indicators in the German language. To gain additional insights into the transferability to other recent crises, we extended our stimuli to the Russian war of aggression against Ukraine in Step 3. While the COVID-19 pandemic as a health-related crisis primarily revolves around public health and safety, tweets regarding the Russian war against Ukraine are fundamentally of a more security-related nature. Despite their differences, both crises share commonalities in their political relevance, the occurrence of common conspiracy theories, the exploitation of the insecure situation by manipulative groups, and their high potential to influence public opinion.

As many other social media platforms, Twitter has consistently adapted its handling of misleading information during the pandemic, reaching from removal or labeling of harmful contents to generic warnings that the information conflicts with public health experts' guidance [106]. While less specific warnings may be time-efficient, users could have problems understanding their reasoning [200]. Recently, Twitter has started the pilot project "Birdwatch" in the USA as a community-based approach [106]. However, at the time of this study, this approach was not yet accessible in Germany. While this work was conducted when the platform was still named Twitter, recent changes have demonstrated an increasing interaction with disinformation sources after accounts controlled by a state are no longer marked as such. This shows a wider relevance of the platform now named X for misleading content in general [475].

7.3.2 Step 1: Quantitative Twitter Study—Deriving Content-Specific Indicators for Misleading Tweets During Crises

Step 1 aims to contribute to the knowledge base on characteristics of misleading tweets during the COVID-19 pandemic as a specific crisis. Thus, we derived characteristics on different levels as potential indicators to assist users in assessing tweets' credibility.

7.3.2.1 Data Collection

To create our sample, we used a Twitter API, scraping a wide range of COVID-19-related tweets using the query *"Coronavirus OR Corona OR Covid OR Coronakrise OR coronadeutschland OR Corona19 OR covid19 OR covid-19 OR #COVID-19"*. We initially collected a total of about 360,000 tweets. The first data collection phase on Twitter took part during a very early stage of the pandemic (27 February 2020 to 6 March 2020) while the second data collection phase on Twitter took part from 14 April 2020 to 22 April 2020. Public life in Germany was still quite ordinary during the first observation period, and a strong focus of reporting was on foreign countries such as China and Italy. Only after the first observation period were contagions detected in all German states, and Chancellor Merkel announced the first Germany-wide lockdown. During the second observation period, more and more COVID-19 cases were recorded in Germany. Germany was still in the first lockdown, and public life had reached a standstill. However, a slow return to public life had already been introduced. The debate about compulsory vaccination was present in the public discourse.

7.3.2.2 Data Processing

To extract a meaningful number of tweets for subsequent examination for indicators, we used Python to process the tweets as follows. (1) We excluded all non-German tweets and eliminated duplicated content like retweets. (2) We generated a random sample to include all potential kinds of misleading information, independent of topic and motivation. (3) We ensured that a sufficient number of misleading tweets was included during a group of researchers' initial review of the tweets. (4) A thorough introduction and a background in political science qualified the researchers to label the tweets with the corresponding codes. A detailed description of the individual steps can be found in the Electronic Supplementary Material (ESM) (see Section A.2.1). To gain a balanced proportion, we randomly excluded some non-misleading tweets from the analysis as they were overrepresented after the final coding. Hence, our final sample for analysis consisted of 2,382 tweets in total, including 50% misleading information such as false information or misleading satire, and 50% tweets that were marked as not misleading. This entails tweets that were labeled as true information, opinions that are not absolute, or satire that is not misleading in the COVID-19 context, both for the sample of the early stage and the second stage of the crisis.

7.3.2.3 Analysis

After extracting a reasonable amount of tweets, we conducted a manual content analysis to thoroughly derive characteristics of misleading in comparison to non-misleading tweets. The content analysis [161] consisted of a manual coding process using *RQDA*. We chose to perform the analysis manually instead of computationally, to not overlook any previously unexpected characteristics and allow for diving deep into the data. A training phase initiated the coding process to obtain a common understanding of the codes. Code categories, codes, and subcodes were developed both deductively and inductively, taking into consideration both theory and the empirical material (see Coding Scheme in Table 7.1). It was an iterative process, enabled by a research group of three. Three researchers with expertise in political science and misleading information, as well as HCI, labeled the tweets according to a pre-established coding scheme by one researcher. Each tweet was at least tagged as misleading or non-misleading. As we are interested in identifying potential indicators for misleading tweets, many non-misleading tweets were not assigned to any of the codes reflecting topics.

We were looking for immediately visible and potentially comprehensible indicators for misleading tweets and thereby differentiated between topical, formal, and rhetorical characteristics of tweets as potential indicators, inspired by related work (see Coding Scheme in Table 7.1 and, for examples, see Table A.3, ESM). With topical indicators, we aim to capture topics that are recurring in misleading posts. Thus, we want to gain insight into what message posts communicate and differentiate lexical or syntactical indicators, such as punctuation, from a post's meaning. We also identify related formal indicators, which may also represent system functions, such as external links or hashtags [518]. Codes capturing these formal and more general content indicators were inspired by the literature on the detection of misleading information [88] and related contemporary ideological narratives [211, 532]. In contrast to topical and rhetorical characteristics and the majority of formal characteristics, the mere existence of hashtags or links is not an indicator of misleading information. Instead, users can conclude a closer look at the hashtags or links contained, e.g., whether they lead to conspiracy theory websites. Specific topical indicators (e.g., "Laboratory Wuhan") were based on the exploration of tweets as were the majority of codes capturing rhetorical indicators of tweets. Thus, regarding the latter, we did not dive into rhetorical structure theory [470] or intend to expose the existence of more abstract underlying rhetorical tactics (e.g., the construction of *echo chambers* [131]). Instead, our focus lies on more basic stylistic means, (e.g., "Exaggeration", "Sarcasm"), which could be identified in the empirical data. We thereby implicitly agree with the assumption that the dissemination of disinformation can operate through underlying rhetorical tactics, but instead focus on identifying the specific stylistic and rhetorical devices employed in the process.

Coders were free to add codes to the existing coding scheme whenever necessary. In that case, the information on the new code was shared with the other coders. In all other cases, after the training phase, the three coders worked independently to achieve a sufficient level of objectivity. Subsequent analyses were based on majority decisions, determining the codings. Intercoder-reliability was tested and reached a value of $\alpha > .7$.

7.3.3 Step 2: Qualitative Online Survey—Autonomous Assessment Strategies of Tweets During Crises

Step 2 extends the discovery of characteristics of misleading tweets during the quantitative Twitter analysis of Step 1 using a user-centered approach. This phase enables us to gain insights into the indicators that individuals already autonomously utilize to evaluate tweet credibility. We created twelve tweets (see Figure A.5; ESM) inspired by real-world examples. Tweet contents such as text and referenced images, links or videos were maintained. However, the original real-world profile names and pictures were changed to include potentially interesting characteristics (e.g., tweets by an unknown fictional doctor) and avoid biases towards well-known personalities or the exposure of real-world Twitter users.

7.3.3.1 Participants

We asked 49 participants of different socio-demographic backgrounds whether they thought statements to be true or false as well as, openly, how they came to their respective conclusions. This was conducted in an online survey format (*SoSci Survey*), with adult participants getting an expense allowance of €25 each. For the analysis, we excluded participants who withdrew prematurely from the study, resulting in $N = 44$ valid data sets. The sample size is in accordance with common user studies in HCI research [81]. Participants were acquired mainly through a university website and Facebook groups. Of the participants, 38% were male; 62% were female. The age was collected in clusters with most participants being between 20 and 29 years old (38%) or older than 54 years (33%). Regarding the highest formal education, five participants had completed vocational training ("Lehre"), two had a (general) certificate of secondary education, four had a vocational diploma, nine had A levels, and 17 had a university diploma. Seven participants made no entry for education. Our goal was to receive a diverse qualitative input, effects of socio-demographic differences were not evaluated. The questionnaire was part of a longer study spanning over seven days, with the fourth day focusing on the issue of online misleading information regarding the COVID-19 pandemic. Our study design did not explicitly introduce the context of misleading information beforehand to minimize a framing effect. The survey answers were collected in open text format.

7.3.3.2 Analysis

We used *RQDA* for a qualitative content analysis of the open-text questions. Consensus coding [606] with discussion to resolve disagreements was conducted by two researchers trained in social sciences, referring to a coding scheme similar to the one of the first quantitative study (for an excerpt of survey answers and their according codes see Table A.5, ESM).

7.3.4 Step 3: Think-aloud Study—Perceived Comprehensibility and Usefulness of Content-specific Indicators

Building on the derived potential indicators during the content analysis in Step 1 and the additional insights on autonomously applied indicators during the online survey in Step 2, we investigated if and how the derived indicators are perceived as comprehensible and useful when assessing tweets. Thus, in Step 3, we get a thorough user perspective on identified indicators as central characteristics to be applicable within a user-centered digital misinformation intervention. Comprehensibility and usefulness are central concepts when gaining first insights into the perceived effectiveness of an intervention in contrast to evaluating the objective and quantitatively measurable effectiveness, for instance, in large-scale field studies or experiments. Comprehensibility refers to the extent to which our proposed indicators are easily understandable in general and in regard to a specific tweet to assess its credibility. On the other hand, perceived usefulness refers to the extent to which our proposed indicators are helpful in achieving the intended goal of enabling users to assess the credibility of a tweet.

7.3.4.1 Stimuli

We included a total of nine tweets as stimuli in our think-aloud study. This includes one tweet with completely correct information (tweet 4) and one tweet, that although having the potential to cause harm, has been professionally checked and consists of correct information as well (i.e., the picture and names mentioned are correct, however the picture is cropped). All remaining seven tweets were officially identified as misleading. For all nine tweets, we have resorted to official fact-checking websites that have performed professional fact-checking on the content mentioned. To allow for comparison with Step 1 and 2, we focused on recent tweets concerning COVID-19 and added three tweets concerning the Russian-Ukranian war as another current crisis. Inspired by the guidelines by Pennycook et al. [421] for selecting misleading and non-misleading tweets we visited official fact-checking websites to first identify suitable topics that are not completely likely to be familiar to participants. We further made sure to select content that had not already been outdated when conducting the study. After identifying recent and relevant topics on the fact-checking website, we searched Twitter for current tweets referring to COVID-19 from both a favourable and critical perspective on official COVID-19 measures, as well as tweets referring to the war from both a pro-Russian and pro-Ukrainian perspective. To avoid biases towards well-known personalities and to focus solely on the tweet content which is our special interest in this study, we blurred profile pictures and profile names of the authors as well as statistics (i.e., number of likes, shares, and retweets). However, when a tweet content included a name or profile picture, we kept the original display as we consider this part of the content itself. Information on the posting date was kept visible as well (see all tweets in Fig. 7.2).

Fig. 7.2 Tweets that were used as stimuli during the think-aloud study. Tweet 4 contains accurate information that has been officially fact-checked by *Correctiv* (https://correctiv.org/faktencheck/2022/06/03/br-tauschte-irrefuehrendes-vorschaubild-in-video-beitrag-ueber-corona-demos-aus/). While tweet 6 has the potential to cause harm, official fact-checking found that the picture and mentioned names are correct (https://correctiv.org/faktencheck/2023/01/04/ja-auf-diesem-foto-steht-ein-ukrainischer-general-vor-einem-stepan-bandera-gemaelde/). All other tweets contain misleading information that was officially disproved

7.3.4.2 Procedure

The study was conducted online using Zoom, where sessions were recorded separately. See Section A.2.2 in the ESM for a detailed description of the study procedure including all items. At first, we assessed demographic information and political attitudes in a questionnaire. We then presented nine real-world tweets. Participants were instructed to keep thinking aloud while performing the following tasks and were told that there was no wrong or right, we were only interested in their individual assessment [163]. We asked the 23 participants about their familiarity with the tweet, its perceived importance, and its perceived credibility. For the two tweets containing reliable information, we asked the participants to imagine that a reliable fact checker (human, organization or algorithm) found the tweet's content to be correct. In contrast, for all seven tweets containing misleading information, we asked the participants to imagine that the reliable fact checker found the tweet to be misleading and therefore, in addition, an algorithm looked for indicators as guidance on evaluating misleading tweets. We then presented a list of indicators next to the given tweet and asked to elaborate on the perceived comprehensibility and usefulness of each indicator. For this, we highlighted the indicators and referred parts of the tweet with color (see Fig. 7.3). Afterwards, participants were asked again to rate the tweet's credibility before the researcher informed participants about the actual truthfulness of the tweet. After all nine tweets were addressed, we asked the participants to summarize their overall impression of the comprehensibility and usefulness of the approach of highlighting indicators for misleading tweets and if it made a difference if decisions on tweets' credibility were performed by an algorithm or by a fact-checking human/organization. We did (not) explicitly introduce the topic of misleading information to avoid biases at the beginning of the study. However, the context of the study became clear very early on.

Fig. 7.3 Example of the stimulus for tweet 9 and the indicator 'negative expressions'. The textbox on the right was translated from German to English

7.3.4.3 Participants

A total of 23 participants took part in our think-aloud study, ranging from 20 to 66 years (*Median* = 32) and covering a diverse set of educational levels (six participants with basic education, ten participants with medium level of education and seven with high level of education). The sample size is in accordance with common user studies in HCI research [81]. We explicitly decided against recruiting solely college students as these are already overrepresented in misinformation studies due to their easy accessibility, and they do not represent the average user. Of the participants, 14 were male, eight were female, and one was of diverse gender. All participants, except for two, stated German to be their native language. To gain information on the social media usage and expertise of our participants, we asked how often they use social media. The majority stated using social media daily ($N = 20$) or several times a week ($N = 3$). More specifically, six stated using Twitter daily, eleven several times a week or weekly, three once or two times a month and six never use Twitter. As our stimuli address misleading and reliable information on the topics of COVID-19 and the Russian war against Ukraine, we further evaluated central general attitudes towards these topics and the general political orientation of our participants. Of the participants, 20 fully or partially agree with COVID-19 measures being useful overall, while three tend to disagree. All participants fully or partially agree with the statement that the coronavirus exists. All participants, despite one, fully agree with Russia waging a war of aggression against Ukraine. The single participant stated to fully disagree. We asked our participants to place themselves on a scale from 1 (extreme left) to 11 (extreme right) regarding politics. Six participants placed themselves on the far left between 1 and 2, nine rather left between 3 and 4, eight in the middle between 5 and 7, and none further right than this. Participants were acquired through the panel provider *Prolific* and got an expense allowance of €12 for an average duration of 60 minutes.

7.3.4.4 Analysis

For data analysis, we automatically transcribed speech to text using *whisper*[1] and manually checked for better quality. We then performed a content analysis with two researchers regarding the dimensions of perceived comprehensibility and usefulness of each indicator as well as more general findings regarding the perceived comprehensibility and usefulness of the whole indicator-based approach. Thus, Step 3 results in a thorough overview with which previously derived indicators are perceived as useful and comprehensible and can be considered suitable to be integrated into a future indicator-based misinformation intervention.

7.4 Results

In the following, we present the results of our threefold study. We start by reporting the identified characteristics of misleading information disseminated on Twitter in the course of the COVID-19 pandemic (RQ1) which we assessed in the quantitative content analysis of tweets as Step 1. Subsequently, we present our findings on which indicators are already being used autonomously by users to assess tweet credibility (Step 2: qualitative online survey) to answer RQ2. Finally, we report how users perceive the comprehensibility and usefulness of identified indicators (Step 3: think-aloud study) to answer RQ3.

[1] https://openai.com/research/whisper

7.4.1 Deriving Content-specific Indicators for Misleading Tweets During Crises

In Step 1, we systematically derived characteristics of misleading tweets in comparison to non-misleading tweets on the topical, formal, and rhetorical levels to create a knowledge base on what misleading information during a specific crisis looks like. Our sample consisted of misleading tweets (N=1,191), which contain claims of holding true information while they are inaccurate. Some of these tweets may have been formulated for satirical reasons, yet they may mislead perceiving users in case they do not clearly reflect satire while referring to virus-related issues. For example, some tweets indicated the healing potential of alcohol: *"In Russia, nobody has come down with COVID-19. So, don't prep yourself with any cans of sausages, of fish, any ready-to-serve meals or water. Buy a lot of vodka!"* (#0046) Because people have believed similar statements with regard to the origin, effect, and options of healing of COVID-19, we categorized such tweets as misleading. The second group of tweets (N=1,077) consists of accurate information, own and thus non-falsifiable experiences or feelings as well as satirical tweets which do not suggest misleading users. Satirical posts (N=114) like *"To be fair, there is nothing more frightening than German bureaucracy. I am not sure whether the #coronavirus wants to deal with it. Viruses also have standards."* (#0779) are assumed to be not misleading as they, again, personalize the COVID-19 virus and put it in an unrelated semantic field (i.e., bureaucracy) to criticize politics.

In the following, we present topical, formal as well as rhetorical indicators of both misleading and non-misleading tweets related to the COVID-19 crisis in Germany (see Table 7.1 for an overview of all indicators).

Table 7.1 The coding scheme is oriented by related scholarly contributions, as indicated by references, as well as tweets (*). Especially topical indicators reflecting instances of relatively recent conspiracy theories were derived directly from the material, as were the majority of codes capturing rhetorical indicators. While respective codes may also be derived from literature, we became attentive to them by looking at the data. Prior literature research helped to categorize the codes and structure them hierarchically

Level 1	Level 2	Level 3
Formal indicators	Emoticons [221, 593] Capitalization [221, 593] Punctuation Marks [221, 593] Hashtags [88, 375] Links [88] Incorrect Spelling Grammar [375, 428]	Sad Hopeful Emoticons [203, 574] Aggressive Emoticons [203, 574] Exclamation Marks [221] Question Marks [221] Incorrect Use Marks [221, 375] Unusual Quotation*
Topical indicators	Media Criticism [533] Restriction of Liberty* 5G [10] Bio Weapons* Laboratory Wuhan* Gates* Real Simulation* Science Denialism [211] Higher Powers [211] Associated Financial Market [540] Islamophobia [376] Antisemitism [211, 532] Traditional Conspiracy	Strong Denial Vaccination [211] Questioning Mandatory Vaccination [211] Homeopathy Influenza* Satan*, God [211] Human Nature* Exclusive Elites [87] Free Mason [532] Prepper AntiGlobalism [532] Nazism* Sovereignty Germany*
Rhetorical indicators	Negative Phrases [486] Rhetorical Question [246] Irony* Satire Sarcasm* Metaphor Neologism* Exaggeration [486] Relativization* Phraseme* Absolute Truth*	
Misleading Non-misleading		

7.4.1.1 Topical Indicators for Misleading Tweets

Regarding the first group of topical indicators, the content analysis revealed that codes of the category of "Traditional Conspiracies" like "Free Mason", "Nazism", "Prepper AntiGlobalism", "Sovereignty Germany" appeared very rarely while codes referring to new, more context-specific themes (e.g., "Restriction of Liberty" or "Gates") were much more prevalent (see Figure A.2, ESM). At the same time, misleading tweets were still characterized by some traditional conspiracist topics like the "Exclusive Elite". For example, one user made clear that

> *"[i]t is about something else, but you will soon learn about it; it is not about the coronavirus. Soon, the hospitals will be overcrowded with children, who will be freed from the underground, where they have been living in underground bunkers for years. Who of you knows about adrenochrome, just google it."* (#1331)

Attesting confidence about having complete, yet exclusive, knowledge of the world is a conventional feature of conspiracy theories, which, in this instance, was put into a contemporary context, referring to adrenochrome (a metabolite which plays an important role for the QAnon conspiracy theory). Related statements may also carry antisemitic content: *"There are some people who have strong doubts about the official version of China regarding the coronavirus (...). There are theories and assumptions that, for example, Soros has something to do with the virus, which would perhaps be very interesting."* (#0271). Further, new trending themes like "Gates" or comparing COVID-19 to conventional influenza regarding its effects (cf. "Homeopathy Influenza") may follow more established science denialist, anti-vaccination narratives:

> *"This is a very big #put-on of the German people! Whether "refugees", whether #corona, whether the new app and the compulsory vaccination with a chip, everything is a plan by criminals like the #WHO #BillGates #RKI #Merkel #Spahn #Rothschild #Rockefeller and their organizations!"* (#1219)

7.4.1.2 Formal Indicators of Misleading Tweets

Regarding formal indicators, we observed a relatively balanced distribution across types of tweets (see Figure A.3, ESM). Links were used in 1,183 tweets, of which 519 were misleading. Although links may make a tweet seem more credible, this may not always be the case. Thus, taking a closer look at the referenced link can lead to a suitable indicator for misleading content (e.g., a link to a conspiracy website). The same applies to hashtags, which on their own are not an indicator of misleading information. Again, however, a nudge to look more closely can lead to a helpful indicator (e.g., hashtag naming a conspiracy theory). Interested in cross-media use and the types of media shared by embedded URLs in tweets, it was revealed that links in misleading tweets were most frequently referring to YouTube videos and images (see Table A.4, ESM). While we did not perform statistical analysis on that particular topic, our findings reveal some first impressions. For instance, 2.56% of non-misleading tweets contained an embedded link to a YouTube video; for misleading tweets, that was the case in 21.58% – such a large difference cannot be seen in any other type of embedded links. Well-known online news sites (Focus, N-TV) were shared to a similar extent across non-misleading and misleading tweets. This may be due to the rather superficial, short-term nature of fast online journalism. Additionally, smaller, less professionally held blogs were used as references. While we derived the share of links that related to URLs containing the word "blog," we did not list the various, relatively unestablished homepages that may account for the greatest share of links in misleading tweets. At the same time, "traditional" conspiracist or misleading websites (Russia Today, Sputnik, Anonymous, pi-news) were not used as much as expected in misleading, but never in non-misleading tweets. Formal indicators common for crisis situations, like capitalization of words or incorrect spelling, were comparatively more present in misleading tweets. Yet, and potentially as the COVID-19 pandemic is no short-term event fueling overwhelmingly highly emotional tweets, misleading tweets were rarely characterized by such formal elements. Looking at misleading tweets, the use of links was comparatively most frequent, suggesting users' interest in legitimization or increased spread of content being important. All formal indicators indicating incorrect language use were more indicative of misleading content than non-misleading, with the use of unusual quotations being solely present in misleading tweets. Thereby, users questioned the existence of the *"virus"* (#1795) or related *"deaths"* (#0024).

7.4.1.3 Rhetorical Indicators for Misleading Tweets

The third group of indicators comprised rhetorical devices that are applied commonly in the context of the creation of misleading information (see Figure A.4, ESM). With respect to negative, absolute language that may intensify conversational interactions, we made out two recurring themes of swear words. First, terms relating to fascism, dictatorship, or the Weimar Republic were used to target the current political system and incumbents (#1398, #1720). Thus, one user notes that *"[t]he Merkel Regime is finally dropping its masks and turning out as an axis-mirrored fascist dictatorship of "Mini-fascists" End the masquerade [sic]! #corona #FRG2020."* (#1668)

Second, stigmatizing political measures and (scientific) public discourse on the COVID-19 pandemic was performed by pathologization. For example, *"hysteria of speculators"* (#1383) was defined as the reason for the outbreak of COVID-19 while support for political measures was achieved because of irrational fears or (creation of) panic (#1749). In sum, 7.81% of misleading tweets contained the words "panic", "hyster*", "insane", or "delusional". Such discursive undertakings were also reflected by the constructions of neologisms like *"Merkill"* (#1170), *"schizophrenic framing-channel"* (#0028), or *"Coronavirus-Hysteria"* (#0108). While "Sarcasm", "Irony", and "Rhetorical Questions" were also, partly predominantly, used in non-misleading tweets, all other rhetorical indicators could most frequently be studied by looking at misleading tweets.

7.4.2 Autonomous Assessment Strategies of Tweets During Crises

To shed more light on the user perspective, the aim of our online survey (OS) in Step 2 was to investigate which content-specific indicators people have already been using autonomously for assessment of tweets, assuming that these are indicators from which they might benefit when highlighted in an assisting tool. We see this step as a complement of the identified characteristics from our quantitative Twitter analysis which allows us to potentially add additional indicators, conform others, or identify particular relevance. While some indicators were autonomously identified by the participants, others, such as specific rhetorical indicators, were not. Adding those indicators to a user-centered assisting tool might have the potential to help users expand their knowledge and encourage media literacy through offering an explanation of "tendentious" (OS02:17) content. In the following, we address topical indicators, formal indicators, and rhetorical indicators successively.

Regarding topical indicators, we can observe that "Science Denialism" indeed played a significant role in our participants identifying tweets as misleading. The absence of scientific references when stating alleged scientific information was emphasized (see Fig. 7.4). The participants were often *"skeptic in general"* (OS02:01) when vaccination statements were combined with COVID-19 topics, especially when scientific references were missing. As instances of "Strong Denial of Vaccination" were also exclusively prevalent in misleading tweets of our second period of observation, it is indicated that users may identify misleading information of a potentially more deliberate nature more easily. In our large sample of tweets for the quantitative study, many were tagged as containing explicit media criticism, referring to the *"mainstream media"*. To our participants, this buzzword constituted a well-known indicator of problematic content (OS13:02). However, some participants assessed tweets by critical journalism as misleading information as well, solely based on the occurrence of the word *"media"* (OS02:01; OS02:03; OS13:04). Hence, further investigations on how to help differentiate between media "criticism" in misleading information and media criticism that is a form of critical journalism may be a valid effort.

While we observed a strikingly large co-occurrence of different conspiracy theories in the same tweets (e.g., "Gates" and "Science Denialism"), participants in our survey did not explicitly reflect on specific conspiracy theories or their combined appearance. Instead, they rather mentioned the dominance of conspiracy theories in general (OS02:05; OS02:06; OS02:07). This suggests that some users may find a more general indication of the "Conspiracy Theory" category satisfactory while highlighting more well-known conspiracist vocabulary or strong statements may also increase comprehensibility. At the same time, evaluation of the answers to the online survey also indicates that it is not necessary to overload users with information on relationships between conspiracist topics and narratives.

When looking at the formal indicators, our quantitative results revealed that, although a relevant system feature, "Links" were not a distinguishing characteristic of misleading content. Participants also considered links in their assessment and found links to traditional news to be a comprehensible indicator of non-misleading content (OS03:07). During our survey, it became clear that the content was decisive. Participants referred to *"#DeepState"* or *"#Simulation Game"* (OS02:08; OS13:09) as an indicator. Thus, respondents stressed their problematic content while also emphasizing that hashtags may be relatively easily identifiable as relevant formal indicators. Some survey participants considered a lack of punctuation or excessive punctuation (e.g., "...") indicative of misleading information (OS02:10; OS02:11; OS02:12; OS02:13). This is consistent with results of the Twitter analysis, which show the prevalence of incorrect spelling or grammar in misleading tweets, and backed by prior research [428].

Regarding rhetorical indicators, while they appear to be significant indicators for misleading information, respondents did not make many statements. Individuals named provoking language and contentious terms as indicators or declared that they *"assess this statement as extremely exaggerated"* (OS03:14). However, they did not evaluate tweets as misleading based on the occurrence of absolute phrases (i.e., "Absolute Truth") or "Relativization". Considering the online survey, it appears difficult to transparently evaluate tweets as misleading, referring to these rhetorical indicators. "Exaggeration", "Absolute Truth", "Relativization", and "Negative Phrases" (including swear words) were valid indicators for misleading tweets in our quantitative analysis. Hence, interested in comprehensible user assistance, future HCI research may benefit from looking deeper into how these rhetorical indicators can be used as indicators for end users, providing a learning effect regarding the identification of misleading information.

Summarizing the findings of our online survey, the strong indication of sources as indicators implies that users may not want to invest too much time into assessing content and rather prefer to quickly evaluate based on obvious indicators. At the same time, regarding each tweet, roughly one-fifth of participants evaluated the tweet incorrectly (see Figure A.6, ESM). These participants, in particular, might benefit from our user-centered approach of giving comprehensible content-specific indicators to encourage media literacy instead of labeling tweets as misleading without explanation. We use the insights gained here on indicators already applied autonomously and not yet used consciously as input for our final Step 3.

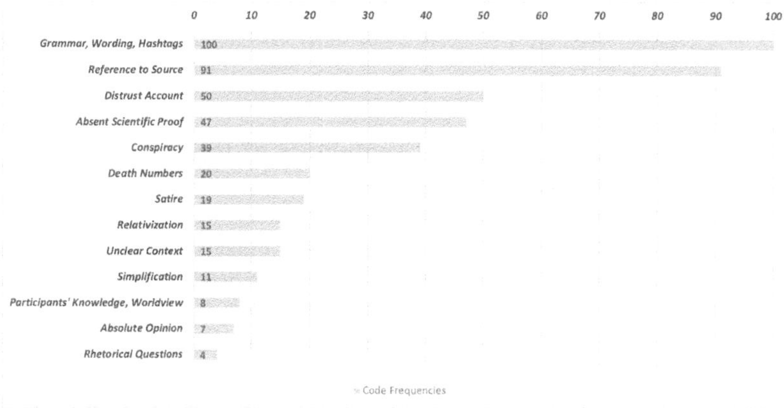

Fig. 7.4 Overview of code frequencies in study 2. For participants, tendentious phrasing was mostly an indicator of misleading information, followed by an absence of references supporting the respective statements or critical opinions regarding sources. Only one participant was convinced by a tweet because the reference suggested proof of accurate information. More specific stylistic or rhetorical indicators were less often relied upon when assessing tweets

7.4.3 Perceived Comprehensibility and Usefulness of Content-specific Indicators

Building on our findings in the two previous steps, we made sure to include tweets containing the following formal indicators: "Emoticons", "Capitalization", "Punctuation Marks", and "Incorrect Spelling Grammar". We decided to include "Hashtags" and "Links" as formal indicators as well, as they were specifically mentioned as helpful in Step 2. For rhetorical indicators, we included "Negative Phrases", "Rhetorical Question", "Exaggeration", "Relativization", and "Absolute Truth". We decided to exclude "Irony", "Satire", "Sarcasm", and "Phraseme" as they did not appear significantly more often in misleading tweets and were not mentioned by participants in Step 2. "Metaphor Neologism" was excluded as it appeared only 14 times in sum within our comprehensive sample in Step 1 and could not be identified for any additional tweets for Step 3. The assessment of rhetorical indicators in Step 3 is of special interest, as they were barely used by participants in Step 2 while being highly relevant. For topical indicators, Step 1 revealed an overabundance of all indicators in misleading tweets. We exemplarily included "Restriction of Liberty" and "Science Denialism" as topical indicators. See our final selection of stimuli in Fig. 7.2 and an example of a stimulus with guidance in Fig. 7.3.

7.4.3.1 Comprehensibility and usefulness of topical indicators

The comprehensibility of the included topical indicators "Restriction of Liberty" and "Science Denialism" was overall very high. For instance, in tweet 1, 20 participants were fully able to explain correctly what "Restriction of Liberty" meant in that particular tweet and in general, indicating a high level of comprehensibility. Regarding the perceived usefulness of the indicators to assess a tweet‘s credibility, the results were also very positive, but with a few drawbacks. For example in tweet 8, 16 participants (fully) agreed with "Restriction of Liberty" being useful, while three participants completely or partly disagreed. Reasons for participants disagreeing with the usefulness of "Restriction of Liberty" was having an inappropriate naming, not being specific enough, or referring to an unsuitable part of the tweet: *"Exactly, and in this case I would even say that if this is recognized as a restriction of freedom, then someone would have to look at the algorithm again."* (P15). For "Science Denialism", some participants referred to this indicator as being the most useful for assessing tweets, while other participants did not agree and stated that it was not useful at all, mostly because they rated the referred part of the tweet as not suitable.

Building on our findings of Step 2, we kept topical indicators very general instead of naming specific conspiracy theories. With a few exceptions, this approach was rated as positive by our participants. As we included a study part where participants were asked to explain in detail how they assessed the tweets' credibility before showing our indicators, we found that many participants autonomously referred to their prior knowledge on a specific topic as central, not yet naming "Science Denialism" or the "Restriction of Liberty" specifically as own criteria. However, when having seen the topical indicators in a previous tweet, participants referred to them autonomously in subsequent tweets, often embedded as part of their prior knowledge on the topic. While we did not aim to quantitatively evaluate efficacy (e.g., to reduce the sharing of misleading information), this indicates a tendency towards a potential learning effect for this short-time situation. Indeed, this tendency for applying indicators to novel content is repeated for other indicators as well.

7.4.3.2 Comprehensibility and usefulness of formal indicators

The overall comprehensibility and perceived usefulness of formal indicators was rather positive with a few exceptions and limitations. All but one participant rated the comprehensibility of "Emotional Emojis" very high and were easily able to explain what this indicator refers to for both tweets it occurred in. Similarly, the usefulness was overall rated very positively with few exceptions, emphasizing the unprofessional character of using highly emotion evoking emojis within a tweet. It is noticeable that participants find this indicator particularly helpful when there are a lot of emojis in the tweet and do not only pay attention to whether the emoji is particularly emotional (e.g., red siren or screaming emoji).

For "Capitalization", all participants were fully able to explain what this indicator stands for, emphasizing a high level of comprehensibility. The vast majority further fully agreed or tended to agree that it is useful to assess the credibility of a tweet.

> *"Exactly, I think that's a good hint. [...] Because capitalization on the Internet is easy to put with shouting in real life. And who must shout, however, in the discussion, has simply no arguments. And so that's simply an exclusion criterion for me when someone writes like that."* (P12)

When evaluating the indicator "Punctuation" we found that the majority of participants again were able to understand what it means and why excessive punctuation (e.g., "!!!") might indicate misleading information. However, some criticized that this may include wrong punctuation as well (e.g., missing commas in a sentence) and emphasized that this can easily be found in non-misleading tweets as well, as not everyone is capable of writing correctly. Again, "Incorrect spelling" was easily comprehensible as an indicator, as all participants were fully able to explain what it means and what it refers to. However, there were mixed results regarding its usefulness, as seven participants for tweet 5 and five participants for tweet 3 fully or partly disagreed with it being a useful indicator to assess a tweet's credibility. Some highlighted the relevance of non-native speakers when considering spelling and punctuation or: *"I wouldn't want to reconcile the content with the spelling. So the credibility doesn't get better because somebody has better spelling. So that would be a fallacy for me. So I can also believe a person who has a worse spelling"* (P4).

While the previous indicators were overall very comprehensible without further explanation, "Hashtags" and "Links" were often not immediately self-explanatory. Many participants were unsure if it highlighted all "Links" and "Hashtags" within a tweet regardless of their content or quality or if it highlighted only suspicious "Links" and "Hashtags". As links to credible sources can be a highly relevant indicator of reliable content, it is crucial to more clearly nudge users to take a closer look at the link's quality and not make the mere appearance of links look suspicious. The same applies to hashtags. Still, the overall concept was rated rather positively and participants were able to understand and explain its meaning as an indicator easily after having seen an example in a previous tweet.

7.4.3.3 Comprehensibility and usefulness of rhetorical indicators

We were particularly interested in rhetorical indicators, as these were identified as highly relevant from existing literature and confirmed by us in Step 1, but were not consciously attended to at all by participants in Step 2. When taking a closer look at "Negative Phrases", our findings reveal a high comprehensibility. For instance, all participants were fully able to explain what this indicator means in tweets 8 and 9, and only two participants slightly struggled to explain it within the other tweets. The usefulness of "Negative Phrases" as an indicator to assess a tweet's credibility was perceived as high as well for 10 to 19 participants. Participants further noted, however, that this depends on the context of the actual tweet, leaving two to six participants rather critical.

> *"Yeah, that is one of the biggest problems here. That that was a strong, strong indicator. [...] Again, it's very emotional. It's designed to get people to react without thinking because of the instant negative associations and the negative impression that it gives, and yeah, it strongly suggests that this is designed to play on people's prejudices rather than make a serious and considered contribution to any genuine debate."* (P17)

Interestingly, participants paid close attention to whether indicators such as "Negative Phrases" occurred within a quote in a tweet. They clearly emphasized that these indicators were not valid to assess a tweet's credibility, as long as this was the case.

Similarly, this is the case for "Rhetorical Questions" as well. Again, while the overall comprehensibility was very high (all participants but one were able to explain the meaning and purpose of the indicator correctly), five participants explicitly claimed low usefulness because the "Rhetorical Questions" were part of a quote. Despite that fact, a majority still rated the usefulness of the indicator high or very high.

> *"And as that comes up more, I see more and more what exactly the algorithm is referring to [...]. That sense of, oh, yeah, it's obviously this. [...] Because there's no serious question, there's no serious answer, and people are left to speculate and come up with more and more and more opinions of their own that have very little basis in fact. So yes, that is indeed a sensible thing to look for as an indicator."* (P17)

For "Exaggeration" as a potential indicator, our findings reveal similarly high comprehensibility with all participants, without exception, being able to correctly explain what it means in the specific tweets. Again, the perceived usefulness was high, as all but three participants agreed or fully agreed that exaggeration represents a useful indicator to assess a tweet's credibility. The same positive feedback applies to "Relativization" and "Absolute Truth": *"So I would rank it more important than, for example, punctuation or capitalization. So the claim to absolute truth is always a bigger red flag for me"* (P13).

7.4.3.4 Inter-individual Differences

To assess differences in *age*, we created a sub-group of participants under 30 years old ($N = 10$) and at least 50 years old ($N = 2$). While this number of participants does not allow statistical conclusions, it revealed some tendencies of differences regarding the comprehensibility and usefulness of certain indicators that would we worth taking a closer look at in future studies. For example, while all younger participants considered "Hashtags" at least rather comprehensible from the beginning, both older participants did not consider it comprehensible from the beginning.

To assess differences in *education*, we created a sub-group of participants with low ($N = 5$) and high education ($N = 10$). For example, slight differences in the comprehensibility of the indicator "Restriction of Liberty" were visible in tweet 1: while this indicator was absolutely comprehensible for all higher educated participants (10 out of 10), it was only rather comprehensible for two out of five lower educated participants. Interestingly, there were no clear differences in the overall assessment of the comprehensibility of the approach. Both education groups considered the approach to be similarly understandable (at least fairly understandable: nine out of ten in the higher education group, six out of six in the lower education group). The usefulness was also seen as rather useful by eight out of ten in the high education group and six out of six in the low education group.

To assess differences in *political orientation* (measured on an eleven-point left-right scale), we created a sub-group of participants with a self-reported strong left political orientation (1 or 2, $N = 6$) and a center political orientation (5, 6, 7, $N = 5$). Overall, there was little difference in views on the comprehensibility and usefulness of the different indicators. Regarding "Science Denialism", three out of five participants found the indicator itself useful but did not consider the example in tweet 2 to be appropriate for this category. No one in the centrist group held this view. Thus, the left-leaning group seemed to have a stricter understanding of what they would consider science denialism. Regarding the "Incorrect Spelling Grammar" in tweet 5, four out of five left-leaning participants did not consider it a useful indicator, while only three out of eight in the middle group did not consider it a useful indicator. The overall ratings of comprehensibility and usefulness did not differ between the two groups.

Apart from demographic characteristics and the resulting differences and similarities in the assessment of comprehensibility and usefulness, we further evaluated differences among participants with diverse attitudes towards COVID-19 and the Russian war against Ukraine, revealing findings on particularly skeptical participants (e.g., because of partly conspiracy-theoretical views). Therefore, we compared those ($N = 3$ and $N = 1$) participants who did not agree with COVID-19 measures being useful or Russia waging a war of aggression against Ukraine (among which was one participant who disagreed with both statements) with the majority that fully or partly agreed ($N = 20$ and $N = 22$). It is not surprising that the participant with general skepticism towards both topics often assessed misleading tweets as rather credible as those address content that fits their own worldview.

> *"And it is also true that there are Nazis in Ukraine. Only they are covered up by the news. [...] That's why I find that somewhere, that it's just being covered up for the German news, but only states that Russia, Russia, Russia is evil. But it is never shown what Ukraine is actually doing."* (P19)

Further, the participant, in general, assesses the majority of individual indicators as not useful and states that they do not help assess the tweet's credibility at all. Instead, the participant states to use their own knowledge on the topic and does not want to receive any additional input from the intervention. Surprisingly, when asked about the general comprehensibility and usefulness of the approach, the participant assesses it as rather comprehensible and useful. Upon further inquiry, it becomes clear that the participant would have liked more explanation of the approach. While these participants' answers revealed rather entrenched views, some of which are influenced by conspiracy theories, resulting in a tendency to reject the proposed intervention, the other participants with skepticism towards COVID-19 measures did not reveal similar tendencies. In summary, our findings qualitatively show strong limitations of even a relatively less patronizing intervention when aiming to address people with opposite views in relation to official fact-checking results.

7.4.3.5 Comprehensibility and Usefulness of the Overall Indicator-based Approach

In addition to the individual indicator types, we were interested in how participants rated the overall approach of displaying indicators next to tweets that have previously been identified as misleading. The overall comprehensibility of the approach was rated as (very) high by 18 participants and rather low by one participant, leaving the rest for undecided ratings. Similarly, the overall perceived usefulness was rated as (very) high by 19 participants and rather low by one participant. A number of interesting criticisms and suggestions for improvement of the approach emerged, for instance, regarding more detailed display of information, improvements of the actual naming of indicators, habituation effects, and resulting neglect of other indicators in the tweet. One participant explicitly pointed out the risk that highlighting problematic parts of tweets will draw attention to them all the more and that this content could be remembered accordingly. Exemplary quotes of these key aspects can be found in the following:

> *"So in principle, I think it makes sense, but I think it's sometimes too abbreviated to know what it means. [...]"* (P1)

> *"The only thing that could be criticized is if certain pre-selections are made, that could perhaps influence the user, because [...] then some people can only focus on what is displayed and then they might ignore the rest."* (P9)

> *"There should be, I think, if then a little bit of work on the terms, what might be meant by that. What I just said, 'claim to absolute truth', is a bit vague, I would say. My problem is always with science that everything is extremely, I say, vaguely expressed. No normal user can understand what is actually meant."* (P18)

> *"I think that this will be very clear for most people. I could imagine that it might help to maybe display a little question mark as a popup, [...] a kind of button that explains exactly what it is about [...]"* (P20)

> *"And for me, for example, colored highlighting does not go far enough. Or even runs the risk of reproducing it. But then it also emphasizes that."* (P15)

While the quantitative effectiveness (e.g., to reduce sharing of misleading information) has been investigated by other research [35], our findings reveal some further mostly qualitative outlooks on expected effects of the indicator-based approach. Specifically, we assessed how credibility ratings changed after participants were confronted with the indicators. Indeed, for all nine tweets, there were changes in credibility rating for the better. For instance, four participants changed their rating for tweet 5 from (rather) credible to not credible after being confronted with the indicators. The most common effect, however, was reinforcing the pre-exposure presumption that a tweet is likely misleading and thus reducing uncertainty.

To sum up our key findings (see also Table 7.2) regarding the comprehensibility and perceived usefulness of our topical, formal, and rhetorical indicators, our findings reveal an overall surprisingly high comprehensibility of potential indicators for misleading tweets and mixed but rather positive perceived usefulness to assess a tweet's credibility. However, to further improve the comprehensibility and usefulness of the approach, working on the actual terms of indicators was suggested by some participants, for instance, by using more specific terms than "Restriction of Liberty" (e.g., P12). Further, especially incorrect spelling and punctuation received some criticism as it is a common way of writing more informally on Twitter and, for instance, non-native speakers may be more prone to making spelling and punctuation mistakes while still creating reliable content. Most of that criticism was relativized when participants reflected on the previous step of pre-filtering misleading information manually or automatically and indicators being only applied to tweets that have already been detected as probably misleading based on other characteristics than our indicators. Overall, participants learned to apply and autonomously name indicators after a while when being confronted with novel tweets. While this is a positive tendency towards a potential learning effect, this was true for true contents as well, where participants looked for indicators and expected the tweet to be false when a reliable tweet contained, for example, a rhetorical question (see tweet 4) - an implication to be thoroughly addressed in subsequent studies on this approach. When comparing the results on topical, formal, and rhetorical indicators, there is a tendency for topical and rhetorical indicators to be assessed as slightly more useful. Our qualitative findings reveal that especially "Spelling", "Punctuation", and "Hashtags" were evaluated less useful in comparison to all other indicators.

Some participants pointed out our concrete steps to further improve the indicator-based approach. For example, some proposed mouse-over effects or pop-ups provide more information and sources for the indicators to increase their usefulness. While this study contributes to the general foundation of indicator-based and user-centered interventions to combat misleading information, other studies in HCI focus on user interface design and how to display relevant information without information overload for end users. Follow-up studies may consider our participant's suggestions for the user interface design to assess the approach under more realistic conditions (e.g., a click-dummy), complementing existing studies [e.g., 51, 350] and evaluating the efficacy of indicator-based interventions as a central condition.

Table 7.2 Key findings regarding the comprehensibility and perceived usefulness of content-based indicators (Step 3)

Key Finding	Explanation
Topical, formal, and rhetorical indicators are generally comprehensible	Majority of participants rate overall comprehensibility of indicators as high (Sect. 7.4.3.5)
Topical formal, and rhetorical indicators are perceived as rather useful	Majority of participants rate overall usefulness of indicators as high (Sect. 7.4.3.5)
Topical and rhetorical indicators are more useful than formal	Topical (e.g. "Science Denialism") and rhetorical (e.g., "Negative Phrases") indicators are assessed as slightly more useful than formal indicators (e.g., P20, Sects. 7.4.3.1 and 7.4.3.5)
Inter-individual differences	Usefulness of indicators is sometimes assessed differently by different user groups and comes with limitations especially for users with entrenched world views (P19, Sect. 7.4.3.4)
Preference of specificity, wording, and level of detail differs	Broader indicators such as "Restriction of Liberty" are partly not considered specific enough (e.g., P15, Sect. 7.4.3.1) and some participants prefer additional explanatory information (P1) or different wording (P18) (see Sect. 7.4.3.5)
Indicators not valid if within a quote	Indicators within a quote are often not considered to be valid (e.g., P23, Sect. 7.4.3.3)
Indicators are learned	Indicators are learned and applied in the course of the experiment (e.g., P14, Sect. 7.4.3.5)
Indicators are mistakenly learned also for truthful tweets	In the case of truthful tweets that still contain the presented indicators, participants sometimes rate the truthfulness of these tweets as very low (e.g., P21, Sect. 7.4.3.5)
User interaction may increase usefulness	The perceived usefulness of the approach might be higher if it worked with mouse-over effects (e.g., P6, Sect. 7.4.3.5)

7.5 Discussion

In this study, we evaluated the comprehensibility and usefulness of indicators to guide users in dealing with misleading information in crises. Investigating German tweets related to COVID-19, we identified misleading as well as non-misleading content. We conducted a content analysis, which entailed a manual coding process, aiming at grasping the diversity of misleading tweets. Based on our sample, we could examine frequencies and co-occurrences among different types of characteristics. Further, we weighted our findings of topical, formal, and rhetorical indicators against the results of an online survey from a user perspective. Building on those findings, we finally assessed the perceived comprehensibility and usefulness of the identified indicators within a qualitative think-aloud study with participants from diverse backgrounds, extending the crisis context of tweets to the Russian war against Ukraine. This allowed for the derivation of user-centric indicators as an important step to facilitate the design of technology interventions as countermeasures against misleading information after (automatic or manual) detection that preserves user agency.

7.5.1 RQ1: What are Topical, Formal, and Rhetorical Characteristics of Misleading Tweets in Crises, with the COVID-19 Pandemic Serving as an Example?

Our study reveals a variety of topics of science denialism and classic conspiracy theories [211], partly adapted to the new context of the COVID-19 pandemic, to be found in misleading tweets. Further, a combination of formal and rhetorical indicators such as "Capitalization", "Exaggeration", "Negative Phrases" is proposed as the best-fitting model for misleading tweets in a balanced sample and adapts to prior work. We also found that misleading tweets in the COVID-19 context tend to contain co-occurrences of multiple conspiracy theories in a single tweet. Allowing for a transnational discourse, our findings reveal several similarities to studies from a U.S. perspective [e.g., 240] with a potential to generalize indicators to the global pandemic or other crises, while other characteristics are specific to the COVID-19 context in Germany (see Table 7.1, esp. "Sovereignty Germany").

7.5.2 RQ2: Which Indicators are used Autonomously to Assess the Credibility of Tweets?

We also follow existing research [e.g., 350, 491] in our interest of encouraging media literacy by giving comprehensible indicators in contrast to many current countermeasures that tend to label or delete content without offering a clear explanation. Thus, we contribute to subsequent countermeasures against misleading information after successful detection. In our survey, we investigated which content-specific indicators are already being used autonomously. It might thus be beneficial to also display these indicators to the user in addition. Autonomous assessment strategies have been partly investigated in other research, for example regarding adolescents assessing TikTok videos [224] or adults assessing different types of content including text, images, and videos [513], revealing capabilities to identify complex but well-known indicators (e.g., lighting in AI-manipulated videos or the source of information) while neglecting others (e.g., emotional sounds). We observed a similar diverging applicability of indicators in our study. Participants focused on simple and easily visible indicators, such as the presence or absence of references, in their assessment of presented tweets. Less explicit indicators, such as rhetorical indicators, were not considered specifically. More sophisticated information, for instance, regarding rhetorical indicators, was not considered thoroughly. Hence, for our final think-aloud study, we specifically included rhetorical indicators to examine whether and how they can be rendered more tangible to end users.

7.5.3 RQ3: What Comprehensible and Useful content-specific Indicators can be Derived with Regard to the Design of Digital Artifacts Assisting users on Social Media in Dealing with Misleading Tweets During Crises?

We evaluated the comprehensibility and perceived usefulness of topical, formal, and rhetorical indicators within the contexts of COVID-19 and the Russian war against Ukraine. In our context, those factors are thought to be crucial in increasing trust [51]. Our findings suggest focusing on easily visible indicators (e.g., "Capitalization") and expanding future research on how to make more complex indicators [470] accessible to end users. Interestingly, when confronted with rather complex indicators (e.g., "Absolute Truth" or "Rhetorical Question"), the majority of participants from different educational backgrounds were able to comprehend its meaning and positively evaluated the usefulness of our proposed indicators to assess tweets' credibility. When comparing the assessments of indicators, our qualitative findings reveal that especially spelling, punctuation, and hashtags were evaluated as less useful. Our participants emphasized the limitation of spelling and punctuation as indicators, as these also commonly occur with non-native speakers and people with dyslexia (see Sect. 7.4.3.2). This complements suggestions by Plotnick et al. [428] to look into how indicators such as grammar can still be used to assess credibility during crises when time for content creation is limited. Some participants emphasized that to make the media literacy approach accessible to a wider group of users, it may be necessary to make indicators as guidance more comprehensible in general. This may imply personalizing digital misinformation interventions to pick up on different levels of media literacy. The specific indicators identified in Study 1 in the COVID-19 context could be readily applied to the additional crisis context of the Russian war of aggression against Ukraine. These more generic indicators instead of very specific ones (see Table 7.1 Level 1 vs. Level 3) were motivated by study 2, in which the preference for more generic indicators slightly stood out, revealing the potential to apply our findings to other (crises) contexts.

While an indicator-based approach, in contrast to giving credibility labels without any explanations, may already counter reactance, it still comes with several limitations regarding rationality and potential biases [62, 188]. Consistent with the understanding of Diaz Ruiz and Nilsson [131] that general belief systems and susceptibility to disinformation are intertwined, our qualitative results emphasize that users who have already formed a firm opinion on a particular topic and are, for example, in a bubble that tends to be based on conspiracy theories and can only be reached to a limited extent even with this less patronizing intervention. Nevertheless, while those participants were generally very critical of single indicators, they nevertheless assessed the overall comprehensibility and usefulness of the approach rather positively. This positive assessment is consistent with other studies encompassing indicators and their potential to enhance media literacy [51, 224, 299]. There is a tendency that such an approach, when giving more explanations, is perceived positively, but with the attitude: "Not for me, rather for others". This is in line with findings of other indicator-based research on misleading information and reflects some extent of overconfidence in end users that has already been investigated [75, 224, 390].

With this work, we contribute to the design of successful countermeasures against misleading information on social media, considering a user's perspective after detection. Our study further proposes the following implications based on the analysis:

7.5.4 Implications for User-oriented Assistance in Dealing with Misleading Information

In light of our findings, we derived four implications that give orientation in the design process of user-oriented platform-mediated measures against misleading information in the context of public crises. The implications were derived by synthesizing the findings from Steps 1, 2, and 3, triangulating insights from our quantitative Twitter analysis, online survey on autonomous assessment strategies, and the user-centered evaluation of indicators' perceived comprehensibility and usefulness.

(1) Consider comprehensive indicator-based interventions to address the need for transparency. The perspectives from our German context in Steps 2 and 3 suggest that users benefit from a comprehensive indicator-based misinformation intervention during crises, revealing an overall very positive assessment of the indicator-based approach. Out of 23 participants, 19 rated its usefulness (very) high, specifically naming its transparency as a reason. In addition, we observed first insights into changes in credibility ratings for the better (e.g., four changes from (rather) credible to not credible for tweet 5). Indeed, research [299] revealed that users prefer transparent countermeasures over binary labeling or deleting without explanations. Providing users with comprehensible indicators implying why a specific content was identified as potentially misleading, is considered a promising approach to avoid the counterproductive socio-psychological effect of reactance and backfire effects [221, 391]. Indicator-based interventions should be designed to comprise a variety of levels. In this study, we demonstrated how content-specific indicators on a topical, formal, and rhetorical level can be derived and evaluated in terms of user-centered applicability, leaving opportunities for future work on additional levels (e.g., creator profile, interactions) as related work has already been partly addressed. Our findings expand existing research on indicators that focus on the U.S. context [626] and address the need for a stronger user perspective [195, 299]. While other approaches, such as [129] have attempted to identify the rationale behind why people are susceptible to believe disinformation and related legitimization strategies, we claim that our approach is also capable of addressing those dimensions indirectly through a set of appropriate indicators. For instance, *cognitive legitimacy* could be addressed through topical indicators that reveal narratives of conspiratorial world views. However, we emphasize that the indicators identified here may not fully represent all of these dimensions.

(2) Keep indicators rather general and simple by default. Our findings suggest that users rather identify common buzzwords and umbrella terms as indicators than consider more complex or specified characteristics (e.g., OS13:02; OS02:05; OS02:06). To avoid information overload and foster productive processes of sense-making [244, 533], it is crucial for user-centered approaches to balance between the benefits of highlighting only general indicators (e.g., "conspiracy theory") for simplicity versus enhancing transparency and trust by highlighting more detailed indicators (e.g., "Free Mason"). While offering rather simple indicators is desirable, whitebox-based assisting tools may benefit from an optional feature of giving detailed indicators for personalization, which is also in line with prior findings [221]. Nevertheless, we believe that it is still worthwhile to develop new strategies to encourage users to reflect on more complex implications such as underlying legitimization strategies [129] and rhetorical tactics [131] to promote media literacy. This is supported by our findings, which show that easily identifiable features such as links or references to sources may, as already indicated by [129], influence user perception of disseminated information.

(3) Focus on visible indicators. When investigating characteristics of misleading tweets regarding suitable indicators to assist users, visibility is a central factor for comprehensibility. Sophisticated approaches for feature detection, for instance, using network analyses [518], often come with a good detection accuracy; yet, respective indicators may often not be suitable as they are not understandable for the end user at first glance, when looking at a tweet. Our findings imply that users may not want to invest too much time when assessing content and prefer to quickly evaluate based on obvious indicators such as topics (*"#DeepState"*; e.g., OS02:08) and formal indicators (lack of/excessive punctuation; e.g., OS02:10). Thus, we consider it a valid approach to highlight comprehensible indicators to the end user which are immediately visible when looking at a misleading tweet and do not require additional investigations like looking at a user's Twitter bio or understanding their network.

(4) Allow for personalization in indicator-based interventions. Our research emphasizes how demographic factors, such as age and education, and individual attitudes, including COVID-19 skepticism, impact user responses to indicators. To accommodate these diverse user profiles, future user-oriented assistance should be adaptable and flexible. Especially when considering the findings on general comprehensibility and perceived usefulness of our approach in Step 3, participants emphasized a diverse need and requirement for additional information and explanation. We specifically aimed to include a diverse group of participants from different sociodemographic backgrounds to reveal differences regarding comprehensibility and perceived usefulness (see Sect. 7.4.3.4). While some prefer very simple and general indicators, others criticized vague expressions (P18) and preferred more detailed explanations and more specific expressions. In addition, the comprehensibility and perceived usefulness of the indicators partly varied to some extent between demographic groups according to age, education, and political orientation (e.g., two out of five lower educated participants did not fully comprehend "Restriction of Liberty" while all ten higher educated participants fully comprehended it). Personalization of indicator-based interventions that allow for different levels of assistance may address that challenge. Indeed, personalization and tailored interventions have already been proven to be promising with regard to susceptibility according to personality types [494] and in other contexts such as cybersecurity [e.g., 146] to encourage compliance. Especially considering that the context of misleading information has a great potential for reactance and rejection due to opposing (political) attitudes, media literacy education measures with little perceived paternalism are desirable and could contribute to effectiveness. While our findings suggest the benefits of personalization, it comes with additional efforts. In particular, while some user differences or preferences can be covered by a simple configuration (e.g., the binary selection of a detailed or basic version of intervention regarding indicators), others, such as (educational) background or political orientation, cannot easily be covered while considering privacy needs. Thus, its feasibility and potential are still to be investigated in future research.

7.5.5 Limitations and Future Work

Our work provides valid insights into the field of HCI and particularly into misleading information on Twitter during a long-term global crisis. However, our study comes with some limitations.

First, we did not include account characteristics such as registration date, number of followers, retweets, or users' age. These are potential indicators for misleading information as well, which becomes clear when looking at Twitter bios containing QAnon references. Yet, we explicitly decided to focus on the content of tweets as one very central and immediately visible component. Concentrating on immediately visible components that do not require visiting an author's profile facilitates the capability to reflect on tweets' accuracy while in a common user scenario on Twitter (e.g., solely scrolling through the timeline after following a hashtag during a crisis). This may, of course, be complemented by nonetheless relevant account information and tweet reactions. Their applicability as easily comprehensible user-centered indicators is still to be evaluated in future studies.

Second, identified indicators for misleading information cannot be generalized easily. While our study shows that indicative features may be apparent to varying extents, overall low frequencies regarding non-misleading content suggest indicators to be plausible. Topical indicators (e.g., "Media Criticism") may be applicable to other situations, and indicators specific to the COVID-19 pandemic (e.g., Wuhan Virus) may be reformulated more abstractly (e.g., as source, responsible actor). Indeed, in our think-aloud study in Step 3, we included several tweets on the Russian war against Ukraine and found that the applied indicators were assessed as useful in that context as well. Still, the sample size of our studies does not allow for a definitive generalization to other crises and contexts, suggesting further (long-term) studies with multiple cases and bigger sample sizes as a valuable next step. It is important to notice that both contexts tend to have a polarizing effect on opinions and often come with entrenched opinions. This, of course, may have had an effect on our user study results. Comparing our sample to interactions in non-disruptive situations allows us to gain detailed insights into the indicators' respective prevalence. Due to the clear focus on German-language tweets, we want to provide an analysis that would allow cross-country comparisons as previous research on misleading information during the COVID-19 pandemic demonstrates a clear focus on the U.S. Connecting to international comparisons of social media discourses, it would prove useful to study thematic "spillovers" that occur in virtual spaces across linguistic or geographical borders. Recognizing that the COVID-19 pandemic was a global crisis, it is noteworthy that not only specific narratives of misleading information may differ between countries, but also the measures that proved effective in addressing this issue.

Third, we decided to balance our sample between misleading and non-misleading tweets to include enough misleading information for valid assumptions. However, the balanced sample comes with limitations. Hence, it is crucial to stress that the findings were based on a balanced sample and may significantly differ in a real-world relation between misleading and non-misleading tweets.

Fourth, user-centered misinformation interventions, including our indicator-based approach, typically rely on preceding manual or automatic detection of misleading information that successfully pre-filters whether a given piece of information is potentially misleading. This is a highly complex task and has been a central research area for years [518, 613], especially within the field of machine learning. Displaying indicators on all tweets per default would result in a non-acceptable number of false positives and is not the aim of this approach. For future work, studies might fully implement and evaluate the combined approach of automatic or manual detection of misleading information and subsequently extracting and displaying indicators within content that has been successfully identified as misleading.

Fifth, while we gained rich qualitative insights regarding the perceived usefulness and comprehensibility of the derived indicators, our work does not constitute a large-scale representative experiment to make conclusions about statistically significant effects. Future studies might apply our findings on user-centered indicators to an online experiment or survey with a larger sample size to allow for comparisons between a control group and experimental groups confronted with the intervention. Related studies typically investigate quantitatively how an intervention affects sharing behavior of misleading information compared to other interventions or a baseline group without interventions. Based on our study, complementing related research that demonstrates controversial findings on whether transparent information and explanations significantly impact efficacy [179, 349], prospective statistical research may build on our insights. The interpretation and comparability of intervention efficacy strongly depend on the research design, and finding a consensus in measurement has been highlighted as central to the development of successful interventions. Future studies might consider existing frameworks for efficacy measurement that have been proposed [193]. Future work might build from our insights by evaluating our derived design implications on a larger scale. For instance, it would be valuable to pick up on a personalized indicator-based approach that considers different requirements and preferences derived in our work (e.g., regarding rather general indicators versus indicators broken down in detail), testing changes in misinformation sharing behavior under different conditions against each other. Personalization is typically based on a suitable way of user segmentation, sometimes realized using psychometric tests like the General Decison-making Style [500] or personality types [494] and should rely on a thorough consideration of privacy needs and weighing additional efforts.

7.6 Conclusion

In this work, we evaluated promising indicators to assist users in dealing with misleading information in crises. Our primary contributions (C) and findings (F) involve an initial (C1) thorough assessment of 2,382 German tweets on Twitter (now X) within a realistic crisis setting to extend the knowledge base of our specific context. This examination allowed us to (F1) identify indicators for misleading information that pertained to both topical (e.g., context-specific themes like "Restriction of Liberty"), formal (e.g., links and hashtags that occur in similar measure in credible tweets but are a valuable cue for misleading information when users being nudged to take a closer look at their content), and rhetorical levels (e.g., claims to "Absolute Truth"). We then (C2) conducted an online survey to explore how users autonomously evaluate a tweet's credibility. Our findings shed light on (F2) a range of indicators that users use independently (e.g., buzzwords like "mainstream media" as a topical indicator), as well as characteristics that they are unaware of (particularly rhetorical indicators of misleading information), revealing the benefit to display them in an assisting tool. Consistently focusing on user-centered design, we (C3) evaluated the perceived comprehensibility and usefulness of our previously identified indicators. We found (F3) that the indicator-based approach is perceived as useful overall within the context of COVID-19 and the Russian war against Ukraine. When presented with tweets, the topical, formal, and rhetorical indicators improve users' perception and evaluation. We gained (F4) positive feedback regarding comprehensibility and perceived usefulness (e.g., for topical indicators when kept rather general, emotional emojis, and capitalization), but also criticism for others (e.g., "Punctuation" being too excluding for non-native speakers). (F5) Besides opportunities, we also identified challenges and limitations of the indicator-based approach for short-text content. This applies especially to users with general skepticism towards official services or even entrenched conspiracy theory views but also concerns indicators not being valid if within a quote or drawing attention to misleading content, even more so when highlighting specific parts.

Digital Resilience in Dealing with Misinformation on Social Media during COVID-19

8

Abstract In crises such as the COVID-19 pandemic, it is crucial to support users when dealing with social media content. Considering digital resilience, we propose a web app based on Social Network Analysis (SNA) to provide an overview of potentially misleading vs. non-misleading content on Twitter, which can be explored by users and enable foundational learning. The latter aims at systematically identifying thematic patterns which may be associated with misleading information. Additionally, it entails reflecting on indicators of misleading tweets which are proposed to approach classification of tweets. Paying special attention to non-expert users of social media, we conducted a two-step *Think Aloud* study for evaluation. While participants valued the opportunity to generate new knowledge and the diversity of the application, qualities such as equality and rapidity may be further improved. However, learning effects outweighed individual costs as all users were able to shift focus onto relevant features, such as hashtags, while readily pointing out content characteristics. Our design artifact connects to learning-oriented interventions regarding the spread of misleading information and tackles information overload by a SNA-based plug-in.

Original Publication *Stefka Schmid, Katrin Hartwig, Robert Cieslinski, and Christian Reuter. "Digital Resilience in Dealing with Misinformation on Social Media during COVID-19 A Web Application to Assist Users in Crises". In:* Information Systems Frontiers *26.2 (2024), pp. 477–499. DOI:*10.1007/s10796-022-10347-5

Supplementary Information The online version contains supplementary material available at https://doi.org/10.1007/978-3-658-47885-8_8.

K. Hartwig, *Navigating Misinformation*, Technology, Peace and Security | Technologie, Frieden und Sicherheit,
https://doi.org/10.1007/978-3-658-47885-8_8

8.1 Introduction

The COVID-19 pandemic has confronted people all over the world with different challenges. The main challenge concerns measures to contain the virus, which include a variety of government-led measures aiming at physical distancing and vaccination [127]. Particularly in the context of public health crises, adequate communication of relevant information is crucial. Believing and trusting inaccurate information may not only result in regressive world views but also careless behavior and physical harm of individuals [192]. Considering the spread of misleading information on social media, the World Health Organization (2020) has referred to the problem as an "infodemic". From the perspective of *Crisis Informatics*, which is dedicated to information systems in disruptive situations [527, 534], increasing digital resilience allows individual, informal actors to better contribute to crisis management. The challenge of misleading information has been identified as a barrier to rely on social media in crises [463]. While misinformation has been perceived as an exogenous issue, more recent works interested in successful collaboration of social media users have noted ideological foundations and (state) actors' involvement [386, 532, 533].

Our work focuses on targeted social media users and proposes to productively encounter misleading information on social media in a context of high uncertainty, need for transparency and higher risk of information overload [167, 289]. Thereby, we assume users with relatively strong media literacy as well as non-expert users to interact with information on social media which is reliable to varying degrees [280]. Looking at empowering ways of assisting users in interacting critically [107, 365], we have developed a prototype of a web app which relies on network analysis. Aware of its relative complexity [174], we presume visualization of networks of tweets to offer a usable overview for people with different media competences and prior knowledge. The tool allows to further investigate tweets as it includes word clouds and lists with comprehensible indicators for misleading content. Aiming for digital resilience of user "crowds" [536], it is important to assist users who individually show different levels of motivation, routines and skills. While general short attention spans need to be considered, we assume that assisting tools enhance learning and comprehension especially when individuals use them consciously [419, 423]. In return for carrying the costs of dedicated time, users may be supported by a relatively transparent web app which aims to train systematic information processing skills. Thus, we are interested in answering the following research questions (RQs):

RQ1: *How can a web app which aims to foster users' ability of systematic information processing be designed?*

RQ2: *How can user learning be effectively enhanced in the context of an "infodemic"?*

Following other inquiries of design science [558], we are interested in problem-solving. Thus, we focus on both the development RQ1 and evaluation of a web app which is usable independently of individual media literacy (RQ2).

We particularly aim to respond to its usability for lay users during crises, for which digital resilience is necessary to overcome and adapt to external shocks in the long-term, including potential information overload [228]. Building upon social media being considered a tool for individuals to successfully navigate stressful situations [107], the web app aims to contribute to digital resilience by empowering users to participate in informed decision-making. More specifically, the web app aims to enhance systematic and analytical processing of information instead of heuristic or intuitive thinking [237, 371]. On Twitter, information quality of posts may be very heterogeneous and thus pose a cognitive burden on social media users. Our aim is to consider a tool which helps to relieve individuals from this burden, counter information avoidance, and enhance critical, systematic thinking. We further understand motivation to be dynamic [243]; it is thus important that the web app itself does not contribute to information overload but that it offers a satisfactory experience to both expert and non-expert users [186, 237].

Ultimately, however, our technical solution carries limitations and should rather be perceived as a complementary approach to other educative endeavors of digital literacy [353]. As noted by Gregor and Hevner [190], design science research contributes to IS research not only by the practical work of developing an artifact but also by offering theoretical insights of different levels of abstraction. Presenting an "exaptation" [190], we rely on existing technology to address a new problem. Communicating our findings (see Sect. 8.6), we discuss results on learning about systematic processing of information on social media to contribute to the knowledge base. Referring to an "infodemic", we extend our field of interest onto contexts which are characterized by a vast amount of disseminated information of different information quality which increases the potential of individual information overload and intuitive, non-analytical thinking.

Following a design research approach, we refer to the conceptualization of resilience in IS research by [228] and identify transparency as an additional design requirement. We present the development of a web app in JavaScript and Flutter, relying on pre-established Twitter data sets and SNA in Python. Based on a two-step evaluation (*Think Aloud* study), we find that independent from their level of media literacy, participants were able to generate a specific network and explore it with regards to thematic trends and potentially relevant indicators. However, users with specific background knowledge put more emphasis on the assessment of "black-box" classification of misleading vs. valid information. Flexibility proved to be advantageous while participants were unequally overwhelmed by both the network visualization and misinformation on the COVID-19 pandemic as an issue. This highlights the relevance to consider different levels of media literacy as well as motivations.

Our work pays special attention to strengthening digital resilience of human-social media information systems and proposes learning-oriented ICT [280]. The web app may be embedded as a plug-in and can be accessed by an external link. We therefore offer an alternative intervention to top-down, overwhelming warnings directly integrated in the UI, by providing substantial information on tweets to individual users who are ready to invest more time for learning.

In the following, we present related work (see Sect. 8.2) which problematizes user-centered measures against misleading information with regard to users' systematic information processing in times of crisis. Subsequently, we introduce our *design science approach* which consists of the formulation of design requirements of a relevant artifact, its development, and an iterative evaluation. We then present the concept and implementation of the web app (see Sect. 8.4) and illustrate the main findings of the two-step evaluation (see Sect. 8.5). This is followed by a discussion of our results (see Sect. 8.6) and complemented by a conclusion (see Sect. 8.7).

8.2 Related Work

8.2.1 Sense-making, Misleading Information and Information Overload in Crises

In disruptive situations, people face a high level of uncertainty as well as the need to exchange necessary information [116, 309]. Processes of sense-making and coping often entail the dissemination of unverified and misleading information on social media [117, 315, 316]. Leavitt and Robinson [316] stress that information visibility, afforded by design, is a crucial factor in users' perception of a crisis on which crowd-based efforts of volunteering, sense-making or emotional support are built on. Although misleading information may be a general problem, its distribution in the context of a global pandemic carries great costs. Science denialism, information overload, avoidance or fatigue may result in the loss of lives due to non-systematic processing or unawareness of relevant information [211, 237].

While validation of information in the context of crises such as earthquakes may be essential [183], social media users rely on information of different quality and also have to cope with uncertainty in long-term disruptive situations such as epidemics [383, 502]. Referring to the MERS outbreak in South Korea in 2015, Yang and Lee [619] emphasize the existence of an "information vacuum" which allows "risk-alarming frames" of media coverage to build on anxiety. As studies on social media use during the COVID-19 pandemic have indicated [9, 274], anxiety is quite common across the globe. However, responses to anxious posts also comprise misleading information [9] whose display without information on reliability inhibits users from productive interactions. Thus, sense-making in crises, which entails systematic processing of information, takes place in an environment where users more easily perceive information overload and information quality is not necessarily assured [289]. Thereby, information quality depends on uncertainty, novelty, ambiguity, and complexity [237]. Information accuracy is an important criterion for information quality [323] and, following Kahn, Strong, and Wang [279], Hosseini et al. [242] relate to different dimensions of information quality, including understandability, when they assess relevant requirements of transparent systems.

As found out by Hong and Kim [237], information overload, defined as "the state of lacking cognitive resources to process information", prevents individuals from systematic processing of information, learning, or critical thinking. Instead, people engage in heuristic processing or intuitive thinking, which relies on trusting external opinions and precludes understanding [237, 371]. Although this may be part of coping with crisis situations, it prohibits users to fully adapt to new contexts. Further, Misra, Roberts, and Rhodes [371] emphasize that information overload is a subjective experience, which is prevalent when "feeling burdened by large amounts of information" which cannot be used efficiently or effectively. This may be the case in emotional crisis situations [244].

8.2.2 User Media Literacy and Resilient Information Systems

Successful, systematic processing of information relies on a resilient human-social media IS and thus a clear, transparent display of information. It is also positively associated with social media users' motivation and their levels of media literacy [237, 303]. The broad concept of media literacy allows to capture not only "an individual's ability to understand printed text and to communicate through print" [303] but refers to knowledge about different media environments, including (lack of) journalistic verification processes and different forms of representation of information and their purposes [353]. Importantly, McDougall [353] points out that media literacy is not equal to digital skills but presumes individuals to assess content on topics they are not familiar with based on their critical thinking skills and understanding of media content. Arguing against simplifying solutionist approaches, tools may refrain from suggesting easy, binary categorization into fake news and the truth but instead, aim to enhance social media users' media literacy and thus processes of analytical or systematic thinking [353].

Various studies have focused on digital resilience [176] and understand resilience as the "capacity of individuals and communities to bounce back" [466] or even 'bounce forward', adapting to external shocks [378]. Depending on information technology as well as human dispositions, an information system can be improved purposefully and share a diverse set of attributes which reflect resilience, such as learning, rapidity, flexibility & diversity, and equality [228]. Considering these attributes, learning is assumed to be foundational [228]. As highlighted by other scholars, it implies users' active engagement which initiates critical thinking [324, 353]. Resilient information systems do not exclude the potential of information overload a priori. Instead, they allow for learning from related experiences, i.e., increasing media literacy [353], and to foster systematic information processing over heuristic thinking.

While Koltay [303] as well as Hong and Kim [237] found media literacy to be a more important factor influencing the potential of information overload, users' socio-economic background may play a role as well [501]. Considering user modelling for enhanced requirement specification, Adikari, McDonald, and Collings [5] point out that users are generally assumed to be able-bodied and may have different needs, goals, and preferences. As with information overload, user experience is also subjective [186, 578]. Besides media literacy, Hong and Kim [237] identify motivation to be essential for learning. However, affected by user experience and changing goals, motivation of both "professionals and amateurs" [303] is a dynamic variable [243, 578]. Motivational factors are an individual's attitude and emotional state, entertainment gratification as well as perceived ease of use [243]. The latter may be defined by the extent of old knowledge that could be used to routinely interact with information technology [381]. However, McDougall [353] emphasizes that while critical media literacy is more likely to foster digital resilience, it should not be framed as an individual responsibility.

8.2.3 Suitability of Existing Approaches

To avoid information overload for social media users, automatic detection efforts have been made [613, 631]. However, accurate detection may be challenging. For example, Shu et al. [518] point out the topical diversity of misleading content and the increasing importance of rhetorical structures and Zhou and Zafarani [631] capture the relevance of different categorization of content. Besides diverse approaches of detection, technical interventions may be beneficial when they differentiate between "malicious" and "normal" users, by assisting the latter to "improve their ability to distinguish fake news" [631].

In this regard, Clayton et al. [104] find that general warnings about media were even counterproductive while more specific warnings show a small positive effect. Further, their study shows that "rated false" tags are more helpful in correcting views on information accuracy than displays of posts being "disputed".

Following inoculation theory, Roozenbeek and van der Linden [469] analyze the effect of an online "serious" game, in which users play the role of creators of misinformation. Although isolated from real-world social media environments, a learning game emphasizes the importance of substantial interventions [469]. Further, Guess et al. [194] emphasize that discernment between false and accurate news can be increased by "teaching people more effective heuristics" instead of relying on quicker, cost-sensitive measures [555] which require users to pay attention to the assessment of singular posts. Lehman and Miller [324] also note that active engagement (in their case microblogging) increases learning effects of individuals. In case users find interaction "simple, fun and educational" [363], they are ready to invest time and increase their media literacy.

In sum, the relevance of learning with regard to interacting with misleading information autonomously is highlighted by different studies, emphasizing the enhancement of media literacy as promising [123, 280, 365]. Using Agent-Based Modelling, Gausen, Luk, and Guo [177] evaluate measures against misinformation, distinguishing between blocking users, as well as behavioral approaches, accuracy flags, and "societal inoculation" [327]. The latter implies that it is necessary to create "mental antibodies" [184] against misinformation by fostering users' skills to assess information accurately.

As related scholarly work on the COVID-19 pandemic has indicated, SNA is assumed to produce generalizable insights into the distribution of content on social media. Thus, it offers grounds on which users can build insights into broader debates and learn about narratives. For example, Ahmed et al. [10] use SNA and present clusters which reflect debates surrounding 5G and COVID-19, including relevant actors on social media. Similarly, Memon and Carley

[356] focus on propagating actors of COVID-19-related misinformation, presenting insights regarding misinformed communities on Twitter. Milani, Weitkamp, and Webb [366] also analyze diffusion networks regarding vaccination debates on social media which focus on polarizing actors. Their model has been implemented by Cui et al. [113]. It assesses content for interested users and allows them to track a post's dissemination using a network graph. While Shao et al. [505] propose Hoaxy which allows to gain insight into the diffusion of misinformation by studying user networks, Yan, Lapata, and Li [618] also include visualized topical groups of tweets based on their semantic proximity. Proposing Check-It, Paschalides et al. [411] combine user analysis of online social networks with a Deep Neural Network for linguistic analysis and cross-checking articles with a list of fact-checking websites. Sharma et al. [509] investigated narratives that were promoted in misleading tweets in English during the COVID-19 crisis in 182 countries. It must be taken into account, that as pointed out by Shu, Bernard, and Liu [517], nodes of networks do not necessarily reflect users.

Further, the display of comprehensible indicators, giving insight into potential criteria for classification, may reduce reactant behavior by social media users as countermeasures are no longer perceived as top-down initiatives [465]. For example, Aswani, Kar, and Ilavarasan [26] and Hartwig and Reuter [221] propose understandable indicators for misleading social media content. Such "white-box" approaches allow users to get an impression of what happens during information processing between in- and output. Results of a survey conducted by Kirchner and Reuter [299] support the need for comprehensible countermeasures in contrast to non-transparent warning signs, with respondents demanding explanations. However, transparency itself can lead to information overload in the absence of information accuracy, clarity, accessibility, or understandability [242]. Therefore, the importance of information quality should not be neglected in transparent information systems.

Various works capture the diverse technical countermeasures both proposed and undertaken by research and social media platforms [111, 177, 617]. Roughly, it is differentiated between deletion, redirection or reduction of visibility of posts, and behavioral approaches tackling users' perception of social media content by warnings or labeling. Looking at existing approaches, there is little focus on introducing surrounding broader debates or giving more specific explanations to labeled content independent from user features (see Table 8.1).

Table 8.1 Examples of existing countermeasures implemented in major social media platforms; own research

Social Media	Countermeasures	Binary Labels	Learning Effect	Top-down Bottom-up	Broader Debate
Twitter [568]	labels	no	yes (explained by user features, further info)	top-down	no
	warnings	yes	yes (explained by user features, further info)	top-down	no
	reduced visibility	n/a	no	top-down	no
Twitter [567]	labels	no	yes (explained by thematic features, further info)	bottom-up	yes
Facebook Meta [360]	labels	no	yes (explained by user features, partly further info)	collab.	no
	warnings	yes	yes (explained by user features, partly further info)	collab.	no
	reduced visibility	n/a	no	top-down	no
Instagram [249]	labels	no	yes (further info)	collab.	no
	warnings	yes	yes (further info)	collab.	no
	reduced visibility	n/a	no	top-down	no
YouTube [147]	labels (on spec. topics)	yes	yes (explained by user features, further info)	top-down	no
	reduced visibility	n/a	no	top-down	no

(continued)

Table 8.1 (continued)

Social Media	Countermeasures	Binary Labels	Learning Effect	Top-down Bottom-up	Broader Debate
Reddit [254]	bans, "quarantine" of subreddits	no	collab.	no	
TikTok [247]	label	no	yes (further info)	collab.	no

Considering prior findings on media literacy, information overload in crises, and resilience as well as currently implemented countermeasures on social media, we aim for the development of a web app which efficiently enhances users' systematic processing of social media content. Thus, we follow technical approaches which focus on "societal inoculation" [177, 184] against misinformation and aim at reducing information overload and fatigue in the context of a pandemic. Aimed at attributes of resilient information systems [228] and transparency implying information quality [242], the proposed intervention invites users to explore thematic patterns of misleading information and specifically considers the costly challenge of information overload.

8.3 Methodology

Considering misleading communication on social media regarding the COVID-19 pandemic and platforms' rather reserved interventions, we find the design and development of a web application which supports long-term adaptation and transformation of problematic phenomena highly relevant. Thus, we conducted a *design science approach* as an appropriate method to create new and innovative artifacts. Based on an identified empirical problem, this approach comprises the design of an artifact as well as an evaluation, aimed at systematic problem-solving [235]. Contributing to IS research with an exaptation [190], we subsequently communicate more abstract findings which can be derived from the instantiated artifact. Connecting to the existing knowledge base on misinformation, information processing and overload as well as digitally resilient information systems, formulated implications refer to both kernel theories about human behavior and mid range theories on design and crisis management [190]. While we do not contribute with insights on the third level of abstraction, i.e., fully developed design theories, the situated implementation (first level) is a contribution in itself and also allows for the discussion of rules with regard to the design of Internet-related socio-technical information systems (second level) [190].

In our case, the artifact is a web app and evaluations took place iteratively in form of two *Think Aloud* studies. The design science approach consists of five steps, namely (a) achieving problem awareness and showing its importance, (b) suggesting a solution's objectives, (c) the web app's design and development as a solution to the problem in a suitable context, (d) an iterative evaluation of the artifact using the *Think Aloud* method, and finally (e) the communication of a conclusion [417].

We follow an iterative evaluation process in which a two-step *Think Aloud* study aims at refining the design of the web app [235]. *Think Aloud* studies are a suitable method to gain rich qualitative and user-centered feedback on problem-solving, using a novel artifact such as our web app. Performing the two studies iteratively has facilitated the refinement of the artifact based on obtained findings [163].

While specific empirical analyses may precede the development of applications [461, 614], our work follows existing studies which have noted the importance of social media for coping with crises and identified related problems of unreliable content and information overload [107, 167, 559]. We also rely on insights of studies on the spread of misinformation, particularly noting users' needs for an explanation as well as the unsustainability of low-effort interventions [299, 423, 469]. Considering disruptive situations in which individuals need to manage external shocks, we are specifically interested in user-assisting tools which enable individuals to learn. Linking the question of strengthening digital resilience to media literacy as a basis for handling crises allows focusing on identified attributes of the former [228]. These help to structure both the development process and evaluation. Thus, our perspective allows to combine efforts of developing problem-solving applications with the generation of insights complementing the IS knowledge base [235]. In the following, we present the concept and implementation of the web app (see Sect. 8.4) and findings of the two-step evaluation (see Sect. 8.5.3) which point out both potentials and limitations determining the process of learning, i.e., increasing digital resilience.

8.4 A Concept for Digital Resilience on Social Media

We propose a web app to provide social media users with an overview of potentially misleading and non-misleading tweets with the aim to encourage media literacy and thus improved digital resilience and also investigated how the latter can be increased using an interactive tool based on SNA. We used the four attributes of *learning*, *diversity & flexibility*, *equality*, and *rapidity*, characterizing digital resilience [228] as a frame of reference for the development process. Derived from the knowledge base (see Sect. 8.2), we present *transparency* as an additional design requirement (see Table 8.2).

Assuming that users can benefit from an intervention which considers specific posts but emphasizes their thematic connections, the SNA-based web app shifts the focus away from singular instances of misleading information onto topical trends which updated are more (or less) prevalent in misleading or valid tweets. We chose this approach to offer a user-assisting tool which enables individuals' to improve their systematic processing skills and discernment of posts' information quality in the context of crisis situations.

Despite varying attention spans and potential information overload, users are encouraged to actively generate network visualizations and focus on classifying information as valid or misleading. Thus, the web app can be considered a tool to enhance crucial learning processes. While this engagement takes time, we aim for a relatively easy process of network generation, highlighting the function while further exploration of tweets and proposed indicators are added to an information system which allows diversity & flexibility. Aware of different user backgrounds, i.e., personal motivation and media literacy, it is crucial to offer various options to generate resources. Especially during the COVID-19 pandemic, which has accelerated digitization, it is relevant to include new users and allow digital resilience to be fostered independent of their level of media literacy.

Table 8.2 Design requirements contributing to digital resilience. Interested in increasing media literacy, we rely on resilience attributes and identified indicators by Heeks and Ospina [228] which imply learning. Transparency was added as another abstract design requirement during initial awareness raising. Thereby, we used findings by related works (see Section 2) which also involved considering *user attributes*

Knowledge Base	Requirements	Indicators
– *user needs: sensemaking* – low info quality: high uncertainty,novel, ambiguous, complex contents in crisis – info visibility is crucial – info quality, including accuracy is necessary in transparency rhetorical structure plays a role	Transparency	accessibility, consistency, certainty, accuracy, understandability, clarity
– training media literacy: systematic instead of heuristic processing – understanding processes of information production based on printed text – reduction of info overload – understanding broader narratives, topical diversity – active involvement – generalizable results on which users can build insights	Learning	capacity-building, new & trad. knowledge, reflective thinking
– *different needs, goals, preferences of users*	Diversity & Flexibility	diff. courses of action, emerging opportunities, adaptable decision making, innovation mechanism
– life-long learning independent from *media literacy and sociodemographic factors* – *expert & non-expert users* – *motivation is dynamic* and should not be decreased	Equality	strengthened competencies, gap reduction, inclusiveness, openness & accountability

(continued)

Table 8.2 (continued)

Knowledge Base	Requirements	Indicators
– user costs (investing time) – ease of use & entertainment – topic is stressful – *motivation is dynamic* and should not be decreased – info avoidance, fatigue are problematic in crisis	Rapidity	rapid resource access, assessment/coordination, mobilization

The backend of the web app was implemented using *Python 3.9* (2020). The *Flask* library builds an HTTP interface that is used to access the frontend. *NetworkX* enables the generation of relatively large graphs and also offers implementations of graph and network algorithms. Most of the UI is written in *Flutter* [185], a framework for *Dart*. The generated network is visualized with the JavaScript library *D3.js* (Data-Driven Documents) version 6 [61]. While the web app was developed using state-of-the-art principles and algorithms (e.g., the CNM algorithm to divide nodes into groups), it comes with potential for the innovative user-centered exploration of a social network in the context of misleading information.

8.4.1 General Concept

The web app interface consists of four main components (see Fig. 8.1).

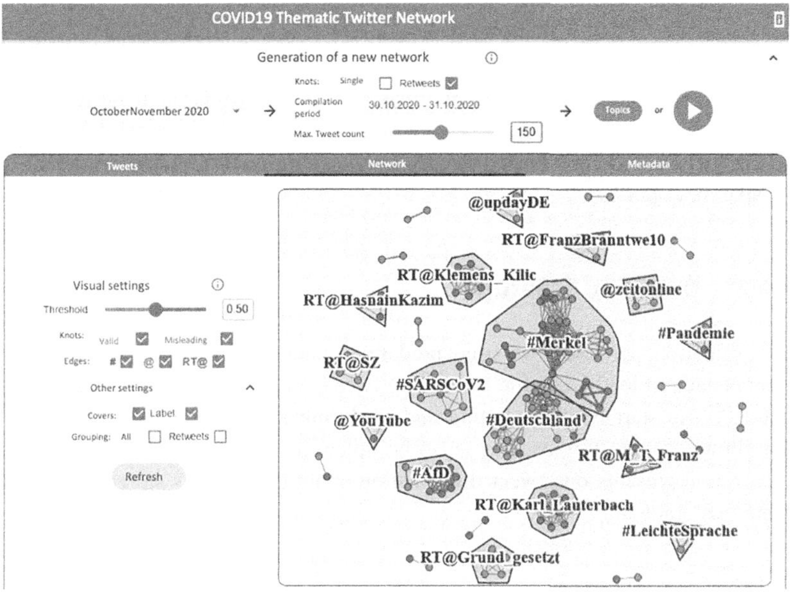

Fig. 8.1 The interface of the web app while the generated network is selected as a visualization component

First, at the top of the interface, there are several options to set up the web app, e.g., by loading a specific data set and making customized configurations before generating a network. Therefore, users' active participation is required, which in turn encourages learning [194, 324, 353]. Second, the core of the web app represents the visualization of a social network for Twitter contents. Third, besides exploring Twitter contents in the visualized social network, the user can switch the data representation towards a list of tweets along with other relevant information. Fourth, the contents of a Twitter data set are visualized as word clouds when selecting the component "metadata". Offering diverse options for exploration, the web app allows flexibility [228]. Users can dive deeper into data and recognize known system features of Twitter, while others preferably rely on the broad thematic overview [237, 303].

All four components of the interface include options for user configurations regarding the specific representation of data to encourage interactive exploration. In the upper area of the interface, users can generate a new network at any time (see Sect. 8.4.2). The lower area either shows a network visualization (see Sect. 8.4.3), a list of all used tweets (see Sect. 8.4.4), or various metadata as word clouds (see Sect. 8.4.5). This ensures the web app can be used by users with different preferences [5]. In all areas, tool tips (text that is displayed with a mouse-over effect), provide users with comprehensible descriptions for individual components, thereby contributing to information quality as a requirement for transparency [242].

8.4.2 Data Set & Setup

For evaluation purposes, we used predefined data sets (see Sect. 8.5.2). Thus, users needed to load a csv-file containing tweets and additional information (separator ";") to start both generating and exploring a network. However, the web app allows for any further data set to be uploaded, as long as it fits the format. Each row represents one tweet. Information about the tweet is stored as a JSON string (see Table 8.3).

Table 8.3 Overview of data set

Column	Keys	Description
Tweet	content	raw tweet text
	enrichedData	optional, list of occurring hashtags
	mentions	optional list of occurring mentions incl.retweet, reply
	retweetTo	Twitter ID of the originating, retweeted tweet
	startTime	posting time, year-month-day-hour:minute:second.msecondZ
	url	URL of the tweet
User	content	optional, raw thext of the account description
	isVerified	whether the account is verified
	numStatuses	number of created statuses (tweets)
	numFollower	number of accounts that follow creating account
Score	scoreBERT	optional, decision of BERT classifier (value betw. 0 and 1)

Our data sets for evaluation contain the output of a trained "black-box" classifier to automatically detect misleading contents based on BERT (Bidirectional Encoder Representations from Transformers), a state-of-the-art NLP model from Google [128]. Its output is a number between 0 (probably not misleading content) and 1 (probably misleading content). The classifier achieved an accuracy of 0.70 (precision: 0.74, recall: 0.66, F_1-score: 0.70). BERT calculations were performed beforehand and its outputs were included into the data sets for evaluation, rather than performing the classification during run-time while using the web app.

To explore Twitter contents using the web app, the user first specifies a setup scenario. Therefore, a predefined source data as csv-file is selected containing tweets and associated relevant information (see Sect. 8.4.2). Subsequently, users can make additional configurations. For instance, a limited time frame for tweets within the data set can be chosen by selecting specific start and end period for when the tweets were created. This information is stored within the csv-file. Further, users can customize the visualization, by deciding whether to include retweets or by choosing to display isolated nodes without connecting edges. Users further have the possibility to define the maximum amount of tweets to be included in the network. Finally, the hashtags and mentions used to connect nodes can be adapted manually, with which individuals can identify terms that are relevant for a specific narrative [316, 509].

In our scenario, hashtags containing "Corona" or "COVID" are not used as connecting edges by default, as the data collections used for evaluation solely contain tweets related to COVID-19 (see Sect. 8.5.2). For all steps of the setup phase, users can access additional information and explanations when moving their mouse over the tool tips. This increases transparency while principally keeping information as concise as possible [242, 323]. To gain information about German Twitter communication regarding the COVID-19 pandemic, users need to click on the "play"-button, resulting in the calculations for three distinct visualizations (social network, list of tweets, word clouds) of the chosen Twitter contents which all offer different possibilities for interactive exploration.

8.4.3 Network

Based on the data, a node models a tweet and its author while undirected edges represent thematic connections between the tweets. These are established by hashtags (#<tag>) or mentions (@<username>) (see Fig. 8.2). A sub-type of mentions used here are "retweets", i.e., forwardings of (foreign) tweets to one's own contacts. These are usually marked in the tweet text ("RT @username>") and can also be clearly identified by the collected metadata. The type of edge depends on the labels (i.e., hashtags, mentions or retweets) used and the number of connections. A retweet is weighted heavier, since hashtags and mentions are often similar in such cases. As mentions always contain a specific user name, a connection between public figures and specific tweets can be established. To model the topics or trends of the created network, the nodes are divided into groups using the CNM algorithm [103]. The most frequent hashtag, mention or retweet, used in the connections is displayed as the entire group's topic.

While the network visualizes relations which do not represent concrete behavior between user accounts, it shifts focus away from personalized interactions onto relevant topics [517, 518, 618]. This builds grounds for enhancing systematic processing of users as they gain more general insights into the dynamic debate [518]. Prioritizing edges (mentions, hashtags, retweets) over nodes reflects the relevance of these system features in affording "information visibility" [316]. Thus, individuals can learn about dynamics and functions of social media which fosters critical media literacy. Network visualization of Twitter communication on the COVID-19 pandemic allows to gain a broad overview. It thereby does not only allow to identify thematic patterns but also to relieve users from anxiety during emotional encounters [9, 274, 619].

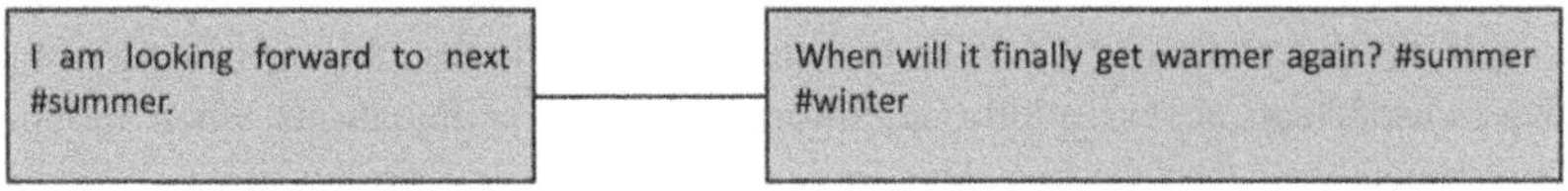

Fig. 8.2 An example of two tweets being connected by hashtags. The authors have been omitted here

The visualization of the network is the core of the web app (see Fig. 8.1). A "classic" representation is used, i.e., nodes are modeled by circles and edges with lines between circles, which are also colored differently according to the available properties, for example "misleading" or "valid" for nodes or "hashtag", "mention", or "retweet" for edges. The visualization can also be configured by the users, for example by hiding certain nodes or edges. The predominant topic (= label) of a group of nodes is displayed in its center.

Within the network component, further configurations can flexibly be made while exploring the data (see Fig. 8.1).

As a focus of our work is to encourage user exploration of Twitter contents regarding misleading versus non-misleading tweets, the output of the BERT-classification of tweets as probably misleading or not misleading is displayed within the network component. To enhance autonomy for advanced users, users can adapt the threshold which adjusts from which score a tweet should be displayed as "misleading". This value can be set at any time, so there is no static binary classification. However, the threshold is set to 0.5 by default, while "valid" nodes are colored green and "misleading" nodes are colored red to present a simplified dichotomous differentiation for lay users in a situation where information quality is low due to its novelty and complexity [237].

A slider or input box can be used to set the threshold for misleading tweets. This ensures users are aware they are dealing with probabilities which technical solutions have to offer and more easily invest time to engage [242, 353, 469, 555].

Several check boxes can be used to hide certain nodes (misleading, not misleading) or edges (hashtag, mention, retweet) depending on user preferences [5]. Related to the grouping, borders or topic labels can be drawn or, for further abstraction, all related tweets can be reduced to one node. The labels represent the most popular labels of the respective group (see Fig. 8.1, right). If the network contains multiple retweets of a particular tweet, these can also be reduced to one node. The web app displays the network graph interactively. Users can drag individual nodes across the canvas by holding down the left mouse button, or single left-click to explore selected tweet data in a new window. When the mouse hovers over a node or edge, some data is displayed in a tool tip, ensuring accuracy and accessibility of information [242]. The mouse wheel enlarges the display and allows a more detailed view of the network.

Depending on the settings above, nodes and edges are colored in certain colors. If preferred, an alternative color scheme according to Wong [608] is used. The thickness of the edges shows the number of connections between two tweets. When tweets are reduced to groups, the display slightly changes. The size of the nodes (= groups) then represents the number of attributed tweets; analogously the thickness of the edges symbolizes the number of connections between two groups. The score of a group is the median of all associated tweet scores. A left mouse click on a node displays all tweets and some metadata of the selected group in the new window.

8.4.4 Tweets

In addition to the network component, the web app offers an exploration of Twitter contents by displaying all tweets as a list along with relevant information. In contrast to the network graph, the display is largely text-based but also allows a certain degree of filtering or sorting. Thereby, one can take a closer look at the tweets used in the network. All relevant node data is displayed for each tweet. If preferred, the tweet can be opened in a new window. In addition, the list can be further customized: A slider sets a minimum and a maximum score of the BERT classifier, retweets can be hidden and a drop-down menu allows the selection of a sorting order. If preferred, the value can be displayed as a percentage instead of the decimal score. In addition to the BERT classification, we detect and display comprehensible indicators for misleading information during runtime within this component; having been inspired by prior research:

1. capitalization [221, 593]
2. frequent punctuation marks [221, 593]
3. frequent and conspicuous emoticons [221, 375]
4. frequent hashtags [26]
5. number of URLs [88]

While such attributes do not offer a holistic explanation for contents' misleading character and are by far not deterministic in this regard, indicators function as points of references for systematic processing of information which is crucial for users [237, 299]. Critical assessment of tweets includes considering highly emotional content and relying on bridging system features, such as hashtags or embedded links [244, 316]. To limit the potential of suggesting indicators to adequately explain BERT scores, we refrain from causal statements and opt for clear and consistent [323] display of indicators independent from score values. While indicators are generally comprehensible, engaging users rely on their prior knowledge about social media features at the beginning [242].

8.4.5 Metadata

The last visualization component displays all detected labels (separated by hashtags and mentions) in word clouds and the labels that occur together are listed. The metadata area component displays various information about the generated network. The size of the text depends on the absolute frequency of the label. Labels that were used only in one tweet or that directly contain "Corona" are also displayed using a checkbox. Further below, connections between labels are shown. Display of connections between topical labels as well as word clouds pose the most abstract and simplified exploration of Twitter data as there is no corresponding presentation by tweets or visualization of thematic relationships. For reproduction purposes, the parameters used in the last setup phase can be viewed in this area.

8.5 Evaluation

In the following, we present our method (see Sect. 8.5.1) as well as the pre-established data sets used for the evaluation (see Sect. 8.5.2). Subsequently, relevant findings considering digital resilience are illustrated (see Sect. 8.5.3).

8.5.1 Method

The web app was evaluated in a two-step *Think Aloud* study [163]. Prior to this, a test run was also carried out with four people which allowed the test leader to prepare and check the recording procedure [229]. Responding to the feedback, a search box for tweets was added. The iterative approach was guided by the aim of improving the web app, paying special attention to the needs of non-expert users. Both evaluative meetings took place on Zoom with 16 participants. Nine people participated in the first round between February 27, 2021 and March 1, 2021 and seven participated in the second evaluation between March 8, 2021 and March 11, 2021. On average, each unit lasted 39:14 minutes or 42:36 minutes, respectively. To avoid bias by prior experience, subjects participated only once in total. Individuals differed in age and motivation. We further distinguished between lay users and experts according to level of media literacy. This in turn is understood as dependent on habitual Twitter use and media consumption (see Electronic Supplementary Material (ESM), Table A.3).

The participants were briefed on the *Think Aloud* method [163] and consented to their session being recorded for the purposes of this scientific study. Further, a friendly and relaxed atmosphere was critical to create a quiet space in which participants could articulate their thoughts freely [163]. The undertaking researcher thus used appropriate language and clarified that participants' thoughts were not categorized as right or wrong. He mostly refrained from interfering to not interrupt participants but promoted the process of "thinking aloud" by asking open questions (see ESM, Table A.7). Since the test persons usually have little or no prior knowledge in the field of networks, an introduction (see ESM, Figure A.7) was handed out and gone through after an initial phase of exploration. Thereby, it was not only ensured that participants had a "general impression of the system" [229] but also indicated differences in use, based on individual background.

As part of the design product evaluation [516], a prefabricated task was presented subsequently, which was carried out by each participant in interaction with the web app (see ESM, Table A.7, 3.). This comprised a guided network creation, identifying a predominant topic, and searching for indicators of misleading information in tweets with "#Merkel". Simultaneously, participants were encouraged to express their thoughts out loud. The researcher also took additional notes regarding possible ambiguities or peculiarities. The transcript and the standardized tasks allow for comparison of usage. Following Fonteyn, Kuipers, and Grobe [163], a conventional interview phase was conducted succeeding the *Think Aloud* phase. The latter allows a direct, consistent insight into participants' thoughts, while the former indicates their retrospective views. Following the meetings, the researcher transcribed the audio material from which we then derived findings regarding digital resilience.

8.5.2 Test Data Set

For the evaluation, users relied on prior collected data sets of three observation periods (see Table 8.4). Therefore, in each of the time periods, the query "Coronavirus OR Corona OR Covid OR Coronakrise OR coronadeutschland OR Corona19 OR covid19 OR covid-19 OR #COVID19" was sent to a Twitter API. The data sets were further restricted to tweets in the German language ("language = de") and thus reduced in size. Limiting text length of retweets to 140 characters, the data files also include respective tweets but may not comprise the entire post. Afterwards, for each tweet a pre-trained BERT classifier calculated a score between 0 and 1 which indicates the accordance with misleading contents (see Sect. 8.4.2). The filtered data, complemented by the BERT-output, was then saved in a specific format (see Sect. 8.4.2).

Table 8.4 Collection of test data sets. Observation spanned over diverse phases of the long-term crisis to capture Twitter communication in the context of different numbers of infected cases and governmental restrictions

Observation Period	Tweets	Crisis Context in Germany
02/27/2020–03/06/2020	156,455	beginning of crisis (infections, public debate)
04/14/2020–04/22/2020	205,198	during first "shut-down", gov. restrictions
10/29/2020–11/27/2020	334,215	during "lock-down light", before second one

8.5.3 Results

Aiming for digital resilience, we structured the analysis of the two sessions according to formulated design requirements (see Table 8.2). Our work is a proposal and offers ground for identification of both potentials and limitations of resilient user-assisting technology.

8.5.3.1 Transparency

Considering attributes of resilience [228], in particular the essential characteristic of *learning*, these may only be prevalent when technical systems are transparent to users and when *transparency* also implies information quality [242]. Accessing the application, some participants (e.g., EVAL-1A) wished for a better introduction. In particular, this was more relevant to individuals who had had less prior engagement into networks (EVAL-1C). This was contrasted by one user who understood the purpose of the network visualization immediately: "*As I have already worked on own data sets and built edges and nodes [...], it is a bit easier for me to understand*" (EVAL-1B). Thereby, they could rely on prior knowledge [228], while other users could only access new knowledge after an expanded introduction to the web app as well as a continuous text description had been provided for overall comprehension [242].

Further, users who were relatively new to social media communication did not change the threshold and adjust probabilities of classifying content as misleading or valid, taking BERT results, when asked explicitly, as "*given*" (EVAL-1C). Only expert users showed awareness regarding the slider and were eager to gain more detailed insight. Although tool tips were often disregarded in both the first and second evaluation, individuals who showed higher interest and felt comfortable with exploring the application could get more insights which was found highly relevant in terms of their perception on the classification of content (EVAL-1A). At the same time, binary coloring supported users who did not dedicate their time to exploring the threshold. This shows the importance of transparency to users' experiences while effectiveness of design depends on their media literacy and information quality, indicating a subjective notion of understandability [242].

8.5.3.2 Learning

Learning is foundational to digital resilience as it builds ground for adapting to disruptive situations [228]. Considering both expert and lay users, we evaluated the web app's usage regarding the potential for capacity-building, opportunities to rely on (new and old) knowledge, as well as systematic, reflective thinking [228, 237, 353]. Presenting networks to users, we aimed at offering individuals with insights into thematic connections between tweets which may be more (or less) prevalent among misleading (or non-misleading) posts. In this regard, all participants were able to acquire new knowledge while all of them were informed, albeit to varying degrees, about the COVID-19 pandemic and aware of the spread of misinformation. For example, EVAL-1B, who mainly relies on conventional media for news consumption and is not active on Twitter, concluded that

> "*one may assess clearly that, here, clusters are established comprising valid and some misleading [tweets; AN] and that there are also sometimes connections between accurate and misleading information [...].*"

Another participant who does not use social media and felt a bit overwhelmed in the beginning could subsequently identify thematic patterns such as on politicians of the German government (EVAL-2L, see ESM, Figure A.8, Figure A.9). This indicates that although individuals may not habitually use social media, overview of Twitter communication by networks allows for systematic processing of information [237]. In particular, description of thematic communities and notification of connections among tweets of varying accuracy reflect active engagement and critical reflection instead of intuitive thinking [194, 228, 371]. However, only expert users (EVAL-1I, EVAL-2M) followed the idea of a misleading-valid continuum [353] and were open to depart from perceiving misinformation as a clearly and easily identifiable issue. They showed awareness regarding the slider and were eager to gain more detailed insight. They also used the tweet tab for better understanding (EVAL-2M). Changed prior to the second evaluation, binary coloring of the slider supported heuristic thinking [237] by lay users, connecting the threshold to the network:

> "*Uh, it says 'valid' and 'misleading'. So red is misleading and, uh, to some degree fake or at least you don't know whether it is true. And when colored green, you may believe it is true [...]*" (EVAL-2P).

Participants also started to evaluate proposed indicators of misleading tweets and autonomously weighted them against user biographies or linguistic content (EVAL-1H, EVAL-2K). Learning to differentiate between misleading and valid tweets was also initiated by individuals who did not focus by themselves on proposed indicators and had less background knowledge about Twitter, pointing out "*dubious terms*", a user's biography and grammatical inconsistencies (EVAL-2J, EVAL-2L). Thus, although active generation of a network proved to be difficult to some subjects, they could expand prior knowledge and gain new insights into potential indicators of misleading information. As such heuristic cues [237] were already presented to users, they could train systematic processing of information and gain an understanding on how narratives on social media are developed [316].

8.5.3.3 Rapidity

In the context of varying attention spans and quick interactions on social media [423], resilient systems are perceived to provide rapid access to resources, i.e. information, and allow for their efficient assessment, coordination and mobilization [186, 228]. In this regard, initial interaction with the web app is crucial which was at times perceived to require too much effort. One user pointed out that one could familiarize oneself with the application within "*5 minutes* and indicated the need for improvement by including a "*30-seconds video*" or a "*short explanatory text*" (EVAL-1E). Thus, we integrated the latter prior to the second evaluation to accelerate the learning process. Generating a network based on selecting a data set and time frame provided some difficulties to lay users. In general, participants needed several minutes to fulfill the task of generating a network. The time required by participants to fulfill this task was reduced by changes to the generation UI between the first and second evaluation (see Table 8.5).

Table 8.5 Duration of a specific network generation in minutes. The median value of the second evaluation is lower than of the first experiment. One value (8:55 min) had to be corrected because of an interruption of 15 minutes without internet access

EVAL-1A-1I	4:43	5:00	5:56	7:13	1:56	7:03	3:44	3:22	4:53
EVAL-2J-2P	2:28	3:20	5:54	3:05	3:08	8:55	10:24		

In terms of individual times, it should be considered that participants were asked to fulfill a standardized task of generating a specific network. This was done to gain more systematic, yet non-representative, results. Certainly, this took more time than passively noting warning signs. However, once they started, participants paid attention and continued with the task, simultaneously articulating their thoughts. As the start, i.e., selection of the data set and time frame as well as labels required at times more effort than anticipated (EVAL-2M), respective improvement is crucial to increase digital resilience.

Based on the first evaluation and the experiment's frame of using predefined data sets, users found the "Upload CSV" button off-putting (EVAL-1E). To ensure user-friendliness, CVS.files can now be uploaded by accessing the drop down menu and selecting "select source file" (see Fig. 8.3 and Fig. 8.4). Further, we included two arrows to illustrate the process and while a pop-up window appeared in the first evaluation, which some perceived as overwhelming (EVAL-1A), the drop-down field is colored yellow and a hint to select a file is presented with an automatically disappearing snack bar at the bottom of the screen. The first evaluation indicated that the steps necessary for successful network creation were not clear, with users showing frustration over attempts to select time periods (but no data set) (EVAL-1E). Transparency, characterized by information quality, is thus crucial for efficient network creation and to avoid information overload [237, 242, 323, 371]. Additionally, the icon representing a chain of connections was replaced by text ("Topics") as it was misunderstood prior the second evaluation (EVAL-1D). While participants during the second evaluation neither articulated positive feedback regarding the "topics" button, nor explicitly focused on it, it did not inhibit network creation. EVAL-2J was, however, able to describe its function retrospectively: "*Well, I can't remember what the window was, but it probably selects individual clusters in the data set*".

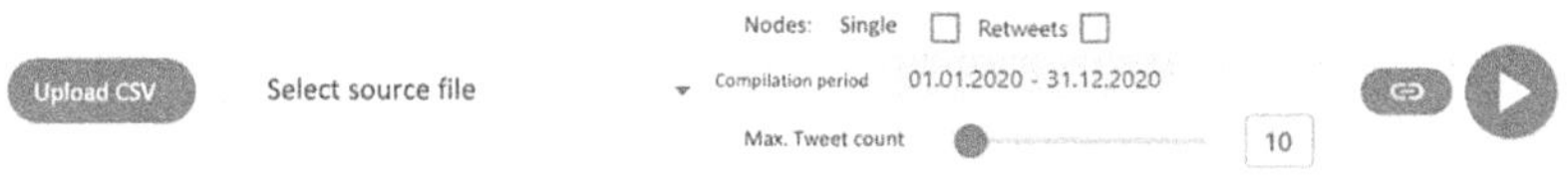

Fig. 8.3 Evaluation 1: If, for example, the play button is pressed without a selected record, a new window appears with a note

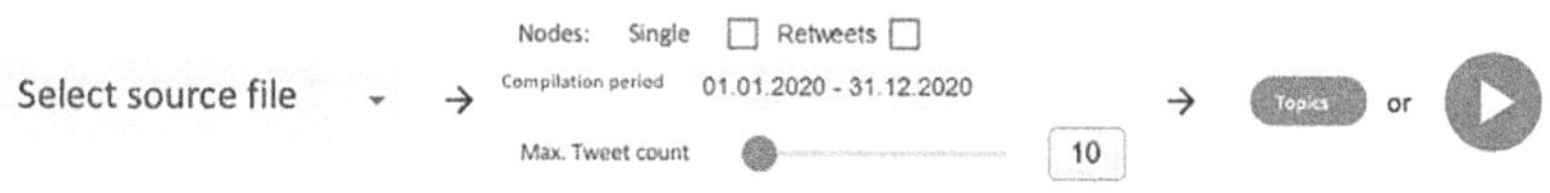

Fig. 8.4 Evaluation 2: Instead of a popup, the drop-down box is colored yellow and a snack bar with a hint is displayed. Arrows reflect the procedural steps for network generation while terms and icons which were perceived irritating in the first evaluation have been eliminated or changed

These changes are particularly relevant to avoid future decreases in motivation as readiness to engage with social media content was sometimes low with subjects being, as expected [9, 237, 274], "*annoyed*" (EVAL-2M) or "*frustrated*" (EVAL-1I) by the pandemic.

8.5.3.4 Diversity & Flexibility

Complementing network visualizations, the web app offers a display of tweet contents which may be more known to users. Thereby, users valued that they can gain more detailed insights into tweets which are part of the generated network (EVAL-1I). Further, some participants (e.g. EVAL-2N) autonomously focused on the word cloud when exploring Twitter communication on the COVID-19 pandemic which emphasizes the importance of an information system's ability to offer various options [228]. While the network component is certainly at the core of our approach, it is relevant to offer additional information to users which may not be satisfied by it. EVAL-2J highlighted that they

> "*definitely [liked] the way of looking at these graphs, meaning the changes one can make to the network visualization. At the same time, [I like] that one can have a closer look right and left at the metadata and tweets. I assume this [the network, AN] is the main function here and I think that's indeed well done.*"

Thus, we offer a more traditional display of lists of tweets which also allows users to look at user characteristics such as the bio and thereby rely on their assessment (EVAL-1E) and focus on specific content and proposed indicators (see Sect. 8.5.3.2). The word cloud also allows gaining an impression that does not consider thematic connections between more or less potentially misleading and non-misleading tweets. However, users (such as EVAL-1G, EVAL-1I or EVAL-2P) identified which topics were important in Twitter communication regarding the COVID-19 pandemic and, engaging with the list of tweets, also expanded their focus onto characteristics of misleading tweets. Based on the two-step evaluation, we interpret users' choices as also depending on their personal preferences [5, 243]. This includes considering the specific crisis context, in which anxiety is relatively prevalent [9]. One individual engaged in heuristic processing [237] in the course of focusing on a specific misleading post:

> "*This [tweet; AN], for example, by this guy is [...]. It does not make sense; it's a pandemic and no conspiracy*" (EVAL-2M).

Offering diverse options, including an overview of sensitive contents proves beneficial due to its level of abstraction.

8.5.3.5 Equality

To make the web app more accessible, we chose conventional colors [228, 243], which users highlighted positively. For example, the "*Twitter-like blue*" (EVAL-1E) or differentiated coloring of misleading (red) and valid (green) information (EVAL-1C) was noted. One participant noted being slightly color blind but that they did not require an alternative color scheme. Another participant with relative high media literacy proposed integrating a dark color mode (EVAL-1A). This was appreciated by one expert user in the second evaluation:

> "*I'm a simple man. I see Dark Theme, I press activate. That's how it is; of course [it's] a 'must have' for any web app*"(EVAL-2K).

Even though it "*took a short moment to find their way around*", operation was "*mostly intuitive*" (EVAL-1G). To reflect a more entertaining environment [243, 363], we also included a "play" button into the generation UI. We further tried to refrain from symbols which may not be well-known and instead relied on text. Similarly, we also reduced the transparent display of technical terms such as "upload CSV" as it caused information overload in the first evaluation (EVAL-1B, EVAL-1C, EVAL-1D, EVAL-1E) [242]. However, the friendly and partially well-known [243] twitter icon was preferred over description of "original tweet" (see Fig. 8.5, Fig. 8.6) which is indicated by positive feedback in the second evaluation (EVAL-2L).

Fig. 8.5 Evaluation 1: The indicators are at the top; the button has a text description

One may also consider that non-expert users are not aware of the meaning of hashtags (#) or mentions (@) (EVAL-2P); yet, as most users who may rely on the web app may have a registered Twitter account and to reduce potential overload, we did not change the interface in this regard. As Twitter language or slang was identified as a barrier to lay users, we displayed the indicators more separately at the bottom of the Section (see Fig. 8.5, Fig. 8.6). Across both evaluations, expert users approached the "threshold" and showed interest in further details.

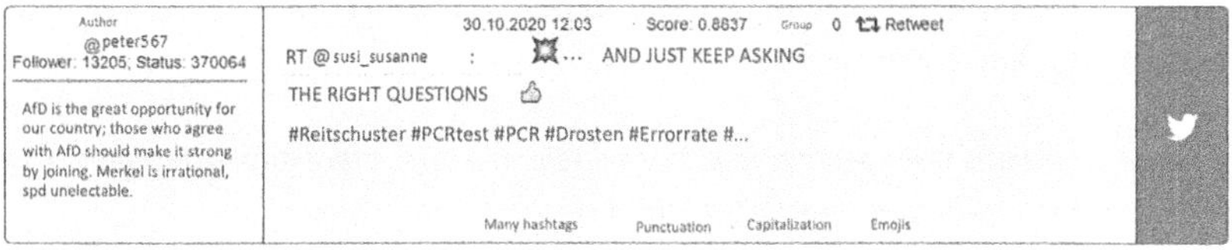

Fig. 8.6 Evaluation 2: The indicators are at the bottom; the button has a Twitter icon

In contrast, lay users rather refrained from exploring the slider freely but kept it in the middle or chose extremes on the continuum (EVAL-1C). In sum, all participating individuals were able to conduct the task of creating a network and thus could increase their competencies. They also showed interest to engage with the topic prospectively.

8.6 Discussion

In the following, we communicate relevant implications with respect to enhancing learning, with the latter being understood to contribute to resilient socio-technical information systems in contexts where a lot of information is disseminated with high frequency and, paired with heterogeneous information quality, may overwhelm individuals (see Sect. 8.6.1). As social media has been increasingly used more broadly, people with different motivations, levels of media literacy, and attitudes use platforms for information retrieval. Insights apply specifically to human-social media information systems but also more generally when humans use search engines for different purposes [303]. Subsequently, we point out limitations and potential future work (see Sect, 8.6.2).

8.6.1 Implications for IS Research on Learning and Resilience in Contexts of Unreliable Information Exchange

Based on our proposed artifact which relies on SNA and the BERT classifier, we can formulate implications which contribute to the knowledge base dedicated to enhancing learning processes when a vast amount of potentially unreliable information is disseminated. Considering design requirements (see Table 8.2), we reflect on our findings with regard to learning, rapidity, diversity & flexibility, transparency, and equality.

(1) Focus on important system features which structure communication permits a deeper understanding of how information is produced.
As identified by [184] and [327], countermeasures which aim at "societal inoculation" are more sustainable compared to interventions which relate to singular posts and require shorter attention spans [419, 423]. Further, as the latter may be perceived by users as categorizing posts simplistically without any explanatory information [299], people may not have the possibility to train their analytical skills as much as they could. As indicated by the evaluation, participants with different backgrounds in media literacy and use were eager to learn about different narratives on social media and patterns which can give orientation to information processing on the Internet. Using a platform's relevant system features as a frame of reference for presenting social media data makes sense as they organize online communication. Emphasis on their structuring function by building the summarizing network on them allows users to reflect on how content is produced and how social media works. To users who have prior knowledge about social media, presentation of social media content by network visualization appears "intuitively" [381]. This allows for further learning processes as cognitive capacities can be used for acquiring more detailed, new knowledge.

(2) "Slowing down" information processing allows to pay attention to posts in the first place but should not decrease user motivation.
Cost-sensitive approaches have been proposed as advantageous [555]. In the context of the high speed of information dissemination on social media, users expect an efficient information retrieval for various purposes. Additionally, platform providers have an interest in satisfying user needs, which makes it difficult to create conditions of foundational learning. However, we designed a web app which affords exploration (in contrast to, e.g., warning symbols) and followed Heeks and Ospina [228] in prioritizing learning as a requirement. This was done because prior literature emphasized the problem of short attention spans and the potential of foundational learning [363, 423]. Additionally, we aimed for rapidity which was also tested for in the evaluation. Non-representative results suggested that, after improving the design, participants were faster. Nevertheless, the iterative evaluation did not indicate that participants of the first round were more impatient and showed decreased motivation as it took longer on average. Understanding motivation to be a dynamic variable [243], it is important to guarantee a satisfactory user experience, even when topics are more serious.

(3) A diverse and flexible system can lead to information overload considering heterogeneous user groups and the subjective dimension of transparency.
Existing studies on interpretability of indicators [221, 593] and transparency in information systems [242, 279] already point out the importance of transparency, implying information quality, to learning. As noted by Hosseini et al. [242], transparency can also lead to information overload, which was validated in the web app's evaluation. Participants with higher media literacy wished for more explanation on the BERT classifier whereas non-expert users were at times overwhelmed by the different interface components. We tried to include these to offer a diverse and flexible design to users considering the objective to resonate with all the heterogeneous backgrounds, user habits and preferences and thus *added* features instead of focusing on how to satisfy intersecting needs and goals. Dynamic networks which allow for zoom-ins on different nodes and clusters already offer a great flexibility which enables users equally to learn and explore information on the different levels of abstraction. However, there is always a trade-off between diversity and transparency when a variety of people use the information system and information quality needs to be assured.

8.6.2 Limitations & Outlook

Similar to Hoaxy [505], the proposed web app operates, with reference to source tweets, in parallel with Twitter. Prospectively, the thematic network may be integrated into the Twitter UI via a plug-in such as TrustyTweet [221]. In this case, future research may set up a different experimental environment and rely on near real-time data from the Twitter API instead of pre-established datasets whose selection required some costly effort by participants. Additional parameters such as custom search terms could be integrated then as well. Further, the two-step evaluation was conducted with a relatively small number of participants and thus offers non-representative findings. As one participant indicated to have slight dyschromatopsia but did not have any barriers in identifying features, complementing works may also consider generalizable results regarding issues of equality, i.e., accessibility and inclusion [228]. For systematic testing, user modelling which considers different motivations and subsequent purposeful sampling of participating users is essential and may complement our illustrative work. Our work contributes to existing research on diverse approaches to counter the spread of misinformation and, following parameters of digital resilience in IS research [228] as well as including transparency [242], emphasizes important considerations for design of user-centered tools that encourage social media users' media literacy. In principle, the web app is context-independent and may rely on near real-time data input. Instead of prior works which have proposed user networks [10, 356, 366], we focus on topics indicated by hashtags, mentions, and retweets. In our view, this implies a plausible shift from personalized accounts to thematic narratives which are to be targeted when dealing with misleading information. However, visualizations may be suggestive, proposing connections to be based on actual interactions between authors of tweets. Displaying relevant labels aims to counteract such impressions. Future works may further investigate relationships among requirements and also consider the impact of individual information processing on collectively used systems.

8.7 Conclusion

The proposed intervention poses a first approach to counter the spread of misinformation with a special focus on increasing digital resilience. Thereby, our design science approach, comprising conceptualization and implementation of a web app as well as a two-step evaluation by *Think Aloud* studies, focuses on the design validation of a resilient information system. In this regard, we consider the relationship between individual human users and the social media platform Twitter and evaluate its potential for digital resilience based on tweets regarding the COVID-19 pandemic in Germany. Considering events such as the "storming of the Reichstag building" [547] in August, 2020, we identify an "infodemic" [611] as a problem which can be approached by offering an assisting tool. Although some participants were overwhelmed at first, the flexible and diverse web app allows for learning. The latter is fundamental to establish digital resilience and aims at shifting user focus onto thematic trends and relevant system functions which help in assessing disseminated information more analytically. Additionally, proposed "white-box" indicators enable reflection about classification of information. Relying on the BERT classifier [128], we do not suggest that information can always be wrong or right depending on the perspective. Rather, the respective threshold aims to emphasize that some information holds more misleading (or more accurate) elements than other and that is is relevant for individual users to assess content critically. Adding to the knowledge base, our approach contributes with a reflection on attributes of resilience as well as on potentials and limitations of information systems aiming to tackle the spread of misinformation.

9 Navigating Misinformation in Voice Messages: Identification of User-Centered Features for Digital Interventions

Abstract Misinformation presents a challenge to democracies, particularly in times of crisis. One way in which misinformation is spread is through voice messages sent via messenger groups, which enable members to share information on a larger scale. Gaining user perspectives on digital misinformation interventions as countermeasure after detection is crucial. In this paper, we extract potential features of misinformation in voice messages from literature, implement them within a program that automatically processes voice messages, and evaluate their perceived usefulness and comprehensibility as user-centered indicators. We propose 35 features extracted from audio files at the character, word, sentence, audio and creator levels to assist (1) private individuals in conducting credibility assessments, (2) government agencies faced with data overload during crises, and (3) researchers seeking to gather features for automatic detection approaches. We conducted a think-aloud study with laypersons ($N = 20$) to provide initial insight into how individuals autonomously assess the credibility of voice messages, as well as which automatically extracted features they find to be clear and convincing indicators of misinformation. Our study provides qualitative and quantitative insights into valuable indicators, particularly when they relate directly to the content or its creator, and uncovers challenges in user interface design.

Original Publication *Katrin Hartwig, Ruslan Sandler, and Christian Reuter. "avigating Misinformation in Voice Messages: Identification of User- Centered Features for Digital Interventions". In:* Risk, Hazards & Crisis in Public Policy *15.2 (2024), pp. 203–235. DOI:* 10.1002/rhc3.12296

Supplementary Information The online version contains supplementary material available at https://doi.org/10.1007/978-3-658-47885-8_9.

K. Hartwig, *Navigating Misinformation*, Technology, Peace and Security | Technologie, Frieden und Sicherheit,
https://doi.org/10.1007/978-3-658-47885-8_9

9.1 Introduction

Recent crises, such as the Russian-Ukrainian conflict and the COVID-19 pandemic, have highlighted the impact of an excess of both reliable and incorrect information, particularly on social media and messaging platforms such as Telegram. Managing this substantial volume of information poses a challenge for governmental policies and societies that are resilient in times of crisis. Indeed, misinformation has the potential to deceive, lead to polarization [325, 395], manipulate elections [283], and thus pose a challenge to democracy. In alignment with previous studies [15, 94, 330, 597], the term misinformation will be employed as a comprehensive umbrella term that encompasses both intentionally fabricated deceptive information ('disinformation' or 'fake news') and unintentionally generated misleading information ('misinformation').

Much research has been conducted on the benefits of social media in disruptive situations, as it functions as a vital information channel [195]. However, content posted on social media platforms (such as TikTok videos or posts on X (formerly Twitter)), and voice messages sent through messaging applications like Telegram or WhatsApp, often contribute to the dissemination of misinformation both during crises and in day-to-day life [384]. Indeed, misinformation in voice messages, particularly within popular messenger apps such as Telegram and WhatsApp, have emerged as prevalent and concerning phenomena [351]. With messengers implementing the feature of sending and receiving voice messages (e.g., in 2013 on WhatsApp [602]), it now is an extensively used feature for a majority of users [351, 603]. Like for text, images and videos, voice messages are a common format to spread misinformation [458]. For example, in April 2020, a WhatsApp voice message sharing misleading information about the anticipated death tolls due to COVID-19 and subsequent medical response in the UK circulated[1]. Indeed, the audible nature of voice messages with its opportunity to demonstrate and evoke emotions through comes with unique implications for misinformation spread [351]. Given the widespread use of the specific format of voice messages and the potential impact on public perception, it is crucial to develop effective countermeasures for maintaining information integrity and fostering a more informed digital society, mitigating the consequences of misleading and false information. The appropriate handling of misinformation in voice messages is significant from various standpoints. Firstly, as messenger users, individuals are presented with a multitude of information and face the task

[1] https://www.manchestereveningnews.co.uk/news/greater-manchester-news/warning-over-hoax-ambulance-voice-18067450

of discerning accurate information from misinformation. Especially in uncertain periods, such as during crisis situations, for example floods, affected persons and helpers seek answers in public messenger groups, thus increasing their susceptibility to fake news. On the other hand, official authorities and decision-makers encounter the challenge of identifying and rectifying rumors and misinformation rapidly, particularly in times of crisis. Technical approaches can assist with managing the abundance of information. The user-centred development and evaluation of technical interventions is noteworthy from both a public policy and Human-Computer Interaction (HCI) research viewpoint.

The ability to share information across messenger groups enhances the pace of information dissemination [122]. The use of speech can convey emotions that may unintentionally cause individuals to propagate false information. Moreover, Wallbridge, Bell, and Lai [590] have illustrated that individuals tend to pay more attention to the method of delivery rather than the content itself. This work adds to existing findings on misinformation in text by emphasizing and addressing the significance of misinformation in voice messages as a prevalent phenomenon, facilitating the education of individuals on characteristics to evaluate the trustworthiness of voice messages. The goal of this study is to expand the present research in public policy by creating a proof-of-concept digital intervention as a technological solution after (manual or automatic) detection with the potential to enhance users' media literacy, and obtaining qualitative and quantitative user-centered feedback on its comprehensibility and usefulness. Addressing users' need for transparency in interventions [299], we concentrate on using comprehensible indicators as cues for misinformation in the evaluation of voice messages. Indicator-based interventions have already been developed and discussed for other modalities, such as text [350]. Hence, we will extend these interventions to voice messages, taking into account their unique attributes such as tonality and speech rate. With voice messages as a common way of communication for users of all ages, we look at it from the perspective of both government agencies and private individuals. We researched the existing literature on characteristics of misinformation in both written and spoken language to gain insight into the attributes of voice messages that could indicate misinformation. Additionally, we created a digital intervention as a proof-of-concept that automatically identifies multiple features in voice messages as possible indicators of misinformation. To gain insight into the potential usefulness of various features as indicators of voice message credibility for users, a user study was conducted using the think-aloud method.

We enhance public policy and HCI research by utilizing the promising existing knowledge of indicator-based interventions, which link media literacy enhancement and technological solutions after automatic or manual detection of misinformation, and applying them to the modality of voice messages as a modality with unique misleading potential. This approach is relevant for the design principles of policy frameworks and can be crucial for public authorities and private users, particularly in relation to public messenger groups. Our study provides initial insights into the perception of the approach by a diverse group of private users. There is potential for future research to extend these findings to practitioners from public authorities. Our main contributions (C) and findings (F) consist of (C1) a systematic identification of voice message features that may serve as potential indicators of misinformation. We (F1) identified 35 specific features on levels of characters, words, sentences, audio, and creators. Additionally, (C2) we developed a proof-of-concept misinformation intervention based on these indicators as a potential media literacy enhancing technological solution and public policy response to the misinformation challenge, and (C3) assessed the perceived comprehensibility and usefulness of numerous features for laypersons listening to voice messages, which revealed that (F2) features directly linked to the content or creator are deemed particularly useful, as opposed to more abstract attributes of the voice message. The study also highlights (F3) the challenges and limitations of the indicator-based approach for voice messages, specifically regarding its presentation to users.

9.2 Related Work

Our research adds to the body of research on digital, user-centered interventions that combat misinformation by promoting transparency and comprehensibility, thereby enhancing media literacy as a public policy response to the misinformation challenge. Related work has already demonstrated the significance of misinformation in social media, particularly during crises. Initial technical support approaches for dealing with this phenomenon have been developed and evaluated. Indicator-based approaches are a particularly promising option for enhancing media literacy. In accordance with other researchers, in this work we define media literacy as the ability to decode, evaluate, analyze, and produce both print and electronic media, that is, to have internalized a sense of "critical autonomy" in dealing with all media [28]. In the following, we discuss the relevance of misinformation in social media for public policy and how it motivates our work (see Sect. 9.2.1). We further review related work that addresses a demand of current research for technological solutions that facilitate media literacy (see Sect. 9.2.2), and literature on features of misinformation regarding both text and audio, serving as a basis for our indicator-based approach (see Sect. 9.2.3). We add to related findings by developing a proof-of-concept misinformation intervention for the assessment of voice messages based on the identified features and evaluating its user-centeredness as a technological solution.

9.2.1 Public Policy Perspective on Misinformation

Social media is indispensable for information exchange during daily life. It comprises providers like Facebook, X (formerly Twitter), Instagram, and TikTok among others, but also messaging apps like Telegram with 700 million and WhatsApp with 2 billion active users monthly in 2023 [538]. Not only in everyday life but also in crises such as floods or pandemics, social media is crucial for both individuals and authorities [102], facilitating attempts to improve safety and security [460]. The research field addressing social media during crises is often referred to as crisis informatics [202, 405] and combines findings from various disciplines such as computer science, HCI, social science and political science. As an important subdiscipline, public policy and administration [236] investigates implementations of policies for public services like emergency services [459]. Indeed, governments and authorities are now increasingly confronted with the usage of information technologies including social media, for example when aiming to engage with citizens during crises. While there are different theories on the relationship between technology and public policy and administration, the sociotechnological theory is most commonly used [451]. It denotes the interrelation of social and technical aspects, meaning that technological progress, organizational needs, and individuals' attitudes shape the utilization of information technology within an organization, thereby impacting social change [451, 459].

While crisis informatics and related research fields have pointed out the advantages and potentials of social media, it comes with several challenges and risks as well [288]. The rapid information access, large amount of data, and (supposedly) anonymity pose a challenge to assess credibility and recognize misinformation. This applies to individuals as well as officials like crisis managers who create and analyze content [248]. For instance, messaging apps with public channels do not only facilitate aid networking during crises but also face criticism for permitting unrestricted spread of conspiracy theories and propaganda [197, 231]. Governments around the world have identified misinformation as a serious threat to national security, prompting calls for regulation of internet technologies [258, 498]. Disinformation campaigns, for instance those launched by Russia and

China in relation to the COVID-19 pandemic [71], impact international diplomatic relations, foreign policy, but also other sectors such as security policy and health policy. As a consequence, governments aim to strengthen confidence in their statements and empower citizens to recognize accurate information [108]. Komendantova, Erokhin, and Albano [305] use the example of the migration discourse to explain the connection between misinformation and policy-making in that misinformation leads to a false public representation of an issue. The general public, who can participate in and influence legislation in various ways, is then prevented from adopting positions grounded in evidence-based decisions due to the inadequate information landscape. Synthesizing knowledge from misinformation research and public policy can be considered a crucial step towards "informed recommendations to [...] public administrators" [108, p.1]. There are three main demands for policy responses to the misinformation challenge in social media identified and discussed in literature, to which our work strongly picks up on: (1) enhance users' media literacy skills, (2) promote technological solutions and self-regulation by platforms, and (3) impose liability laws on platform operators to remove false or misleading content [355, 498, 549]. Our work aims to specifically offer support to public policy approaches targeting media literacy and platform-specific technological solutions by investigating an integration of both. To achieve this, we apply an indicator-based approach that uses characteristics of misinformation in voice messages to assist users in making informed decisions about credibility. In the following, we give an overview on related research on indicators of misinformation that highly motivated our work.

9.2.2 Indicator-based Misinformation Interventions

Developing effective measures to combat misinformation is challenging, both from a political perspective as well as from an HCI perspective. Non-punitive policy interventions that enhance users' media literacy skills as technological solutions that may be applied as self-regulation by platforms [355, 498, 549] are deemed more appropriate than policy approaches that are mainly limited to sanctions and carry the risk of collateral censorship because platform operators are incentivized to excessively delete content to avoid penalties [37]. This is relevant also from a user perspective as users strongly prefer explainable handling of misinformation over simply deleting or flagging content without explanations [299]. Due to its user-centered and potentially media literacy enhancing nature, this approach may be incentivized through policy in a liberal democracy and has been emphasized as a necessary step to "prevent overzealous regulation of speech platforms by the government" [355, p.288] for instance in India. While it is insufficient if taken separately as a focus on purely technical solutions neglects societal factors [355], digital misinformation interventions are developed to complement critical journalism and media literacy education in schools. Technological solutions can take various forms [214] and differ in their suitability to address the demands for policies encouraging media literacy and laws on platforms, including digital interventions that display a warning or correction next to disputed content without hiding the misinformation itself [49], default nudges to promote critical thinking for all social media posts [52], or machine-learning based approaches that automatically identify misinformation and delete or hide it [518]. In this work, we pick up on the idea of post-detection decisions from a user perspective or an approach detached from detection (e.g., via default nudge to encourage reflection). Research suggests that transparency is crucial when aiming to establish trust among users in digital interventions [299] and minimize reactance or other backfire effects [391]—this strongly motivated our work.

Addressing policies to encourage media literacy, efforts have been made to develop digital interventions encompassing the user-centered display of indicators as comprehensible characteristics of misinformation to offer a guidance in autonomous assessment strategies while promoting critical thinking and trust [51]. For instance, Bhuiyan et al. [51] take up the idea of indicators by investigating their utilization like information about the author of social media posts from the perspective of journalists and news consumers. Indicators for the credibility of social media posts have been investigated regarding utility and user preferences for visualization, revealing positive feedback as long as the display of indicators was rather simple [188]. Indeed, research has demonstrated that indicators as user feedback can enable the development of own assessment skills [491], reduce uncertainty [188], and fit pre-existing mental models and practices of users [513]. The vast majority of research has focused on textual content like posts on X or Facebook [214], some extend to images and videos [49, 224, 513]. For example, Sherman, Stokes, and Redmiles [513] identified the source of information as key indicator when assessing different types of content, including fake videos. Hartwig et al. [224] similarly examine the perceived usefulness and comprehensibility of misinformation indicators in TikTok videos within a simulated smartphone app. Messaging apps like WhatsApp and Telegram offer the possibility to share voice messages in private conversations and (public) channels, encompassing an additional prevalent modality of information that nonetheless is confronted with misinformation. Research in (indicator-based) interventions has examined that context less exhaustively, however providing some related studies that guided our approach. For instance, Burgoon et al. [78] use the audio modality to identify characteristics in deceptive speech from discourse, examining both asynchronous (e.g., text chat) and synchronous (e.g., face-to-face) forms of communication, and Maros et al. [347] analyzed misinformation in voice messages and identified indicators we build upon. In the following, we will present research that tackles features of misinformation in various modalities and led to the categorization of characteristics of voice messages guiding our approach (see Section A.4, Electronic Supplementary Material (ESM)).

9.2.3 Features of Misinformation in Text and Audio

To develop indicator-based misinformation interventions aiming to encourage media literacy within a technical solution after manual or automatic pre-filtering, it is necessary to form a knowledge base on typical characteristics of misinformation. Our work focuses on the context of misinformation in voice messages—thus in the following we discuss related research on features of the audio modality that informed our work. As the research landscape of textual content is by far more exhaustive and may be applicable to voice messages when using their transcript, we will further consider insights on textual features. Characteristics of misinformation discussed in related work helped us derive an overview of those potentially applicable to voice messages (see Section A.4, ESM) as a foundation for our technical implementation.

9.2.3.1 Text features

Considering text features for misinformation detection also involves exploring related research areas [518]. These include authorship recognition [1, 6, 630], deception detection [6, 158, 470], clickbait detection [89, 434, 518], and hyperpartisanship detection [433].

For instance, people spreading misinformation might alter their writing style to avoid attribution. Authorship recognition and deception detection analyze style changes to uncover the original author [630]. Zheng et al. [630] classify writing-style features into lexical, syntactic, structural, and content-specific categories for authorship recognition. They seek features that remain consistent across texts from the same writer. Building on that and along with insights from deception and authorship studies [6, 68, 78, 208] investigate deceptive writing style. The study demonstrates that deceit detection benefits from content-related and non-content-related aspects, including authorship mimicry and masking. Feng, Banerjee, and Choi [158] concentrate on deception detection, particularly in product reviews.

Their study employs features like "term frequency—inverse document frequency" for word uni- and bigrams, as well as tf-idf of shallow and deep syntax (utilizing Part-of-Speech tags and Probabilistic Context Free Grammar, PCFG). Both word features and deep syntax aid deception detection, with optimal performance when combined. The connection between authorship recognition and misinformation detection stems from the notion that misinformation spreaders possess distinct writing styles from those sharing genuine information. This suggests treating misinformation detection as an authorship recognition task to differentiate these types of spreaders. Hyperpartisan detection, closely linked to misinformation research, deals with one-sided messages omitting pertinent information [433]. Potthast et al. [433] examine whether stylometry can reliably distinguish hyperpartisan from mainstream news, and explore its potential for misinformation detection. They find hyperpartisan articles are shorter on average, with similar style between left- and right-wing documents. However, stylometry alone struggles to consistently identify misinformation.

Clickbait manipulates readers into clicking on links and reading articles, constituting a form of deception [98, p. 15]. Chakraborty et al. [89] analyze features distinguishing clickbait titles from non-clickbait ones. They note that clickbait titles generally exhibit longer syntactic dependencies, attributed to higher grammatical complexity. Additionally, clickbaits commonly contain words with "very positive" sentiments, like "awe-inspiring". In contrast, Potthast et al. [434] assess clickbaits in tweets across three feature categories: the teaser message, linked web page, and associated meta information. They identify meaningful features such as tweet-wide sentiment polarity, stop-words-to-words ratio, and whether the tweet starts with a number.

9.2.3.2 Audio features

The comprehensive research on features of misinformation in text stands in contrast to a less examined research field of audio misinformation. While there is research that includes the audio track of videos in the detection of misinformation [504], this research uses black-box approaches, and is therefore not applicable to extract potentially explainable features. Others that use audio modality to identify features in deceptive speech, focus on an interrogation scenario [e.g., 78, 326]. Burgoon et al. [78] aim to detect deception from discourse, where they examine both asynchronous (e.g., text chat) and synchronous (e.g., face-to-face) forms of communication. The researchers divided indicators into four types: *quantity*, *vocabulary complexity*, *grammatical complexity*, and *specificity and expressiveness*. They concluded that "deceivers do utilize language differently than truth tellers" [78, p. 91]. Furthermore, it was observed that indicators may vary according to the mode of communication: participants employed fewer conjunctions when using text chat than when using audio chat. Nevertheless, the research team remarked that the participants had limited time to prepare for synchronous communication. As a result, it is suggested that given ample preparation time, the participants may have used language differently. This observation highlights a significant contrast between synchronous discourse and the dissemination of misinformation through voice messages. Voice messages are unidirectional, providing the spreaders with sufficient time to prepare and rehearse the delivery before dispersal.

Furthermore, this observation is supported by the research of Maros et al. [347]—who analyzed misinformation in voice messages—as they found indicators that contradict the findings of Burgoon et al. [78]. One such example is the number of words spoken by deceivers: While Burgoon et al. [78] found that deceivers said "less than truth tellers" [78, p. 94], Maros et al. [347] found that voice messages containing misinformation were longer on average [347, p. 9]. Another difference is that, while Burgoon et al. [78] found that deceivers use less emotional language [78, pp. 94–95], Maros et al. [347] found that voice messages with misinformation tend to contain more negative emotions [347, p. 9].

While many features can be derived from text as transcribed voice messages, others are based on the audio modality. Research has found that speech rate and sound volume may be used as an indicator when assessing credibility of audio material [351]—an insight we build on in our user study.

We have collated the characteristics highlighted in previous research pertaining to the identification of misinformation in text and speech tasks, which are outlined in the ESM (see Section A.4). The properties that can be deduced from voice messages are categorized into five levels: character, word, sentence, audio, and creator level. In this study, we facilitate research into the suitability of these characteristics for voice messages by providing a digital intervention that automatically extracts them from audio material. We then present initial findings on their comprehensibility and perceived usefulness.

9.2.4 Research Gap

Governments around the world have identified misinformation as a serious threat to national security, prompting calls for regulation of internet technologies [258, 498]. This study advances research in public policy regarding a user-centered perspective on misinformation interventions. It accomplishes this by identifying potential indicators of misinformation in voice messages and integrating them into an automated program, aiding further studies on indicator-based interventions. We address two gaps:

Firstly, while misinformation investigations often focus on written content, the relevance of voice messages in messaging apps (e.g., Telegram, WhatsApp) during recent crises highlights the need to consider them as substantial sources of information. For instance, Telegram groups have been significant in propagating conspiracy content [231] and misinformation propagated via mass-messages on WhatsApp are perceived to have functioned as a catalysator for lynchings in India [355]. This emphasizes the necessity to expand the user-centered technological perspective (*1st gap*).

Secondly, interventions combating digital misinformation have shown promise when providing transparent indicators or explanations [299]. Overcoming reactance and meeting users' need for comprehensibility have been underscored [299, 377, 391]. While indicators for text-based content have been partially explored [31, 171, 350, 406], their application to voice messages and audio material is incomplete *(2nd gap)*. While existing research examines misinformation in speech, mostly in impromptu discourse (Sect. 9.2.3.2), we expand on studies analyzing misinformation characteristics in voice messages [347, 348] from a technical and user-centered standpoint, including a proof-of-concept. Our prototype solution is extended by qualitative findings (and supplementary quantitative insights) to demonstrate technology's potential in managing the overabundance of information, researchers, crisis managers, and messenger users face.

We address the research gaps by answering the following research questions:

RQ1: *How can an automated tool be designed to extract features of misinformation in voice messages, supporting public policy efforts for technical solutions that enhance media literacy?*

RQ2: *How do users evaluate the comprehensibility and usefulness of the automatically computed indicators when assessing the credibility of voice messages?*

Our contribution comprises two main aspects. In Step 1, we created an extendable automated program that calculates potential indicators of misinformation from voice messages using features identified in relevant literature (see Sect. 9.3). Subsequently in Step 2, we offer preliminary insights into user-perceived comprehensibility and usefulness of these indicators through a think-aloud study (see Sect. 9.4).

9.3 Step 1: Automatically Extracting Features of Misinformation in Voice Messages

Technical approaches that support crises authorities, crisis management organizations, laypersons, and researchers in dealing with misinformation can contribute to media literacy development, complementing training and professional journalistic work and greater crisis resilience within society. This is especially the case if the solutions use transparent and understandable explanations or indicators. We refer to *features* as specific characteristics of voice messages that have a value during the analysis. For instance, one can analyze and compare features like 'number of unique words' between multiple voice messages. In contrast, we refer to *indicators* as specific features of voice messages that are explainable to end users and indicate that a voice message contains misinformation (e.g., 'The voice message contains common buzzwords related to conspiracy theories').

We designed and implemented a program using *Python* as the backend for such a technical approach. In the future, a user-centred frontend and, for example, an integration within a smartphone app or desktop application could expand the backend. The program receives voice messages and processes them to output features and explanations, as potential indicators for misinformation. For now, we provide Command Line Interface (CLI) functionality. In total, the envisioned usage scenarios are as follows:

1. A layperson wants to analyze a voice message in terms of credibility. This requires functionality that generates indicators for authentic information and for misinformation when given an audio file.
2. A crisis management organization or authority needs to analyze a large dataset (e.g., to find indicators for misinformation and identify trends during a crisis). This requires human-readable output, and functionality that supports the analysis.
3. A researcher needs to incorporate features for a detection model. This requires functionality that, given audio files, generates features that may be relevant to distinguish between authentic information and misinformation. Furthermore, the functionality should be easily extendable and maintainable to accommodate any future application upgrades by researchers.

The program accepts following types of input data:

- a single audio file to extract features for a downstream detection model from voice messages or other audio material
- a single text file (as previously transcribed audio file) to extract features for a downstream detection model from transcribed voice messages or other text material
- a directory that contains a dataset of text files (as previously transcribed audio files) to allow researchers to analyze its features for detection or indicators for end users
- previously extracted features to save time and compute during analysis of datasets

Given the input data, the program uses rule-based and traditional statistical approaches to compute indicators of misinformation to facilitate explainability. This allows users (e.g., laypersons or crises authorities) to check the credibility of voice messages and understand whether—and how—they might be designed to deceive. We generate an explanation for each feature on the server to make them easier to understand for end users.

9.3.1 Data Processing

When audio input is being processed, the program first performs speech-to-text by using *Whisper* [446]. This step is skipped when text input is being processed Then, given (extracted) text, *spaCy* [238] performs preprocessing including tokenization, lemmatization, Part-of-Speech (PoS) tagging, and dependency parsing, which is required to compute word- and sentence-level features. In contrast, character-level features are computed directly on plain text. Thereafter, the computed features are stored in a human-readable .JSON file. Lastly, if features are regarded as comprehensible indicators of misinformation and explanations are necessary, the program extracts these indicators from the generated instance. In case of dataset input in contrast to single files, the processing elaborated above is executed for all files in the dataset to create one instance per file. After all instances are created, we store the instances in a *Dataset* instance and compute averaged features. The outputs as human-readable .JSON files may be opened (e.g., in a browser) for analysis by professional members of crises authorities or researchers. They can then be sent to a detection model for training or testing, or to a digital application as a whole (e.g., a smartphone app) to be displayed to end users. You can find the data flow of the program in Fig. 9.1.

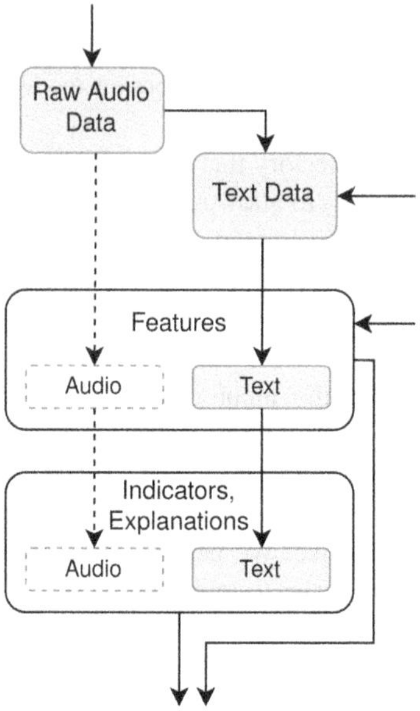

Fig. 9.1 The program's data data flow. While the solid lines represent current data flow, the dotted lines and boxes represent data flow the program may be extended by in follow-up research

9.3.2 Computed Features

The choice of features we compute in our program is motivated by previous research we summarized in Sect. 9.2. You can find the list of features in Table A.4, ESM. While we included a broad variety of features on character, word, and sentence level, this selection is not exhaustive and may be extended based on a systematic literature review, especially on the explicit speech modality that may imbue emotions. Existing research indicates that speech rate and sound volume may compose a relevant attribute of voice messages when assessing its credibility [351]. While these features are not yet included in the prototype implementation, they were part of our user study. For the implemented feature of 'keywords for a specific topic', we specified this for our user study based on existing research. Specifically, we included whether there was a 'claim to be an expert' (e.g., 'I have researched that!') [351] or a 'call for action' (e.g., 'Share this voice message!')

[347, 351]. Additionally, we considered the existing research on the relevance of creator's profiles and thus included whether a voice message was sent by a 'suspicious profile (e.g., name, description of the creator)' which might be detectable via keywords or the excessive usage of emojis.

Like Potthast et al. [433, 434], we use the *General Inquirer* dictionary to compute features that depend on linguistic word categories (e.g., words that express uncertainty). The categories that are used to analyze a file or dataset can be specified in the configuration file. For instance, words related to certainty or uncertainty were derived from the categories 'SureLw' and 'If', words related to affect were derived from the categories 'AffTot', 'NegAff', and 'PosAff', and pronouns were derived from 'Self', 'Our', and 'You'[2]

9.4 Step 2: A User Perspective on the Indicator-based Approach

Having implemented a program that allows to automatically extract a plethora of features from voice messages as potential comprehensible indicators to assist users in dealing with misleading information, as a next step we gained initial qualitative and quantitative insights into the comprehensibility and perceived usefulness of indicator-based approach. In the following, we describe the study design and derived results.

9.4.1 Study Design

We conducted a user study (pre-study $N = 5$, main study $N = 15$) using the think-aloud method, where participants were asked to express their thoughts aloud while listening to audio material and answering questions or performing tasks [163]. Participants were assured that there is no right or wrong, we are rather interested in their subjective perceptions. In said study we aimed to gain insights into (1) which aspects participants autonomously pay attention to when assessing the credibility of voice messages without any assistance, (2) if the participants' perception towards the information's credibility changes when we additionally display automatically computed features from our developed python program,

[2] for further information regarding the categories, see https://web.archive.org/web/20171007184158/http://www.wjh.harvard.edu/~inquirer/homecat

and (3) which features the participants perceive as comprehensible or convincing indicators for misinformation when assessing the credibility of voice messages.

9.4.1.1 Selection of stimuli

The stimuli in our user study consist of four voice messages and corresponding features. See Table A.4 (ESM) for a detailed description.

Voice messages. A systematic approach was used to identify stimuli. In the absence of accessible databases of voice messages, official fact-checking websites were first checked to determine which topics were currently circulating and had been officially debunked. Then, open telegram groups were searched for voice messages on these topics. Two real-world voice messages could be identified this way. We made sure to include different political perspectives, however only politically right-wing misinformation was found in the exploratory search in messenger groups. Thus, we supplemented our stimuli with two artificially generated voice messages. For this purpose, corresponding misinformation in the form of other modalities (Twitter post, Youtube video) was identified on fact-checking sites and, based on this, voice messages were generated by a researcher. Typical characteristics of voice messages were taken into account, such as a length of at least 1:45 minutes [348].

Features. We generated the features as potential indicators for misinformation and their explanations as follows: (1) We used *Whisper* [446] to transcribe the four audio files we collected or generated. (2) Given the transcribed text for each of the audio files, we used our python program's *get-features* command to extract features, such as "number of words", and "number of words that express uncertainty". (3) This creates four *.JSON* files that contain features grouped by character, word, and sentence level. Additionally, we included features on audio and creator level that are known from literature and are not (yet) included into our python program. These features were extracted manually. We did not include all computable features in our user study, as especially for those on character level it is foreseeable that this is not suitable for reporting to private individuals but is more suitable for practitioners who deal with the subject in depth (e.g., researchers developing a detection method). It is crucial to emphasize that the features are not defining for misinformation but can appear and be computed for reliable information as well. As in other indicator-based misinformation interventions, the approach relies on a preceding manual or automatic detection of misinformation as pre-filtering (see Sect. 9.2.2). For a full list of features we presented in the study, see Table A.4, ESM.

9.4.1.2 Participants

In total, we recruited twenty participants (pre-study $N = 5$, main study $N = 15$). As we only used the data of our main study for explicit evaluation, we present the participant's characteristics of the main study with $N = 15$ in the following. The participants' age ranges from 19 to 63 years (*Median* = 32) and covers a diverse set of educational levels (e.g., high school diploma, professional training, university degree). We explicitly decided against recruiting solely college students as these are already overrepresented in misinformation studies due to their easy accessibility and they do not represent the average user. Of the fifteen participants, seven were male, six were female and two were of diverse gender. All but two participants reported German as their native language. Twelve participants reported to receive voice messages at least once a month, and four participants report they might have received misinformation per voice message before. Nine participants stated to currently use at least one messenger group with more than 50 members. As two of our stimuli address misinformation and reliable information on the topic of the Russian war against Ukraine, we further evaluated central general attitudes towards these topics and a general political orientation of our participants. All but one participant fully agree with Russia waging a war of aggression against Ukraine. We asked our participants to place themselves on a scale from 1 (extreme left) to 11 (extreme right) regarding politics. Three participants placed themselves on the far left between 1 and 2, three rather left between 3 and 4, nine in the center between 5 and 7, and none further right than this. Participants were acquired through the panel provider *Prolific* and got an expense allowance of €9 for an average duration of 45 minutes.

9.4.2 Study Procedure

The one-on-one think-aloud sessions were conducted online via *Zoom* in summer 2023. (1) After a brief introduction and obtaining written consent, the participants were (2) forwarded to an online survey for collecting demographics and general usage behavior of messengers. Afterwards, (3) the researcher subsequently displayed four voice messages in random order while sharing the screen. Participants were able to see the original interface of a Telegram group where the voice messages are displayed. After each voice message, the researcher began audio recording the session and the participants were verbally asked to rate the credibility of the voice message. Participants were instructed to keep thinking aloud and were told that there was no wrong or right, and that we were only interested in their individual assessment [163]. (4) The researcher then subsequently displayed the calculated features and asked the participants to rate its comprehensibility and usefulness the assess the voice message's credibility giving explanations. Then, participants were asked to (5) rate the credibility of the voice message again. The researcher clarified the origin and truthfulness of the voice message. Finally, (6) participants were asked to give an overall evaluation of the features and the general approach.

9.4.3 Analysis

Step 2 generated rich qualitative data and complementing quantitative survey ratings on a scale from one to five for comprehensibility and perceived usefulness. This allows for a triangulation of data, combining quantitative survey ratings with qualitative explanations. We transcribed the audio records using Whisper [446], followed by a manual revision and anonymization of participants' responses. We analyzed the quantitative survey items calculating descriptive statistics (*mean*, *median*, *frequencies*) and clustered the verbally given explanations according to the referenced features thematically using *MaxQDA* where a qualitative content analysis was conducted. Codes refer to the individual features and were created prior to the analysis. Additional assessment strategies of participants were added iteratively as novel codes when they emerged during coding.

9.4.4 Results

We report how a diverse set of laypersons as messenger users assess the credibility of voice messages autonomously, and how they assess the comprehensibility and perceived usefulness of a plethora of features.

9.4.4.1 Autonomous Assessments

While users have many assessment strategies regarding layout, interactions and reactions for other modalities (e.g., textual social media posts), they seem to rely mostly on their knowledge when assessing voice messages and only partly relying on characteristics like the usage of specific words or tonality. Specifically, all 15 participants stated to at least partially use their *existing knowledge or gut feeling* when assessing the voice messages in our study. Two participants explicitly stated to consider if the *speaker uses swear words or other negative expressions.* Particularly when *conspiracy theories* were addressed in the voice message, participants were very capable of emphasizing how those keywords (e.g., 'BRD GmbH' as a common conspiracy theory in Germany) influenced their assessment of the voice message. Some described how the voice message seems *contrived and disjointed* and, thus, indicated to be misinformation. Others stated how they differentiated between *personal opinions and facts* when listening to a voice message, for instance via wordings like 'I think'. Only one participant stated to consider the *tonality* of a voice message when assessing its credibility, evaluating if it sounded angry or very emotional in another way. Overall, the majority of participants were very capable of correctly assessing the voice messages credibility for voice messages 1 and 2 where the content was clearly misinformation. Many were unsure when assessing voice messages 3 and 4, which were containing both genuine and wrong aspects of information. We found that three participants corrected their credibility assessment for these messages for the better after being confronted with our indicators, as these nudged a more thorough critical reflection of the content. This effect might have appeared as well, when autonomously reflecting on the content without the indicators for a similar amount of time. Thus, we cannot derive reliable implications regarding a learning effect at this point of the study.

9.4.4.2 Perceived comprehensibility and usefulness of features

Our participants rated the comprehensibility and usefulness on a scale from 1 (not comprehensible/useful at all), over 2 (rather not comprehensible/useful) and 3 (rather comprehensible/useful) to 4 (very comprehensible/useful) and gave additional qualitative explanations and reasons.

Word level. We included several features at word level that can be automatically detected using our prototype. While all features at word level were generally perceived as comprehensible (*mean* > 3.0 for all features) and participants were mostly able to correctly describe what they meant, the perceived usefulness varied among more abstract and more content-related features. On the abstract part, the *number of words* was perceived as rather not useful (*mean* = 2.0) for several reasons. Most prominently, participants criticized that the number of words or corresponding length of a voice message has nothing to do with the content of the message and can be considered a personal trait. While some tried to make sense of it (*"I guess [...] then it's probably to indicate that the lady doesn't have too much information."* (P05, over 56 years, male)), the majority was confident that this feature was not very suitable at first sight.

When taking a closer look at word level features that tend to be less abstract, the perceived usefulness is overall higher.

Some features at word level can be directly related to the content which increased the perceived usefulness. This applies to *verbs that refer to a belief (e.g., I think) (mean* = 2.9; *"Everyone can believe anything." (P03, 26–30 years, female)*) and *words that express uncertainty or certainty* (*mean* = 2.5):

> *"Because this claim to be in sole possession of the truth is already an important characteristic of conspiracy theorists. In that respect, it's also a good unique indicator almost. So if you don't notice all the other things, for me that would definitely be something where I would say: No, I don't think so."* (P09, 36 to 45 years, diverse)

On an even more concrete level, *suspicious words referring to conspiracy theories* were very well received (*mean* = 3.8) in terms of usefulness to assess a voice message's credibility. This is not surprising, as many are very familiar with common conspiracy theories due to (social) media and state it that voice message 2 (see Table A.4, ESM) is like playing Bingo due to the many keywords that occur referring to suspicious content. Similarly, many perceived a *claim to be an expert* (*mean* = 3.1) and *calls to action* (*mean* = 3.0) as rather useful: *"Because that is also related to the claim of truth and that, well, is always a sure sign for me."* (P09, 36 to 45 years, diverse), however always with a need to differentiate between rightful expert claims and calls to action.

Sentence level. Again, for features on sentence level, the usefulness was dependent on how abstract or concrete a feature was received. A high *number of adjectives and adverbs* is perceived as a way of figuratively embellishing a situation to give the impression that the author experienced it personally. This can be used to mislead intentionally, however it was perceived as rather not useful (*mean* = 2.4) due to its abstract nature. The participants struggled with the feature *number of first-person pronoun usage* and spent some effort and time to make sense of it: *"Somehow he may have experienced something that practically just happened, but [...] may not objectively describe some situation"* (P01, 30–35 years, male) However, the feature was overall perceives as rather not useful as well (*mean* = 2.2). Similarly, the *length of sentences* (*mean* = 2.2) and on a broader level, the *vocabulary richness* was perceived as rather not useful (*mean* = 2.1). However, that was very much related to the expression of features. The vocabulary richness of all four voice messages was moderate, which prevents a meaningful comparison within the study. Some expressed it might be useful when there were more extreme expressions of vocabulary richness:

> *"I think the vocabulary used, if it is particularly rich, could perhaps indicate that the person is very well educated, and thus might also provide an indication of truthfulness, but with an average vocabulary, I wouldn't say that's indicative of truthfulness."* (P10, 19–25 years, female)

Similarly, when assessing the usefulness of *number of words with a positive or negative sentiment*, it depended if the expression of the feature was clearly positive or negative, or if it tended to be balanced (*mean* = 2.6). Some participants were very aware of the potential to intentionally use emotional language to mislead and influence: *"Because, sure, using fear you can convince people."* (P03, 26–30 years, female)

Audio level. Both *speech rate* (*mean* = 2.2) and *sound volume* (*mean* = 1.9) were perceived as rather not useful when assessing voice messages. Participants stated that this is a very individual factor, where some tend to speak fast or low and others do the opposite. Nevertheless, some emphasized the misleading potential of intentionally controlling how fast or loud you speak: *"because I think that the, how shall I say, that with the slow speaking speed is also supposed to get across in the message that there is a deep emotional touch somehow."* (P05, over 56 years, male; referring to voice message 1)

Creator level. As expected due to existing research on other modalities or social media platforms, the feature of a *suspicious profile (e.g., name, description of creator)* was perceived as useful (*mean* = 3.7). Indeed, several participants autonomously referred to the suspicious name of the creator when listening to the voice message (and seeing the name of the creator in voice message 2): *"Because of the many emojis in the name and those red punctuation marks and so on. That is almost common knowledge."* (P09, 36 to 45 years, diverse)

9.4.4.3 Overall assessment of the indicator-based approach

Our findings give insights into the overall applicability of showing features of voice messages as indicators for misinformation. Our participants emphasized that it is important to consider the features as a whole rather than as stand-alone components. They noted that these features only make sense when perceived together, in order to obtain a full picture: *"It is comprehensible, but I think it is useful only in combination with all the other indicators"* (P09, 36 to 45 years, diverse) In contrast, this participant also highlighted specific features as a unique indicator (see *words that express uncertainty or certainty*). Furthermore, upon closer examination of all levels, perceived usefulness is closely related to its connection to the content or creator. Participants found features more useful when they could easily recognize its relation to the content. Conversely, more abstract features received less positive feedback.

In contrast to digital misinformation interventions for text, image or even video, voice messages come with the unique attribute of not having any visual indicators. This leads to a particular challenge for interventions which also was highlighted in our study: Receivers of voice messages listen to the message (often only once) and afterwards there are no visual reference points to remember when confronted with a list of indicators. Our study highlights that after listening to the message and gaining feedback (e.g., long sentences), people often do not remember and cannot verify the indicators without the effort to listen to the message again: *"I thought I had understood it very well, but now I realize how quickly you can forget what you have just listened to."* (P03, 26–30 years, female) This is very different to other modalities, where intervention designers can easily place the indicator display directly at the place of occurrence.

9.5 Discussion

In this study we derived a structured overview on features of voice messages as potential indicators for misinformation in voice messages to develop a prototype of a digital intervention and evaluate the perceived comprehensibility and usefulness of computed features for layperson when listening to voice messages. Thereby, we extend existing research on indicator-based misinformation interventions for text [350] and research in misinformation in voice messages [347, 351]. This allows for in-depth findings on the suitability of features on the character, word, sentence, audio, and creator levels as potential user-centered indicators.

9.5.1 RQ1: How can an automated tool be designed to extract features of misinformation in voice messages, supporting public policy efforts for technical solutions that enhance media literacy?

Governments, authorities, and individuals are confronted with an overabundance of both true information and misinformation on social media, during crises and in everyday life. This includes not only tweets on Twitter/X, TikTok videos, or Facebook posts, but also content on messaging apps like Telegram and WhatsApp with public channels, where the distribution of misleading voice messages has shown to be a common problem in recent years [458]. Research has highlighted the serious threat of misinformation and disinformation governments are confronted with, emphasized needs for regulation of online content [258, 498], and the necessity to find ways to empower citizens to distinguish misleading from credible information [108] (see Sect. 9.2.1). Indeed, public policy research aims to define implications for policy makers when investigating the effects of misinformation and its mitigation, including suggestions to maintain fact-checking and flagging efforts [131] and applying non-censoring techniques like labeling and limiting algorithmic promotion [129]. When looking into demands for policy responses to misinformation on social media, there are multiple ways to contribute to these necessities via user-centered technology. As Medeiros and Singh [355] and Tambini [549] among others emphasize, encouraging media literacy can be considered central to counteract the effects of misinformation online. We pick up on that idea by proposing a technological media literacy approach for the user-centered presentation of misinformation indicators in voice messages. Research to combat misinformation on social media while encouraging learning effects has mostly focused on text-based content, especially on Twitter/X, Facebook, and more generic news platforms. Our study addresses the context of voice messages with its unique potentials to mislead (e.g., via emotional tonality) [351] and may offer support to public policy efforts targeting technological solutions for social media platforms and the empowerment of citizens [108].

With the development of a backend solution for a prototype that automates the detection and extraction of features of voice messages as potential indicators for misinformation, this work contributes to informed information handling, particularly in messenger channels like Telegram [231, 351]. The prototype receives voice messages or other audio files, transcribes them, and analyzes them for character-, word-, and sentence-level features. Additionally, the prototype facilitates an extension for audio and creator-level features for future work. By doing so, we add to the public policy response to the misinformation challenge, presenting a potential technological solution to enhance users' media literacy and encourage informed decisions on voice messages' credibility. We envision usage scenarios at both the individual and government levels: (1) Private users of messenger services can have their voice messages (e.g., in large public channels) analyzed and receive a comprehensive output on attributes of the message. This allows them to reflect on the content and make a more informed assessment independently. (2) The intervention encompasses a potential part of a policy toolkit that places a greater emphasis on self-efficacy in dealing with misinformation as opposed to highly regulatory approaches by governments that amount to censorship. Additionally, on a micro-level perspective, people working in crisis management or government agencies may utilize the tool to automatically analyze large amounts of data and quickly identify problematic content, such as conspiracy theories, by specifying specific keywords. This support in monitoring and analysis can supplement manual procedures and extend existing dashboard approaches to classic social media platforms like Facebook and X (formerly Twitter). (3) Researchers in detection methods can selectively extract features from large amounts of data to expand and optimize their detection techniques.

The study examined the extracted indicators from the user's perspective with respect to Scenario 1. As a next step, we recommend conducting more comprehensive investigations from the perspective of government agencies in future work. It is essential to collect user interface requirements to expand our backend component, as our study identified unique challenges (e.g., on the placement of indicators) specific to voice messaging in comparison to other modalities, including text and images, where indicator-based interventions have already been addressed [31, 171, 350].

9.5.2 RQ2: How do users evaluate the comprehensibility and usefulness of the automatically computed indicators when assessing the credibility of voice messages?

When assessing the comprehensibility and perceived usefulness of features to assess a voice message's credibility, we received a mostly positive feedback especially on features that refer to the content itself. Indeed, focusing mostly on the content instead of tonality was the main autonomous assessment strategy even before being confronted with our features, sometimes resulting in a certain gut feeling. However, this was very much dependent on the expression of this characteristic. If it was strongly pronounced and deviated more from the average, it was rated all the more useful.

All features received a comprehensibility rating of *mean* > 3.0 which indicates that the wordings of our features as potential indicators were perceived as rather suitable. Indeed, when taking a closer look at our participants' verbal answers and attempts to explain what each feature means, they were overall very capable of doing so. We found that features closely related to the content of the voice message (e.g., suspicious conspiracy theory keywords, emotionally loaded words, and words referring to the speaker's belief like 'I think that ...') were particularly rated as comprehensible. Also, the feature related to a suspicious profile (e.g., name, description of the creator) was considered particularly comprehensible among the participants. Regarding features on a more abstract level (e.g., number of words, length of sentences, number of first-person pronoun usage), some participants struggled with comprehending what the feature means.

We can observe a similar tendency regarding perceived usefulness. Features that relate directly to the content of the voice message or the creator were rated particularly useful. The suspicious profile (e.g., name, description of creator) and suspicious keywords about conspiracy theories are rated as the most useful feature. Furthermore, other keyword-based features like the claim to be an expert or the call to action [351] were very positively received. On the other hand, the abstract features like number of words, sentence length, number of first-person pronoun usage and the percentage of adjectives and adverbs were perceived as rather not useful when assessing a voice message's credibility. While other research on voice messages [351] and short-videos [226] have emphasized the importance of affectively charged content as misinformation attribute, the features used in our study were too abstract to apply that message in a user-centered way. Similarly, this applies for ascertaining credibility of misinformation in voice messages as "eyewitness, expert, or insider" [351, p.5] which was strongly the case in voice message 1 (see Table A.4, ESM) where we expected the number of first-person pronouns to be a comprehensible indicator for that context—an assumption that was fulfilled only for a few participants. Regarding the audio-based features like speech rate and volume, participants again did not perceive them as useful features, extending findings of related research [351].

Overall, our findings emphasize the importance to combine multiple features from different levels to receive a suitable and useful full picture. However, the indicator-based approach comes with a challenge when applied to voice messages in contrast to other content like text [350], image, or video [226], as there are no visual anchors users can relate to and verify when indicators are displayed. This is a challenge that may be further addressed in future HCI research, having a direct influence on how indicators can be presented to end-users.

9.5.3 Limitations & Future Work

Our contribution comes with some limitations and potentials for future work.

First, the proposed digital intervention is yet to be extended by a frontend part that displays the extracted features as potential indicators for misinformation in a user-centered manner. Regarding voice messages, this is not a trivial task, as the modality comes with limited possibilities for visual anchoring of indicators where they occur in the message. This is unique to audio material, as in text, images, and even video content there are more intuitive ways of highlighting indicators visually immediately at their place of occurrence. Further, user needs for displaying the extracted features may significantly vary between the different usage scenarios of private messenger users, practitioners in crisis management, and researchers dealing with detection approaches themselves—a challenge to be addressed in future work.

Second, our indicator-based approach relies on preceding manual or automatic detection of misinformation. User-centered misinformation interventions typically rely on prior successful determination of whether a given piece of information is misleading. This is by no means a trivial task and has been a focal subject of research in recent years [518, 613], particularly within the field of machine learning. Default display of indicators on all voice messages would result in a not acceptable number of false positives and, thus, is not the aim of this approach. Future work could evaluate the fully implemented approach for voice messages including (1) an automatic or manual detection of misinformation and (2) extracting and displaying indicators for content that has been successfully identified as misleading.

Third, evaluating our approach with a larger real-world sample of voice messages is crucial to make valid assumption about its applicability. While for other content like Tweets or news articles there are publicly available research data sets in multiple languages, this is to the best of our knowledge not the case for voice messages. We therefore applied a systematic and convenient approach of looking for officially debunked misinformation and actively identifying corresponding voice messages in Telegram groups. Due to an overabundance of misinformation in one specific political direction, we chose to extend our real-world sample with two artificially generated voice messages addressing misinformation from an opposite political perspective. This might have influenced the results of our user study, however we did not find any differences in our results when comparing the participants' answers on real-world and artificial messages in comparison.

Fourth, the appropriateness of indicator-based approaches as media literacy interventions is debated, as it depends on rationality [62] and might create false confidence [75]. Further, media literacy has been criticized for only being one of many components within the complex information space [75, 226]. Therefore, there are calls to "rethinking media literacy in the age of platforms" [75, p.17]. However, studies also show positive outcomes of media literacy interventions on critical thinking and behavior change regarding misinformation behavior [530, 607]—an optimistic view we draw on in this study. Our approach is only one of many possible solutions and (like other types of interventions that have already been researched) can only be one of many contributions. For example, the approach has the disadvantage that the features of misinformation in voice messages are characteristics that change over time and with the (technical) progress of disinformation campaigns and require adaptation. With the ever-changing information landscape, the skills and strategies of disinformation creators also change. A combination of different components, areas of responsibility and types of intervention in the interplay of critical journalism, media literacy training and the use of the latest technologies (including AI-based technologies) is therefore essential.

9.6 Conclusion

We developed a prototype as proof-of-concept of an indicator-based digital misinformation intervention for voice messages and conducted a user-centered evaluation of the perceived comprehensibility and usefulness of computational features for laypersons as users of messengers. Thereby, we gained insights into how indicator-based interventions that have been shown to be promising for text-based content can be applied to voice messages as a modality highly relevant in the context of misinformation. We adopt the user perspective of diverse users of messengers like Telegram and Whatsapp. The approach encompasses a public policy response [549] providing a foundation for technical solutions to enhance users' media literacy skills, placing greater emphasis on self-efficacy in dealing with misinformation as opposed to regulatory approaches by governments that amount to censorship. While we identified three usage scenarios of the proposed indicator-based intervention ((1) laypersons that want to assess the credibility of voice messages in crises or in everyday life, (2) practitioners in crisis management organizations or authorities that need to analyze large datasets, e.g., during emerging crises, and (3) researchers requiring features for a detection model), we exemplarily conducted the user study regarding the first scenario. We generated novel insights into the opportunities and limitations of features on character, word, sentence, audio, and creator level as potential indicators for misinformation in voice messages. Our work is relevant to the public policy domain as it addresses demands for public policy responses to the misinformation challenge promoting technological media literacy solutions and preventing overzealous censoring regulations by governments [355]. The challenge lies in controlling the deluge of credible and misleading information. It is crucial to consider the subtleties of voice messaging being a significant yet problematic information channel. We propose an indicator-based misinformation intervention for voice messages to partly address the need. This provides decision makers with a tool to navigate the complex landscape of information dissemination, particularly during crises. Our main contributions are (C1) giving a structured overview on features of voice messages as potential indicators for misinformation, (C2) developing an indicator-based misinformation intervention as proof-of-concept, and (C3) evaluating the perceived comprehensibility and usefulness of a plethora of features to assess the credibility regarding the scenario of layperson listening to voice messages.

From Adolescents' Eyes: Assessing an Indicator-Based Intervention to Combat Misinformation on TikTok

10

Abstract Misinformation poses a recurrent challenge for video-sharing platforms (VSPs) like TikTok. Obtaining user perspectives on digital interventions addressing the need for transparency (e.g., through indicators) is essential. This article offers a thorough examination of the comprehensibility, usefulness, and limitations of an indicator-based intervention from an adolescents' perspective. This study ($N = 39$; aged 13–16 years) comprised two qualitative steps: (1) focus group discussions and (2) think-aloud sessions, where participants engaged with a smartphone-app for TikTok. The results offer new insights into how video-based indicators can assist adolescents' assessments. The intervention received positive feedback, especially for its transparency, and could be applicable to new content. This paper sheds light on how adolescents are expected to be experts while also being prone to video-based misinformation, with limited understanding of an intervention's limitations. By adopting teenagers' perspectives, we contribute to HCI research and provide new insights into the opportunities and limitations of interventions for VSPs.

Original Publication *Katrin Hartwig, Tom Biselli, Franziska Schneider, and Christian Reuter. "From Adolescents' Eyes: Assessing an Indicator-Based Intervention to Combat Misinformation on TikTok". In:* Proceedings of the CHI Conference on Human Factors in Computing Systems. *ACM, 2024, pp. 1–20. DOI:* 10.1145/3613904.3642264

Supplementary Information The online version contains supplementary material available at https://doi.org/10.1007/978-3-658-47885-8_10.

K. Hartwig, *Navigating Misinformation*, Technology, Peace and Security | Technologie, Frieden und Sicherheit,
https://doi.org/10.1007/978-3-658-47885-8_10

10.1 Introduction

Video-sharing platforms (VSPs) like TikTok, which was originally popular for content such as entertaining lip-sync and dance videos for teenagers, is now increasingly filled with questionable, sometimes dangerous and misleading content designed to appeal to and influence people of all ages. The serious consequences of misleading information on VSPs are particularly evident in crises such as the Russian war of aggression against Ukraine [452]. Not only intentionally misleading information can cause great damage, but also false information that is spread unintentionally. 'Misinformation' is often used as an umbrella term that includes both intentionally misleading information (i.e., "disinformation" or "fake news") and unintentionally misleading information (i.e., "misinformation") [15, 97, 597]—we will use the term accordingly.

Human-computer interaction (HCI) research has recently focused on VSPs such as TikTok, YouTube, and Twitch, not only in terms of misinformation, but also in contexts such as mental health or social participation [369]. The question of the transferability of existing findings regarding text-based platforms such as Twitter, where much of the research to date has taken place, to video-based platforms, is highly relevant from an HCI perspective. Misinformation and ways to support users in dealing with it have been researched for many years. More recently, first in-depth insights into misinformation on VSPs have been gained, revealing their relevance as a breeding ground for misinformation [16] and an extensive range of layers (e.g., video, audio, captions) offered by VSPs [387] having a potential to mislead. Research has delved into how adolescents assess credibility in video content, including considerations of comment sections on platforms like TikTok [226]. While existing HCI research emphasizes the relevance and challenges of misinformation on VSPs, we will fill the research gap of considering the user perspective on how to address that type of content with its multimodal potentials to mislead. Even after years of research on digital misinformation interventions and controversial discussions on their suitability, the field is far from a decisive solution. Therefore, further user-centered exploration of specific needs, capabilities, and perceptions is relevant.

This paper aims to extend existing findings with qualitative user-centered insights to sharpen digital misinformation interventions as one building block for dealing with misinformation, to further delineate their possibilities and limitations, and thus to contribute to misinformation research from an HCI perspective. Previous research suggests that transparency plays a significant role in establishing trust among users in digital interventions [299] and minimizing reactance or other backfire effects [391]. To address the preference for transparency (i.e., for "approaches providing [users] with an opportunity for informed decisions" [299, p. 14], for instance using explanations on why a content was labeled as misinformation), we focus on an indicator-based approach that presents comprehensible and user-centered cues for evaluating video-based information. We define an *indicator-based misinformation intervention* as a digital countermeasure that uses identifiable characteristics (indicators) of misinformation to assess and convey the credibility of online content. These indicators (e.g., attention-grabbing layouts or conspiracy theory endorsements) are presented to users for both immediate feedback and educational purposes, enabling them to develop skills in assessing the credibility of content. Typically, these interventions involve highlighting identified indicators within the content (e.g., using color-code [350]). Indicator-based approaches have been developed and controversially discussed in existing research, especially in relation to text [350]. Our aim is to extend these findings to video-based indicators such as edited images, emotion evoking sounds or facial expressions, and refuting reactions to the video. With VSPs such as TikTok as a central platform especially for adolescents, we are particularly focusing on them as potential users who are, on the one hand, regarded by many as highly capable digital natives, but, on the other hand, as particularly vulnerable by others. We conducted a twofold study with 39 adolescents between the ages of 13 and 16 years. First, we used a mixed-methods approach with focus groups and individual surveys to assess existing strategies and the comprehensibility of video-based indicators as a potential extension (Step 1). Based on the findings, an indicator-based prototype smartphone app was developed as a misinformation intervention for TikTok which was qualitatively evaluated in individual think-aloud sessions (Step 2). The sequential studies provide an opportunity for triangulation of data and in-depth findings.

We advance misinformation research by applying existing knowledge about indicator-based interventions to the very current modality of short-videos. In doing so, we adopt the user perspective of adolescents as a particularly relevant user group. Our core contributions (C) and findings (F) are first (C1) evaluating the comprehensibility and perceived usefulness of video-based indicators. In doing so, we (F1) found how indicators on diverse levels (e.g., layout, profile, interactions) expand and confirm the perceptions of adolescents, and (F2) how particular characteristics of manipulated videos are applied intuitively as indicators. We then (C2) implemented and qualitatively evaluated a smartphone app prototype as an indicator-based misinformation intervention and received (F3) an overall positive feedback on the approach and identified participants' substantial understanding of its features. Furthermore, we identified insights into (F4) potentials for transferability of the extended knowledge and skills. Finally (C3), we assessed opportunities, challenges, and limitations of the indicator-based approach and identified (F5) transparency of the indicators as central reason for the intervention's positive assessment, but also (F6) its limitations, such as adolescents' blind trust in the tool and the lack of realistic concerns.

10.2 Related Work

Our work contributes to the design of user-centered digital misinformation interventions for VSPs like TikTok. We discuss related work, addressing how social media is faced with an overabundance of information, and the role of VSPs compared to primarily text- or image-based platforms (see Sect. 10.2.1). We shed light on digital misinformation interventions as one possibility to combat the effects of misinformation (see Sect. 10.2.2), delve into indicator-based misinformation interventions (see Sect. 10.2.3), and give an outline of the role of adolescents compared to others in terms of their susceptibility to misinformation (see Sect. 10.2.4). Finally, we summarize the resulting research gaps and present research questions to address them (see Sect. 10.2.5).

10.2.1 Misinformation on Video-Sharing Platforms

Misinformation research continuously generates new questions due to advancing technological possibilities, such as artificial intelligence and novel multimodal platforms. Social media in general has long been a central arena for the spread of misinformation in this regard, particularly in times of crisis [130, 533, 535]. While text-based platforms have dominated the social media landscape in the past, VSPs have recently gained relevance, transforming TikTok into one of the most successful platforms in the world, especially among younger people. HCI research on VSPs particularly addresses online communities like video game streamers and social participation [39]. This also extends to positive aspects of the platform, such as fostering creativity [393], inspiring playful technology [140], or connecting communities [369, 487]. However, recent events have demonstrated the vulnerability of VSPs to misinformation and other harmful content [16, 42, 338, 387]. In this regard, studies investigate how recommendations affect cross-platform sharing [76], who creates misinformation on TikTok and how users respond to them [16]. Beyond misinformation, research examines, e.g., how dark patterns are applied on TikTok and other social media platforms [367], how VSPs expose creators to hate and harassment [556], and how young people perceive digital safety, including harassment, financial fraud, but also misinformation and deep fakes [165]. Research emphasizes the extensive range of data (e.g., video, audio, descriptions, comments) offered by VSPs [387], which could potentially mislead users. Our work address these misleading potentials through the use of multimodal indicators. A specific relevance appears to apply to comment sections [387], especially among young users, when searching for and evaluating online information [226]—a finding that our work builds on. While there are different approaches to combating misinformation—for example, by teaching media literacy in schools or by supporting the work of professional journalists—one way to help deal with the overabundance of information online is to develop digital misinformation interventions.

10.2.2 Digital Misinformation Interventions

Developing user-centered approaches to combat misinformation poses a challenge within the HCI research community. To address this issue, digital misinformation interventions are employed to assist users in processing online misinformation, expanding on efforts in education and journalism. The term 'digital misinformation intervention' has already been established by several researchers [35, 479, 480]. These interventions vary widely in their primary goal and encompass, for instance, soliciting user feedback after automatic detection or nudges to reduce sharing. While many countermeasures target automatic detection [518, 613] (e.g., through machine learning), studies with a stronger HCI focus explore post-detection decisions or aspects detached from detection (e.g., default nudges for reflection). User-centered misinformation interventions "go beyond a purely algorithmic back-end solution and exert a direct influence on the user in the form of information presentation or withholding" [214, p. 2]. For example, there are approaches that provide a correction of misinformation by displaying a link to a fact-checking website, debunking videos, or corrections within the comment section [24, 49, 349]. Other work suggests providing a binary label to mark content as false [40], however that approach has shown to be less accepted by users when transparent explanations were missing [299]. Trust and distrust play a significant role when aiming to design effective interventions [24]. While some of the studies report promising initial results [84], e.g., to reduce sharing or engagement with misinformation, others demonstrate how common interventions have limited effects in isolation and are more successful when combined (e.g., suspending algorithmic amplification combined with a nudge) [35].

Efforts have been made to offer comprehensible indicators that guide the autonomous evaluation of problematic content, promoting critical thinking or media literacy. In this paper, in line with other researchers, we define media literacy as the ability to decode, evaluate, analyze, and produce both print and electronic media, that is, to have internalized a sense of "critical autonomy" in dealing with all media [28]. However, media literacy has received criticism for being only one of many components of the complex information space [75, 226], for relying on rationality [62], and for creating a false sense of confidence [75], prompting calls to "rethinking media literacy in the age of platforms" [75, p. 17]. Nonetheless, studies indicate that media literacy education can promote critical thinking and alter behavior [268, 280, 600]—also regarding misinformation [530] and especially among younger people [607]. In our study, we draw on this optimistic view.

10.2.3 Indicator-Based Misinformation Interventions

Studies highlight the significance of transparency and comprehensibility in interventions, as they foster trust [188, 299] and can thus be considered a crucial requirement when designing user-centered countermeasures [491]—an insight that guides our approach. Indeed, users prefer to understand why content is marked as misinformation as opposed to receiving binary classifications [299]. To address this, HCI research explores indicator-based interventions regarding misinformation and related phenomena (e.g., propaganda) [299, 350, 491]. For example, Bhuiyan et al. [51] examined how indicators such as information about the author, can be utilized to promote trust, from the perspectives of both news consumers and journalists. More generally, trustworthiness indicators have been evaluated in terms of perceived utility and visualization preferences in social media posts, revealing positive perceptions and a preference for simple visualizations of indicators to reduce the cognitive load [188]. Indicators as immediate user feedback enable users to understand characteristics of misinformation and encourage the development of own assessment skills [491]. While current research primarily focuses on textual content, some insights extend to images and videos. For instance, Sherman, Stokes, and Redmiles [513] identified the source of information as key indicator when assessing different types of content, including fake videos. Although users generally appreciated displaying indicators as a message, favoring simplicity and clarity, the study identified a problem with the overgeneralization of indicators. Research indicates promising advantages (e.g., indicators reducing uncertainty [188], and aligning with users' pre-existing mental models and practices [513]) and potential positive effects on trust [51, 299], perceived utility [188], and the development of autonomous assessment skills [491]. Nevertheless, concerns and challenges persist, with some users (and especially adolescents) still perceiving interventions as patronizing, underscoring the challenge of striking the right tone [226], and other users rejecting indicator-based interventions when their development was not transparent or their design was biased [188]. Our research delves into the nuances of misinformation mitigation and recognizes that while not universally applicable, indicator-based interventions are a promising step towards addressing user needs and preferences.

Identifying comprehensible and useful indicators for credibility assessment is crucial to the development of indicator-based misinformation interventions. Research has primarily explored indicators of misinformation or related phenomena in textual content, often with an emphasis on automatic detection (e.g., to use them as features for training misinformation detection algorithms) [518, 613] and in some cases with a user-centered or social science focus [350, 491, 574]. While text-based indicators (e.g., linguistic characteristics of propaganda techniques such as exaggeration or loaded language [350]) might be partly applicable to other modalities, user-centered indicators regarding video content still require more exhaustive investigation, complementing research that has gained initial insights. Some studies specifically identified or evaluated selected video indicators. This includes the relevance of comment sections with critical reactions of other users, or videos evoking strong emotions [226, 387]. Research emphasizes loaded language or attention-grabbing layout as indicators on social media, including VSPs [607], or highlights the potential of filters and voice changers to mislead [165]. Others more generally gained insights into how users navigate VSPs and tackled indicators for misinformation only briefly, for instance by taking a closer look at hashtag topics indicating misinformation on TikTok [338], or by taking very specific perspectives like characteristics of misinformation in videos on urological topics [392, 615]. Other research tackles misinformation and its indicators on VSPs indirectly, investigating aspects that are closely related to the modality like emotion recognition in videos [50]. Niu et al. [387] provide an overview on multiple components and layers of videos (e.g., sound, profile, reactions) with the potential to be misleading—a foundation we build upon to derive specific indicators for VSPs (e.g., outdated or mismatching sound, facial expression evokes strong feelings). VSPs like TikTok offer a variety of patterns and indicators that have not yet been thoroughly evaluated from a user perspective, hence the motivation for our work. The systematic overview of derived indicators and its sources can be found in Table A.5 (Electronic Supplementary

Material (ESM)). We further summarize the procedure of selecting our indicators in Sect. 10.3.6.

10.2.4 Adolescents as Vulnerable Users

Evidence indicates that susceptibility to misinformation on social media varies among users. Both individual attributes (e.g., decision-making style [34], cognitive ability [124]), and demographic factors (e.g., age [273], language skill [473]) play a role. Young users have been extensively studied due to their exposure to media literacy education, growing up with unique information sources while appearing to have a short attention span, and their status as 'digital natives', which grants them familiarity with new media such as VSPs and technologies like as deepfakes and filters.

Research diverges on the digital and media literacy of young people, with some demonstrating strong literacy of 18–24 year-olds in specific areas, such as health misinformation [157]. The fact, that young individuals belonging to 'Gen Z' generally have the capacity for a significant degree of digital literacy and a basic desire to seek information from different sources was supported by a WHO report based on a survey across 24 countries [580].

However, it is crucial to note that other studies display tendencies to prematurely share misinformation among adolescents under the age of 18 [232, 233]. Limited critical thinking skills and overconfidence contribute to their susceptibility [408]. Lower self-efficacy [401] and limited concern for information trustworthiness among 11–13 year-olds [137, 342] are associated with increased misinformation sharing. Adolescents tend to overestimate their ability to assess digital information accurately [390, 408, 430, 635]. In-depth qualitative work on the information practices of 13–24 year old adolescents and young adults suggests that they do not conceptualize information and online information processing as isolated and narrow processes, but as broad and situated in a social context [226]. This highlights the role of social cues and peer influence in adolescent's online information handling—a conceptualisation from which new vulnerabilities can potentially emerge.

These vulnerabilities, combined with high exposure to online misinformation due to excessive online activity, underscore the vulnerability of young adolescents. Their limited critical thinking skills, responsiveness to social cues, and limited digital literacy emphasize the special need for tailored attention and support in order to critically evaluate online information.

10.2.5 Research Gap

This study advances HCI research on misinformation interventions through the identification and assessment of user-centered indicators for misinformation on VSPs like TikTok. Based on this, we developed a prototype smartphone app designed as a digital misinformation intervention for adolescents. The study addresses gaps in the following areas:

1st gap: Focus on VSPs. The vast majority of research on social media misinformation centers on text-based content. However, recent crises such as the Russian war of aggression against Ukraine and the COVID-19 pandemic have shown that VSPs like TikTok are a highly relevant breeding ground for misinformation [16]. Existing HCI research highlights the significance of this challenge and the need for further research on how to address this specific type of content.

2nd gap: User-Centered Interventions. User-centered countermeasures that promote media literacy are better received by users when provided with comprehensible explanations [299]. Researchers have emphasized the importance of finding interventions that counteract reactance and take into account end-users' needs for comprehensibility [299, 391]. This can be achieved by providing comprehensible indicators of misinformation as user feedback—an approach that has already been partially addressed for text-based content [31, 171, 221, 350], but requires further development from a user perspective [195, 299]. Video-based content, with its multimodality, is interesting in that it presents its own unique opportunities for misinformation [387], which in turn require new strategies for user feedback (e.g., regarding emotional music, filters, and false context of sounds).

3rd gap: Adolescent Vulnerability. Adolescents are particularly susceptible to misinformation, and research suggests potentially harmful overconfidence [75, 481]. Addressing the needs of adolescents in dealing with misinformation on VSPs like TikTok requires further investigation aligning with their tone and style preferences [226], building on HCI research that has already produced significant findings [226, 387] and by involving adolescents in a design process from the beginning.

Considering related studies and combining the resulting gaps, our overarching goal is to answer the following research questions:

RQ1: *How do adolescents evaluate misinformation indicators in terms of comprehensibility and usefulness?*

RQ2: *How can indicators be applied to a smartphone app as a digital misinformation intervention to assist adolescents on TikTok?*

RQ3: How do adolescents perceive opportunities and limitations of an indicator-based digital misinformation intervention?

10.3 Methodology

In the following, we provide details on our participants, ethical considerations, research method, analysis, and stimuli. To answer our research questions, we conducted a twofold study with 39 adolescents aged 13 to 16 in Germany in the summer of 2023. Our study consists of two steps of data collection (see Fig. 10.1): focus groups (Step 1) and individual think-aloud Sessions (Step 2), involving a different set of participants.

In Step 1, our goal was to gain insight into which indicators adolescents autonomously use to assess misinformation on TikTok, and into how they evaluate established indicators in terms of comprehensibility and perceived usefulness. This allowed us to identify a set of potential user-centered indicators for VSPs, and for TikTok in particular, thus primarily targeting RQ1. We chose to implement focus groups in order to gain a variety of insights from the interactions and discussions between participants [314]. Interactive focus groups facilitate the exchange of ideas and the exploration of different viewpoints, which corresponds to our overall aim of identifying the most promising indicators for adolescent users.

For Step 2, we developed a smartphone app prototype as a digital misinformation intervention based on the previously in Step 1 established indicators, and evaluated how adolescents interacted with it, rated its usefulness, and identified challenges and opportunities—primarily to address RQ2 and RQ3. We thereby employed the think-aloud method [329] to gain rich insights into user perceptions during real-time use, rather than retrospectively [478], which is a method widely used in HCI and misinformation studies [380, 408, 478]. In contrast to focus groups, it provides a more individualised and introspective insight into participants' thought processes and decision making as they interact with the prototype independently. By combining both focus group (Step 1) and think-aloud (Step 2) methods, our aim was to gain both depth and breadth of insight and to answer the research questions from a variety of perspectives. Overall, both methods provide rich insights and complement each other to facilitate an understanding of users' needs, behavior, and experiences.

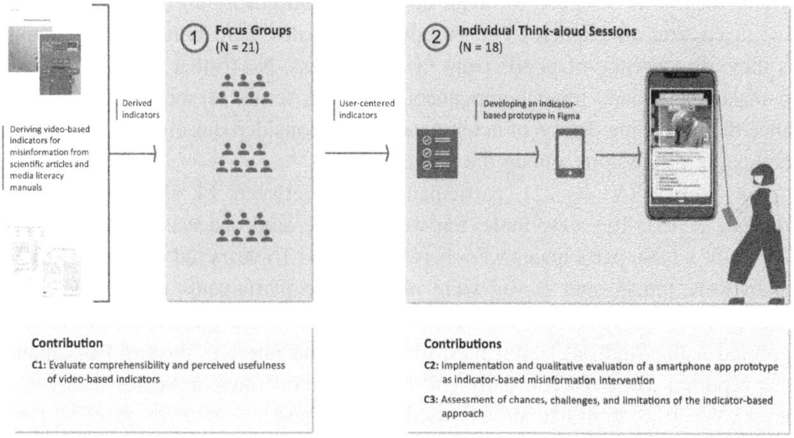

Fig. 10.1 Flow of our twofold study design demonstrating key contributions of each step

10.3.1 Recruitment and Participants

We carried out Step 1 with an entire school class and Step 2 with a separate school class in addition to a cohort of four additional female participants from a youth center to ensure reasonable gender parity. Recruitment was mediated through contacts with a local digital media education center—which, however, had not previously conducted any thematically relevant training courses with the specific pupils in our study. The majority ($N = 38$) of the participants were students at an integrated comprehensive school, i.e., they are mixed in terms of their intended school leaving qualifications (e.g., Abitur, Mittlere Reife, Hauptschulabschluss in Germany), and one additional student came from a secondary high-school ('Hauptschule'). The overarching inclusion criterion was the age range of 13–16 years to gain insights specifically from young adolescents. In addition, the choice of school for the recruitment of participants ensured the coverage of diverse levels of education and learning abilities. The specific classes (one for Step 1, another for Step 2) were chosen on the basis of time availability within a project week at the date of data collection. All pupils attending class at the time of data collection were included in the study and no specific exclusion criteria were applied. For the focus groups (Step 1), the participants were divided into three age- and gender-balanced groups with the help of the teacher. This arrangement was designed to break up cliques and ensure an open and productive discussion atmosphere. The groups as a wholes were rather homogeneous, as they all consisted of pupils from the same class. No further steps were taken to make the groups more homogeneous, e.g., in terms of social media usage, since the remaining degree of heterogeneity was considered useful for stimulating interesting discussions among the participants.

For Step 1 ($N = 21$), participants were between 14 and 16 years old (*median* = 15). Ten were male, ten were female, and one was non-binary. For Step 2 ($N = 18$), participants were between 13 and 16 years old (*median* = 13.5), seven were female and eleven were male. Since participants interacted with a smartphone in Step 2, we also documented their usual operating system. Six reported using Android, seven iOS, four both, and one was unsure. Participants also reported using TikTok daily ($N = 25$), several times a week ($N = 5$), or never ($N = 9$), but all had already used it in the past.

10.3.2 Ethical Considerations

When involving minors, ethical planning and conduct are crucial. Accordingly, the positive voting of the university's ethics committee (IRB Number EK 47/2023) was obtained in advance. Stimuli and study procedures were coordinated with a local media education center and teachers to include interdisciplinary expertise. The content shown was adapted to the age of the participants, excluding any violent or disturbing images (e.g., war, explosions) and politically biased messages. The videos were presented within a TikTok simulation to control the selected content. The truthfulness of the stimuli was clarified immediately after the study and the adolescents were provided with additional written information. Explicit informed consent (in age-appropriate language) for participation and data use was obtained in advance from both parents and youth, in accordance with the permission of our institutional ethics review board. Participants received a €15 book voucher as compensation.

10.3.3 Step 1: Mixed-Methods User Study with Focus Group Discussions

10.3.3.1 Mixed-Methods Approach

Individual paper-based surveys provided insights into autonomous assessment strategies immediately after watching a TikTok video. Focus groups were chosen as they enable discussions among participants and generally provide a broad range of opinions [314]. They are frequently applied in HCI in general [314] and particularly in the context of misinformation [165]. Thereby, in-depth explanations were generated, which was crucial for assessing participants' comprehension of certain indicators of misinformation, for identifying which indicators require improved explanations, and for understanding our participants' general needs. This mixed-methods approach allowed participants the necessary space to think individually before engaging in the group discussions. By triangulating the data, we gained deep insights into our young participants' responses.

10.3.3.2 Study Procedure

Three parallel focus group sessions of approximately equal size were each led by a member of the research team, following a structured guideline with standardized language. Each session lasted 120 minutes. After a brief introduction, written informed consent was obtained and demographic information was collected before audio-recording began. Participants sequentially viewed five TikTok videos, each following a systematic procedure (see Section A.5.1, ESM for details): (1) Participants answered general questions individually on a paper questionnaire regarding the familiarity of the video and the credibility assessment. (2) The focus group received multiple indicators for misinformation and discussed their understanding and usefulness (see Table A.5, ESM for sources and representation of indicators). Finally, researchers provided clarification on the videos' truthfulness. One of the five videos contained only factual content. Thus, no cues for misinformation were displayed. (3) After following the procedure for all five videos, we presented three additional indicators that were not included in the selected videos but still seemed promising to evaluate based on our literature review. Again, participants discussed their comprehensibility and usefulness. (4) Afterwards, the focus groups engaged in general questions about the indicators overall. The study procedure was identical for all three groups, i.e. all participants watched the same videos (in a different order) and were asked the same questions. The general aim of the focus groups was to obtain a variety of perspectives. For this reason, we decided to conduct three separate focus group discussions rather than one large group in order to create a constructive atmosphere and reduce the influence of very dominant individuals. The range of three focus groups—also in contrast to only one large focus group—allowed us to have in-depth discussions that provided different insights from a variety of participants while maintaining a sufficient level of control over the dynamics of the conversation. Subsequently, all three researchers reported similar experiences in their focus groups.

10.3.4 Step 2: Individual Think-Aloud Sessions

We designed and evaluated a smartphone app prototype (see Fig. 10.2) as a digital misinformation intervention for TikTok, incorporating and refining the indicators from Step 1 (see Table A.5, ESM for representations in Step 1 and 2).

10.3.4.1 Developing a Figma Prototype

We designed the smartphone app (= 'Misinfo-App') as a prototype using *Figma*[1], where TikTok videos are embedded in a simulated 'ForYou page' as can be found on TikTok. Different views of our Figma prototype design can be seen in Fig. 10.2. The prototype corresponds to a realistic click dummy with a simulated detection. To disrupt the everyday user experience, the app was integrated into the prototype as an *active intervention*: From the simulated ForYou page, a video can be shared with the Misinfo-App using the share button built into TikTok. The user is then taken to an overview of the detection results (see Fig. 10.2b and 10.2f). If misinformation was successfully detected, the corresponding indicators are listed there. By swiping, the user can switch to the detailed descriptions (see Fig. 10.2a, 10.2d, and 10.2e). Some indicators are associated with appropriate highlighting of the referenced items in the video (e.g., display of the comments (Fig. 10.2e) or the profile for the video). An arrow leads the user back to the ForYou page.

[1] https://figma.com

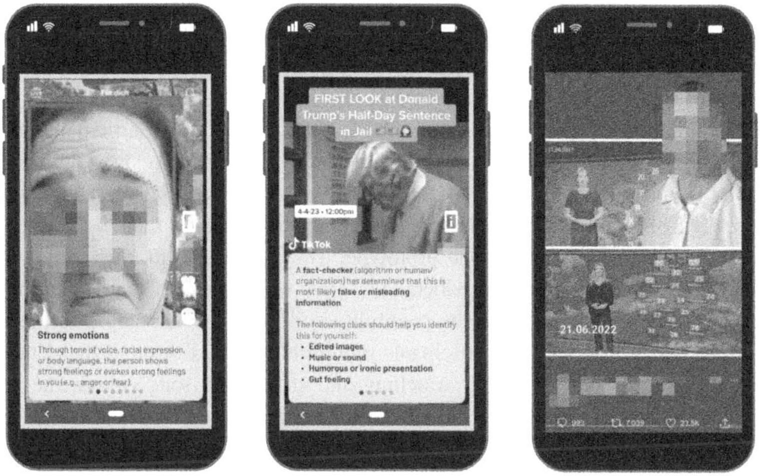

(a) Video Pets (misinformation)

(b) Video Trump (deep fake)

(c) Video Weather (misinformation)

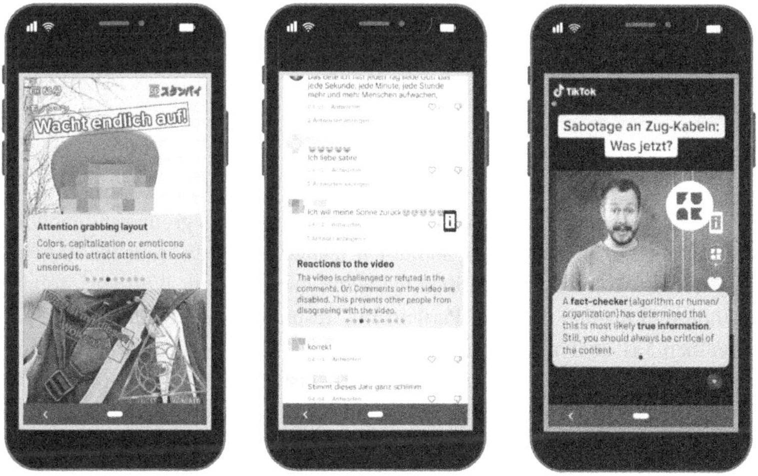

(d) Video Clouds Part I (satire)

(e) Video Clouds Part II (satire)

(f) Video Trains (true news)

Fig. 10.2 Overview on the stimuli videos we used in Step 1 (all) and Step 2 (all but c) demonstrating several views of the smartphone app's UI within our Figma Prototype. Faces of private persons were pixelated for publication. UI elements were translated from German to English

10.3.4.2 Study Procedure

The one-on-one think-aloud sessions were also conducted by three researchers, following a pre-defined systematic guideline. Each session lasted 40 minutes. (1) After giving a brief introduction and obtaining written informed consent, a short simulation of the think-aloud method was demonstrated. Participants were generally instructed to "think aloud" while they were using the smartphone. We recorded audio and screen touch to evaluate our participants' interaction with the app. (2) Participants were instructed to imagine being on their ForYou page on TikTok, using our smartphone prototype. They were able to click and swipe everywhere while watching the videos. (3) They were asked to share each video with our Misinfo-App to check its credibility and to look at the app's output for as long as they wanted. All participants thoroughly swiped through the detailed indicator descriptions. (4) After watching and sharing all four videos, the participants switched from our prototype to an online survey to fill out the SUS and a demographics questionnaire. To accommodate the young age of the participants, the language of the SUS was slightly modified. We adopted the wording of the study by Putnam et al. [442] and slightly reformulated all items that have a grade level >6 in the original (see Section A.5.3, ESM). (5) Finally, we asked open-ended questions to gain deeper insights into the overall indicator-based approach, its usefulness, and any concerns about an underlying algorithm. A detailed description of the study procedure can be found in Section A.5.2 (ESM).

10.3.5 Analysis

Steps 1 and 2 generated rich qualitative data from the focus groups and individual think-aloud sessions, as well as additional quantitative data from the individual survey items. We locally transcribed the audio material with *Whisper*[2], followed by a thorough manual revision. Participants' responses were anonymized.

We then (a) analyzed the quantitative survey items calculating descriptive statistics (*mean, median, frequencies*) and (b) clustered the free-text question thematically according to the individually mentioned assessment strategies from Step 1. We (c) employed thematic analysis for the focus group discussions (Step 1) and think-aloud sessions (Step 2), which is common for this type of study [66, 201] and which serves to identify and interpret thematic patterns within qualitative data [66]. Overarching themes were thus derived from a set of pre-defined codes that were created prior to the analysis and were iteratively added as new

[2] https://github.com/openai/whisper

topics emerged during coding. Two coders independently applied the codebook (see Table A.12, ESM) to the data and any ambiguities were discussed to achieve a consensus. Inter-coder reliability, measured by Cohen's kappa score, demonstrated substantial agreement ($k > 0.61$) across all iterations. The themes as thematic clusters derived from our codes form the basis for our results reports in Sects. 10.4.1, 10.4.2, and 10.4.3.

10.3.6 Stimuli

The stimuli for Steps 1 and 2 were selected and revised in consultation with the local digital media education center, teachers (especially with regard to simple language, age appropriateness, and representative mapping of content on the adolescents' ForYou page), and the ethics committee. They comprise real-world TikTok videos and indicators of misinformation that could help assess content on VSPs. The videos and indicators are similar in Step 1 and Step 2 (see Table A.5 (ESM) for representations in both steps).

Videos. An overview of all included videos can be found in Fig. 10.2 and Table A.10 (ESM) provides a detailed description of the content of the videos. TikTok videos were selected as stimuli following a systematic process. (1) We consulted official fact-checking websites to identify current topics that have been officially debunked which covered a wide range of misinformation. Either these sites already linked to specific recent TikTok videos, or we manually searched TikTok for these videos, referring back to the officially debunked content. (2) We excluded videos that were not age appropriate. For all videos, we chose those that were most suitable for the age group and were approved by the digital media education center and teachers involved in the study. (3) We made sure to include diverse forms of misinformation. While misinformation types can take various forms [285], we chose our selection of videos to represent a realistic sampling of diverse forms of misinformation on VSPs without a claim to completeness: deep fakes (Fig. 10.2b; manipulated video showing Donald Trump in prison), satire with potential to mislead, as well as conspiracy theories (Fig. 10.2d/10.2e; a TikToker making fun of common conspiracy theories by pointing to clouds that block out the sun and referring to the pharmaceutical industry being responsible), and commenting on and sharing misinformation from other users that the person sharing it believes to be true (Fig. 10.2a; a TikToker expresses anger at the alleged call by Fridays for Future to ban pets over 10kg). We included one video containing factual information (Fig. 10.2f; a presenter reports on the sabotage of railway cables). The same videos were used in Steps 1 and 2 with the exception of Video Weather (Fig. 10.2c; a TikToker comments on an alleged manipulative attempt by media to overstate global warming) which was omitted in Step 2, due to time constraints and because the video was found to be too complex for the age group to understand the content in Step 1.

Indicators. In accordance with other research [387], we differentiate the following layers of short-videos on VSPs: spoken words, video, sound and music, layout, content and message, profile, reactions and interactions, and the overall impression. Characteristics within these layers might represent valuable indicators for misinformation, e.g., a sound that evokes strong feelings. To systematically identify a comprehensive selection of indicators for misinformation on TikTok, we (1) used these layers as reference. We (2) conducted a literature review on misinformation in VSPs to identify indicators within the layers, examining which characteristics have already been captured and/or scientifically investigated as potential indicators of misinformation in videos (see Sect. 10.2.3). To compensate for the gap of scientific publications, we (3) consulted non-scientific official media literacy manuals [43, 250, 499] and fact-checking websites [110] that specifically address TikTok and its particular features. This was (4) supplemented by literature on mainly text-based indicators that may also be suitable for video-based platforms (e.g., attention-grabbing layout) and have been evaluated in the context of misinformation and other related concepts like propaganda [350], conspiracy theories [211, 574], or political extremism [203]. An overview of indicators integrated in our approach can be found in Table A.5, ESM.

10.4 Results

In Sect. 10.4.1, we report how adolescents assess the comprehensibility and usefulness of indicators as cues for video-based misinformation on TikTok derived from Step 1. Building on those findings, we developed a Figma prototype of a smartphone app in Step 2 to evaluate how indicators can be applied as part of a digital misinformation intervention (see Sect. 10.4.2). Thereby, we gained insights into how adolescents perceive opportunities for and limitations of the proposed indicator-based approach (see Sect. 10.4.3). We present the results as themes obtained from our thematic analysis. Our core contributions and findings are summarized in Table 10.1.

10.4.1 Assessment of Indicators for Video-Based Misinformation

For all indicators derived in Sect. 10.3.6, we evaluated their comprehensibility and perceived usefulness through focus group discussions in Step 1.

10.4.1.1 Video

Adolescents are familiar with AI-generated *manipulated images* (see Fig. 10.2b), enabling them to identify indicative details and features such as inconsistent lighting conditions. While the information is considered easily understandable, its usefulness is considered limited due to perceived obviousness

> *"I think you can see it anyway, because the photos that are just generated, there is just such an artificial intelligence [...]. There were also photos with the Pope where he was wearing MontClaire jackets and things like that. And these are pictures like that."* (focus group 2)

Similarly, *using filters* (e.g., on voice or face) are a common feature on TikTok and well-known. Opinions on the necessity of acknowledging filter usage vary: *"Many older people are just fooled. I know it myself, I can give examples, but anyway, I think it makes sense, it's good."* (focus group 1) Although participants were able to explain the meaning of *facial expressions and body language evoking strong feelings* (see Fig. 10.2a), its usefulness to assess a TikTok's credibility received mixed feedback. *"So I don't think it's very useful, because you can be calm, you can be cool, but you can still lie. "* (focus group 2) When presented the rather complex indicator of the *origin or time stamp of a video being outdated*, participants struggled with the understanding of its meaning regarding potential misleading content. However, this indicator was included in Video Weather (see Fig. 10.2c), which participants struggled to understand overall due to its complexity. Hence, this might have strongly affected the comprehensibility of the indicator in this context.

10.4.1.2 Spoken Words

Among the young participants, understanding how *tone of voice can evoke strong feelings* and can signal misinformation was present: *"So he raised his voice a little bit to make you think that he was scared himself. I think he wanted that to be more believable."* (focus group 2) Nonetheless, as with facial expressions, participants rated this as not necessarily a useful cue since truth on an emotional topic can be told with a raised voice, too.

10.4.1.3 Sound & Music

Interestingly, while evoking strong feelings through facial expressions or a speaker's voice received mixed feedback, *emotional sound or music* were deemed comprehensible and useful indicators. Participants successfully applied the indicator to both funny / entertaining music, identifying satirical videos, and sad music, identifying an intention to mislead. Regarding sound, the *outdated origin of sound* received positive feedback with participants correctly grasping its role as an indicator:

> *"I also find it quite helpful that when you hear gunshots or something in a video. And then the video is not about gunshots, I think it is right that you know that it is not part of the video. For example, if I'm watching a video about the Paris riots and suddenly I hear a bomb explosion, that doesn't fit in either."* (focus group 2)

10.4.1.4 Layout

Regarding layout as an indicator for misinformation, *attention-grabbing layout through color, capitalization, emoticons* (Fig. 10.2d) is viewed as readily understandable indicator of unseriousness, up to the creator's discretion. *Spelling or grammatical errors* on the other hand, also suggesting being nonserious, are dependent on creators' language skills, if not from official sources. These are not necessarily evaluated as related to the credibility of the content, but by some participants considered relevant:

> *"Well, I think it comes across as really nonserious when you don't have punctuation and a comma and nothing at all in that whole sentence. I think a serious source would never want to just run off in a sentence like that, so fast that everybody gets kind of upset about it."* (focus group 2)

More critically, two out of three focus groups explicitly addressed the issue of dyslexia: "*So, rather useful [...] but that could be that someone with dyslexia wrote this.*" (focus group 3)

10.4.1.5 Content & Message

Some potential indicators refer to specific characteristics of a video's modality. Others, regarding content and message, are applicable to all types of information. For example, any social media post containing *conspiracy theories* (e.g., using hashtags like #simulationtheory) has the potential to mislead. For TikTok videos, our participants were able to explain what this indicator means and rated its usefulness rather positively. Similarly, stating an *opinion without sources* was assessed as comprehensible and useful indicator. It is important to note that adolescents are particularly familiar with this type of indicator as it is a common way of teaching how to assess information in schools.

10.4.1.6 Profile

A hint to the profile of a creator was perceived as useful means to assess if a video might contain misinformation and was actively applied after watching the videos. Particularly, participants applied this measure to check if a video was meant to be satirical, originating from a comedy account. However, they were not aware of satirical content's potential for misdirection (e.g., hiding radical right-wing messages through satire). Overall, themes, questionable content patterns, and the creator's apparent expertise were reviewed by examining the profile, with familiarity fostering trust in known sources (e.g., official news profiles) and skepticism toward unfamiliar ones.

10.4.1.7 Reactions & Interactions

Reactions and interactions, including if *the content is questioned or refuted in the comment section* (see Fig. 10.2e), are pivotal for adolescents when assessing video credibility. This strategy, has been highlighted in previous research [226] and was reaffirmed in our study. Participants proactively examined comments to assess overall sentiment and to identify potential misinformation or satire. They viewed comments critically, particularly when other users used irony and did not explicitly voice their disagreement: *"They don't really explain it to him, they just make him look stupid."* (focus group 3) Conversely, *deactivation of comments* was considered a useful indicator of misinformation: *"Where it's so blatant or something, and the comments are off. Then people usually don't accept any criticism at all. And then you can't see in the comments if it's true or not. [...] I always find the comments disabled to be such a red flag."* (focus group 2)

10.4.1.8 Overall Impression

Encouraging critical reflection of a video's credibility is advocated by media literacy resources and applies to VSPs as well. Unlike objective indicators, this is subjective and based on individual reflection. The idea of this individual indicator to exhibit it as a default nudge: "Listen to yourself. What does your gut say?" and it yielded differing participant responses. Whether a video *evoked strong positive or negative feelings* was rated as rather not useful, especially because the theme of evoking feelings had already been addressed more specifically regarding body language and tonality, and was perceived somewhat redundant. Nonetheless, participants rated this nudge to self-reflection positively when it was phrased more generically as *gut feeling: Overall, the story is hard to believe*: *"I think you can tell right away that it's a little bit nonserious, and that automatically makes you think that the person might be talking garbage."* (focus group 1)

10.4.1.9 Autonomous Assessment Strategies

In the vast majority of cases in Step 1 (86%), participants stated that they had not previously known the videos which they were shown. Nonetheless, they often correctly assessed their truthful content before the video-specific indicators were discussed. In the best case (regarding Video Clouds, see 10.2d), 17 out of 21 participants correctly identified a video as containing misinformation, the remaining four at least identified it correctly in tendency ("rather not true"). In other cases (regarding Video Weather, see 10.2c), however, the assessment was more difficult: only nine participants tended to assess the content correctly, six tended to assess the content incorrectly and an additional six made no statement.

Despite these varying results, it is evident that the participants are, to some extent, capable of making their own correct assessments. In our study, we found that they did so by applying several assessment strategies autonomously. The adolescents were very successful in recognizing if a video was meant to be *humorous or ironic* and emphasized that potentially harmless satire and irony may be misleading as well, when not identified as such. Further, some participants particularly recognized sarcasm or irony as means to mislead:

> *"The irony, as if it is completely clear that all the animals are being brought in and so it somehow shows that she believes all the crap and thinks that she is the enlightened one who knows everything and is so much smarter than us because she knows all the conspiracy theories."* (focus group 2 regarding Video Pet, see Fig. 10.2a)

While this type of indicator was originally not included in our study, we included it in Step 2 as part of 'Content & Message' as it was mentioned autonomously multiple times (see Table A.5, ESM). Moreover, participants applied a strategy of *comparison with existing knowledge*, where they contrasted the given information with what they had already known about the topic, for instance, based on news reports or common sense. Similarly, they frequently displayed attempts to autonomously *identify reasoning or alternative logical explanations*, as in: *"So rather not, because where should these pictures come from, because in prison you're not allowed to have mobile phones. That's why no."* (focus group 1 regarding Video Trump, see Fig. 10.2b).

Besides, participants autonomously paid attention to the content of the *video description*, in particular with regard to the hashtags therein or whether any sources were provided. They also rated the *number of likes and followers*, as well as the *general appearance* of a person in the video as a sign of the reliability. Conversely, participants displayed a variety of strategies to recognize credible information, ranging from the naming of sources and provision of further video evidence, to their general assessment of an account's appearance. Particularly the notoriety or the presence of a verification badge were perceived as useful indicators.

10.4.2 Applicability of the Indicator-Based Approach

We evaluated how the indicators from Step 1 can be applied to a smartphone app as a digital misinformation intervention. We incorporated feedback from Step 1 to refine the indicators in Step 2, where we optimized their arrangement based on the input from adolescents, resulting in a more streamlined and less redundant set of indicators (see Table A.5, ESM for details). Specifically, an indicator related to humorous or ironic content was introduced based on this feedback, and indicators (e.g., related to strong emotions or wrong context) were merged to eliminate redundancy. These refined indicators then informed the development and evaluation of a TikTok-based smartphone app prototype aimed at mitigating digital misinformation. We derived findings regarding the usability and suitability of our indicator-based approach in Step 2.

10.4.2.1 Usability of the Misinformation Intervention

SUS and qualitative results on perceived usefulness. Our smartphone app as misinformation intervention yielded a positive average SUS score of 81.7. This favorable outcome aligns with the qualitative feedback on the usefulness and handling of the app. One of the most frequent positive comments was that the use of the familiar share button and the swiping within the results display is intuitive and very straightforward. The recording of screen touches also confirms this: our participants were able to independently navigate from the ForYou page to the Misinfo-app and within it right from the start. There were critical complaints about occasional loading problems of the app, which can be attributed to Figma's limited possibilities for prototype display with a lot of video material an overlays. In addition, several suggestions for improvement were made, which are presented below.

Barriers of usage and mitigation. We collected insights into what hinders adolescents from using the evaluated Misinfo-app and found that factors strongly differ between individuals. While the majority supported the use of the share button and the individual and detailed display of the indicators, three out of 18 participants found it too cumbersome and wished for a more compact display or a direct integration as a plugin for TikTok to reduce the *effort*: *"Maybe by not having to share it so much and then it just being there, for example, by me being able to press anything on TikTok and then it being displayed that way."* (P16, female, aged 13–14). Moreover, two criticized the app for being too *complex* for some users, as the feedback texts might be overwhelming for users who have a short concentration span, dislike reading or are illiterate: *"But for people like me who don't like to read, it's just rather more convenient to have: 'This is fake. Period.' [...] Young peoples' patience is also very short. [...] Or even my [anonymized], when she downloads the app, she can't read."* (P03, male, aged 15–16). This highlights the diverse experiences of young people in their immediate environment and their ability to empathize with potential other users. Conversely, two participants stressed that, for many videos on their ForYou page, checking for credibility is just *not relevant* because it mainly consists of entertainment videos, and they are rarely confronted with more serious content. Besides, especially the detailed feedback on single indicators using mainly text-based explanations was criticized by a small number of participants and might be a relevant factor when applied on a larger sample of adolescents. Some participants independently proposed a potential mitigation strategy using *personalization*.

10.4.2.2 Proposed User Groups.

We asked our young participants why or why not they would like to use the Misinfo-App on a regular basis and who else they would consider as potential users.

Adolescents as users. Participants liked the idea of the app having a means to transparently check for misinformation in occasions where they are uncertain about the veracity of the content. They especially emphasized positive social effects of being well informed: *"So I don't go to school and say, 'Hey, have you heard this?' And it's not even true."* (P12, male, aged 13–14) or *"I would use it. Because I didn't go to school once because of some weather thing [...] on Tiktok and that was a lie."* (P09, male, aged 13–14). Moreover, it became clear that receiving feedback from the app would not be relevant for the majority of videos, but rather for specific cases: *"So an app like this would actually be helpful. I don't see these videos too often, but when I do, I always research them and want to know if they're true or not."* (P09, male, aged 13–14)

Other users and other platforms. We were additionally interested in whom they thought would benefit from the app and found that our participants thought that older or very young people would benefit the most. Through personal experiences, the adolescents exhibited a heightened awareness of older adults' susceptibility to misinformation. Several participants stated, for example, that they take on the role of fact-checkers for older relatives in their families, especially regarding video-based content. These aspects were brought up in both the individual think-aloud sessions and in the focus group discussions: *"Yes, definitely. I would recommend it to my grandma, my little sister. Those are the people that should be using it."* (P02, female, aged 15–16) or *"Yes, especially with my grandparents, who very often, very, very often fall for fake news."* (P03, male, aged 15–16) or

> *"I can imagine that for Facebook because there are a lot of older people. For example, when the earthquake happened in Turkey, there was a lot of fake news, my [anonymized] told me that this earthquake was made by America and the country is about to be invaded. [...] As an example, I've seen it, but a lot of people believe it, I know it."* (Focus group 1)

The emphasis on relevance, especially for older individuals, also suggests the potential applicability of the indicator-based approach to other platforms, such as Facebook, where older people spend more time on social media.

Responsibility and confidence. The high responsibility of adolescents as fact checkers, e.g., for older relatives, regarding VSPs is certainly understandable and often justified based on the accumulated experiences and intuitive upbringing with these technologies. However, our results regarding how participants rated the truthfulness of TikTok videos also show limitations of those capabilities. For example, the adolescents initially made incorrect judgments in some cases without the support of the app. Six out of 18 participants assessed the truthfulness of at least one of the videos incorrectly before being exposed to the Misinfo-App results. This is not surprising, as VSP's misinformation can be misleading in multiple and often intricate ways. Some level of overconfidence when evaluating one's own credibility assessment skills is very common and reflected in our study.

10.4.3 Perceived Opportunities and Limitations of the Indicator-Based Approach

Our results of Step 2 revealed several insights into how adolescents perceive benefits and limitations of the proposed indicator-based intervention for VSPs. Below, we outline our main findings.

10.4.3.1 Expected Learning and Applicability

The goal of our qualitative study design was to collect in-depth insights into adolescents' perceptions and needs. The effectiveness of digital misinformation interventions, e.g., to reduce sharing of social media posts, has been investigated in other research [35]. In contrast, our findings provide further qualitative outlooks on expected learning effects or applicability of knowledge and behavior learned when introducing an indicator-based intervention.

Applying indicators. We observed how participants quickly adapted the displayed indicators to novel TikTok videos: *"Okay, now that I read through this [the indicators], I realize that his arguments don't make any sense. So no sources either, just babbling something."* (P01, male, aged 15–16). In the further course of the study with the same participant, it became apparent that this indicator is now autonomously added for critical reflection: *"Above all, the cases are sources, I've already thought of that in my head, so it's just kind of written in random text, without any sense or anything, without sources or at least anything. So that's just obviously fake."* (P01, male, aged 15–16).

Changes in credibility assessment. Moreover, two aspects could be identified with regard to changes in participants' assessment of information credibility. On the one hand, there were several instances in which participants changed or consolidated their assessment of the information presented with the help of the indicators in Step 2. Referring to the attention-grabbing layout as an indicator, for example, one participant commented as follows:

> *" I didn't understand why the colors and such have something to do with it, but now when I think about it, that everything is written so big and especially in red, because red is such a... Red is such a sign of... I don't know how to explain it, for something important. And until now I never noticed it before, but now it makes sense somehow."* (P07, female, aged 13–14).

On the other hand, participants also found the indicators useful for practice and as a reminder to be more able to identify disinformation on their own in the future, even without app support: *"I would say that these tips are needed. So that in the future, even without this app, or so that even without this app, people can really identify for themselves that it's fake news."* (P05, male, aged 13–14)

10.4.3.2 Transparency

The proposed approach of the Misinfo-App is based on the concept of transparently giving insights via user-centered indicators. We were interested in the adolescents' perceptions and (mis)conceptions of this approach and gained in-depth findings.

(Blind) trust in the algorithm. Out of 18 participants in Step 2, twelve explicitly stated to fully trust in the (simulated) algorithm's decision on whether a TikTok video is misleading or not. When asked for reasons, some participants expressed a general belief in computers being more intelligent or reliable than humans:

> *"Yes, so of course we trust because we are on the smartphone a lot and it has everything to do with computers. And that's why I trust it because computers are usually smarter than people. So of course I trust, I use my cell phone 24/7 and it has everything to do with a computer. And that's why, yes, of course."* (P07, female, aged 13–14)

While misconceptions of algorithms and AI-systems in particular have already been studied (e.g., revealing the *automation bias* [543]), our participants' trust in the application is also influenced by how stimuli and detection results were presented. To focus on our main goals and adhere to strict time constraints, we decided to not intentionally include false positives or false negatives, which would inevitably arise in a realistic environment of a detection app from time to time. The provision of indicators was an important element in building trust and giving adolescents a sense of autonomy in their assessments, demonstrating the great potential of our indicator-based approach: *"And there are also reasons. If you just say, yeah, it's fake, you wouldn't believe it so quickly because there are no reasons and I need reasons to know whether it's fake or not."* (P11, female, aged 13–14) Indeed, many adolescents preferred independent thinking and viewed the app as a supplementary aid: *"But for me, I would also rely more on myself and just search it again to see if I really know. After a while I would trust the app more."* (P03, male, aged 15–16)

Envisioning consequences of mistakes. As a related concept to trust in the (simulated) algorithm, we gained insight into how participants were able to envision the consequences of the Misinfo-App making mistakes. Again, this was biased by our study design which did not include mistakes of the app. Nevertheless, we can derive a general impression that our adolescent participants were overall aware of the app, algorithms, or more specifically AI systems making mistakes: *"Yes, every AI makes a few mistakes. ChatGPT has also made a few mistakes."* (P03, male, aged 15–16) In contrast, four out of 18 participants were sure that such systems would never make mistakes but would always function perfectly and a lot more precise than humans. When brainstorming about possible consequences of the app making errors, all 18 adolescents were confident that it would not have severe consequences. We then informed our participants about the possibility of the system to make incorrect decisions and the resulting consequences. All in all, this again demonstrates the advantage of the indicator-based approach over purely algorithmic detection, where errors would be even more serious.

Indicators versus binary feedback. In general, providing indicators as comprehensible and useful explanations why a video contains misinformation has been received largely positively and seems to encourage trust and knowledge applicability to other content or platforms. Indeed, when asked more specifically, the majority of participants liked the approach and considered it as a key component of the digital intervention. Their overall impression can be summarized as: *"I would like to know why it is Fake News, that's why I would have some tips about it."* (P10, male, aged 13–14) Only one participant did not like the effort and text-heavy presentation and argued for binary labeling as fake or not, or at least for an option to toggle additional information.

Table 10.1 Core contributions and findings of the focus group discussions (Step 1) and think-aloud sessions (Step 2)

Core Contributions	Core Findings
C1: Evaluate comprehensibility and perceived usefulness of video-based indicators	F1: Indicators on diverse levels (the video itself, layout, content and message, profiles, reactions and interactions, and overall impression) expand and confirm the perceptions of adolescents
	F2: Particularly characteristics of manipulated videos are used intuitively as indicators by adolescents
C2: Implementation and qualitative evaluation of a smartphone app prototype as indicator-based misinformation intervention	F3: Overall positive feedback on the indicator-based application and good understanding of its features
	F4: Potentials for transferability of the extended knowledge and skills
C3: Assessment of opportunities, challenges, and limitations of the indicator-based approach	F5: Transparency of the indicators for misinformation are a central reason for the intervention's positive assessment
	F6: The indicator-based approach comes with limitations like adolescents blindly trusting in the simulated algorithm and a lack of realistic concerns about the robustness of the intervention

10.5 Discussion

In this study, we evaluated how comprehensible and useful indicators for video-based misinformation can be applied to a smartphone app as a digital intervention to assist adolescents. In a twofold study design, we first conducted a mixed-method user study including focus group discussions with adolescents and individual surveys to examine existing strategies in dealing with video-based misinformation, complementing existing research [226], and to gain novel insights into the comprehensibility and perceived usefulness of indicators derived from literature. Building on those findings, we then developed an indicator-based smartphone app prototype, and evaluated the applicability, opportunities and limitations of this approach in individual think-aloud sessions with adolescents. This allowed for the examination of in-depth findings on adolescents' needs and expectations. In addition, we derived overarching design implications for user-centered, transparent digital misinformation interventions for adolescents from the following detailed responses to the research questions (see Table 10.2).

Table 10.2 Design implications for user-centered countermeasures targeting adolescents in dealing with misinformation

Design Implication	Explanation
(1) Promote self-reflection	Encourage users to critically evaluate content based on their own gut instincts and indicators. Encourage self-reflection by including prompts such as "What does your gut say?" to motivate users to assess their feelings in terms of a stimuli's credibility
(2) Use comprehensive, transparent misinformation indicators	Design interventions that provide comprehensive indicators across different components of video content (sound, layout, interactions, etc.) to allow for a well-founded assessment
(3) Allow for personalization	Allow users to customize the intervention according to their preferences to allow for individual circumstances, abilities and needs. This includes users who might be overwhelmed by text-heavy explanations and who require simple visual cues as well

(continued)

Table 10.2 (continued)

Design Implication	Explanation
(4) Mitigate overconfidence	Recognize that young users may be overconfident in their ability to identify misinformation and design interventions to promote self-awareness and humility when in content evaluation
(5) Consider user trust and clarify algorithm limitations	Mitigate blind trust in algorithms by establishing a clear understanding of an algorithm's limitations and potential errors. Encourage users to use the indicators as a tool to develop their own skills in evaluating information, thereby promoting a sense of autonomy

10.5.1 RQ1: How do Adolescents Evaluate Misinformation Indicators in Terms of Comprehensibility and Usefulness?

We found that our young participants were very capable of comprehending various types of indicators regarding different components of a TikTok video. They assessed the video itself, the tonality of spoken words, emotional sound and music of a video, the layout, content and message, the profile, reactions and interactions with a video, and the overall impression which is mainly based on the adolescents' individual gut feeling (design implication (1), Table 10.2). It was noteworthy how adeptly the adolescents were able to identify satirical content by referencing the comment section, the creator's profile, or by relating the video to the music accompanying it. As individuals who have grown up with advances in AI, our participants were very intuitive and natural in handling and recognizing manipulated images. Indeed, they were often able to autonomously name characteristics of manipulation such as video exposure or wrong details which they used as heuristics to assess a video's credibility. These findings support the notion that adolescents and young adults, as digital natives, have a certain level of digital literacy and ability to use information competently, as suggested by some studies [157, 580].

Our findings extend and confirm existing research on trust heuristics regarding VSPs content that have been clustered as "convenience, aesthetics, and tone" by Hassoun et al. [226]. Further, similarly to the findings of previous research

[226], our participants particularly emphasized a high relevance of familiar creators for credibility assessments, and in general revealed some socially-oriented concerns on what would happen if they did not recognize misinformation and shared it among their peers, reporting individual experiences. The potential of using source reliability as a key indicator for evaluating online information was also demonstrated in a randomized controlled trial, where an intervention aimed at enhancing this ability improved the students' capacity to distinguish between reliable and less reliable sources [425]. Thus, it is not surprising that our participants regularly opened the comments section of TikTok videos before deciding if the content is misleading or not.

While existing adolescent evaluation strategies in relation to the content of VSPs have been studied qualitatively [226], we build on these findings by supporting the need for shared sense-making through indicators in the domain of 'reactions & interactions'. However, we stressed that these indicators should always be used in conjunction but with others that extend the view on characteristics which are detached from other users' assessments and potential manipulations.

10.5.2 RQ2: How can Indicators be Applied to a Smartphone App as a Digital Misinformation Intervention to Assist Adolescents on TikTok?

Building on those findings, we developed a prototype smartphone app as an indicator-based misinformation intervention and gained in-depth insights in individual user study sessions. Overall, the comprehensive approach of integrating misinformation indicators across different components of a video was found useful by our participants (design implication (2), Table 10.2). The majority of participants saw value in the approach for themselves or other TikTok users of their age, especially because of the transparent indicators and the freedom of choice that this preserves. In contrast to more technical approaches that use machine learning methods to automatically detect misinformation [e.g., 518, 613], this approach enables transparency and allows for self-deliberation, and may be a suitable extension to follow after automatic detection via machine learning. The positive assessment of the adolescents is consistent with other research, suggesting that users generally prefer comprehensible explanations to binary labelling approaches [299]. In this context, comprehensibility is thought to play an important role in building trust [51]. The positive assessment of this approach aiming at improving media literacy is consistent with other studies that display the potential of media literacy as a means of combating misinformation on social media [530, 607]. However, when aiming to improve media literacy, it is important to remember that this relies on the rationality of the individual [62]. Particularly for younger people, "online information processing is fundamentally a social practice" [226], which can potentially conflict with rational analysis of information accuracy.

Furthermore, our approach is not entirely immune to the problem that young people may feel patronized by interventions aiming at enhancing media literacy and that their digital skills are not taken into account—an issue that has been raised in previous research [226]. In our findings, this is most evident in the additional user groups suggested. For example, some of our participants see older or very young people as users who could particularly benefit from the approach, rather than themselves because of their already developed skills in identifying misinformation. The varying skill levels observed in adolescents, alongside their reported different preferences (e.g., regarding the amount of text-heavy explanations of the indicators) also suggest that incorporating personalization into digital interventions may alleviate such concerns (design implication (3), Table 10.2)—an approach that has shown promise in other areas of critical information handling [17, 146].

Moreover, our results displayed the balancing act that young people face in their role as judges of misinformation, especially with regard to video-based content. On the one hand, because they are 'digital natives', they are called upon as experts, especially by older relatives, to clarify the trustworthiness of videos. Our qualitative insights regarding the evaluation of TikTok videos indeed reflect a certain competence, especially with regard to manipulated videos. On the other hand, our study shows that in some cases even young people who are confident in their abilities are not able to recognize misinformation on their own and may benefit from (technical) support due to the complex and multifaceted misleading potential of the videos. The concept of 'overconfidence' in assessing media content may therefore also apply to adolescents in our context, as they repeatedly stated that they saw potential in such an indicator-based app, especially for other people (design implication (4), Table 10.2). Such overconfidence and general overestimation of digital skills in adolescents and young adults has been explored in other studies [75, 390, 430, 635] and is partly explained by a lack of critical thinking [408]. Furthermore, the level of self-efficacy in adolescents was suggested as a crucial factor for both checking sources and refraining from sharing unreliable information [401]. Overall, our study indicates significant potential for mutual learning by co-designing studies with young people, which can also benefit other user groups such as older adults.

10.5.3 RQ3: How do Adolescents Perceive Opportunities and Limitations of an Indicator-Based Digital Misinformation Intervention?

Our qualitative findings reveal opportunities for an indicator-based approach to provide useful support in dealing with misinformation. However, they also reveal limitations and challenges. For example, our approach shows potential for an automation bias that encourages blind trust in (e.g., AI-based) algorithms [543]. Only few of our participants showed healthy skepticism about the capabilities and limitations of the technical tool. In particular, it was noted that there was little understanding of the serious consequences of an incorrect output from the application. While this potential for blind trust in automation has been suggested [543], other studies have also shown skepticism towards automated systems [23, 312]. Thus, the clear lack of skepticism in the sample of adolescents in our study suggests a potentially high level of vulnerability (design implication (5), Table 10.2). This makes it all the more important to use transparent approaches when helping adolescents to deal with misinformation, rather than relying solely on automatic detection with subsequent binary labelling [e.g., 40]. An indicator-based approach can, to some extent, counteract this risk by further challenging the participants' own thinking. Our qualitative findings offer initial insights into potential opportunities for the applicability of the indicators to new videos and the transferability of knowledge and skills to platforms beyond TikTok.

10.5.4 Limitations and Potential for Future Work

Our study has some limitations and potential for future work. *First*, quantitative evaluation of the effectiveness of the indicator-based approach in real-world usage is still needed to investigate if there is a long-term learning effect of applying indicators and if this approach prevents misinformation sharing. It is not the aim of our study to identify statistically significant effects, such as reducing the sharing of misinformation. This has been addressed in other studies [35, 76] and remains worthy of investigation in the context of video-based content for future research. It will be interesting to monitor whether the integration of the app via the share button is accepted by users in everyday life compared to a direct integration within TikTok, and how the approach is perceived in a practical usage scenario where the app's output is occasionally incorrect, and the app's consultation is useful for a limited number of videos. *Second*, evaluating the approach with a larger sample of videos might provide additional insights and could be combined with the suggested quantitative study design for future research. *Third,* our prototype is based on a simulation regarding the detection of misinformation and the detection of the individual indicators. Digital misinformation interventions often rely on a successful prior decision (automatic or manual) as to whether a piece of information contains misinformation. Prefiltering whether a social media post contains misinformation is a non-trivial task and a problem that has been the subject of research in recent years [518, 613], especially in the field of machine learning. Displaying our indicators by default on all videos would lead to an excessive number of false positives and is therefore not appropriate. Future work could evaluate the approach fully implemented with state-of-the-art detection techniques for pre-filtering. In our user studies, we observed a group of young people with different levels of learning abilities, ranging from very young adolescents (13 years old) to 16 year-old adolescents. By including an entire school class in combination with participants from a youth center, biases towards a special interest in technology could be well balanced and the sample size is also well in line with the usual measures of qualitative studies [81]. Nevertheless, it would be interesting for future studies to—*fourth*—include a larger sample, e.g., to expand the age range of the subgroups. The use of TikTok is restricted to individuals over the age of 13, but there is no defined upper limit. People in older age groups are discovering TikTok for themselves, wich makes them a particularly interesting group of participants for studies.

10.6 Conclusion

Through our mixed-methods user study with 39 adolescents, we qualitatively evaluated the potential and limitations of an indicator-based misinformation intervention for TikTok as a core VSP.

We advance misinformation research by applying existing knowledge about indicators and controversially discussed media literacy interventions to the very current modality of short-videos. Thereby, we adopt the user perspective of adolescents as a particularly relevant user group. We generate novel insights into the opportunities, challenges, and limitations of indicator-based misinformation interventions. Our core contributions and findings are (C1) evaluating the comprehensibility and perceived usefulness of video-based indicators to (F1) find how indicators on diverse levels (e.g., the video itself, profiles, and interactions) expand and confirm the perceptions of adolescents, and (F2) highlight how the identification of characteristics of manipulated videos is an intuitive approach pursued by adolescents. Further, we (C2) implemented and evaluated a smartphone app prototype as indicator-based misinformation and received (F3) an overall positive feedback on the approach and a substantive understanding of its features, and identified (F4) potentials of transferability to new content. Lastly, we (C3) assessed opportunities, challenges, and limitations of the indicator-based approach and found that (F5) it was particularly well received in terms of transparency, while it (F6) carries limitations, such as missing realistic concerns about the robustness of the intervention. On a social level, our findings underscore the balancing act of adolescents being regarded as digital experts to evaluate social media content, but also being vulnerable to the complex landscape of (video-based) misinformation.

Mitigating Misinformation Sharing on Social Media through Personalized Nudging

11

Abstract The ongoing challenge of misinformation on social media motivates ongoing efforts to find effective countermeasures. In this study, we evaluated the potential of personalized nudging to reduce the sharing of misinformation on social media, as personalized support has been successfully applied in other areas of critical information handling. In an online experiment ($N = 396$) exposing users to social media posts, we assessed the degree of misinformation sharing between groups receiving (1) no nudges, (2) non-personalized nudges, and (3) personalized nudges. Personalization was based on three psychometric dimensions—general decision-making style, consideration of future consequences, and need for cognition—to assign the most appropriate nudge from a pool of five nudges. The results showed significant differences ($p < .05$) between all three groups, with the personalized nudge group sharing the least misinformation. Detailed analyses at the nudge level revealed that one nudge was universally effective and two nudges were effective only in their personalized form. The results generally confirm the potential of personalization, although the effect is limited in scope. These findings shed light on the nuanced results of nudging

Original Publication *Tom Biselli, Katrin Hartwig, and Christian Reuter. "Mitigating Misinformation Sharing on Social Media through Personalised Nudging". In:* Proceedings of the ACM: Human-Computer Interaction (PACM): Computer-Supported Cooperative Work and Social Computing (CSCW). *ACM, 2025. DOI:* 10.1145/3711034 (Minor edits in the numbering of the subsections have been made to the original publication for the readability of this dissertation.)

Supplementary Information The online version contains supplementary material available at https://doi.org/10.1007/978-3-658-47885-8_11.

K. Hartwig, *Navigating Misinformation*, Technology, Peace and Security | Technologie, Frieden und Sicherheit,
https://doi.org/10.1007/978-3-658-47885-8_11

studies, highlight the benefits of personalization, and raise ethical considerations regarding the privacy implications of personalization and those inherent in nudges.

11.1 Introduction

The spread of misleading information has become increasingly common in recent years, with prominent cases such as the 2016 US presidential election or the COVID-19 pandemic [422]. A major obstacle in mitigating the spread of misinformation is the immediate accessibility and sharability of false information on social media. This poses a challenge for the content moderation of social media sites, which often relies on automatic detection of false information [581]. In general, the harmful effects of misinformation arise not only from deliberate deception, but also from the unintentional dissemination of misleading information. In line with other researchers [15, 97, 330, 597] we use the term *misinformation* as an umbrella term that encompasses both intentionally created misleading information, often referred to as *disinformation* or, more rarely in the academic context, *fake news*, and unintentionally created misleading information.

Human-computer interaction (HCI) research has focused on ways to support users in dealing with misinformation and reducing its spread for many years. Research considers misinformation and related phenomena like rumors a crucial concern of Computer Supported Cooperative Work (CSCW) as they *"function to undermine the integrity of the information space and reduce human agency by overwhelming our capacity to make sense of information"* [533, p. 2]. HCI and CSCW research approaches misinformation from different angles, including the analysis of influencing factors on misinformation spread [97], sharing behavior of social media users [336], and the user-centered development of digital misinformation interventions (e.g., displaying warnings or corrections) [214]. However, the field is far from a definitive solution. Therefore, further research on effective ways to combat misinformation is relevant from an HCI and CSCW perspective.

Alongside automated detection approaches and fully transparent user intervention, the concept of *nudging* has become increasingly relevant for reducing misinformation spread on social media. A nudge can be defined as "any aspect of the choice architecture that changes people's behavior in a predictable way, without prohibiting any options or significantly changing their economic incentives" [85]. Nudges have been shown to be effective in a range of areas including health [554], online security [284], shopping [155], productivity [313] and many others. Importantly, is has also been successfully applied in the context of misinformation [418, 581]. However, other studies have shown limited effectiveness of nudging [e.g., 25, 173]. In other areas of online privacy and security, the potential for personalized as opposed to one-size-fits-all approaches has shown promise in addressing differences in user characteristics [414]. Thus, while nudging has shown promise as an intervention to reduce the sharing of misinformation, its inconsistent success prompts research into strategies to improve its effectiveness. Our aim in this study is, therefore, to apply the personalization approach to nudging in the context of misinformation in order to assess its potential to reduce the spread of misinformation on social media.

To test this, we designed an online experiment with generic social media posts and measured the likelihood of sharing posts containing misinformation in a group with (1) no nudges, (2) non-personalized nudges, and (3) personalized nudges. The personalization of a pool of five nudges was based on users' decision-making style, their need for cognition (NFC) and consideration of future consequences (CFC) to achieve an accurate mapping of user characteristics. Our results underline the superiority of personalized over non-personalized nudges as a cost-effective and user-centric measure to effectively reduce the spread of misinformation. Thereby, we expand promising findings on personalization from studies in other areas such as cybersecurity. We complement and extend existing HCI research on digital misinformation interventions by applying the concept of personalized nudges to the misinformation context, addressing the research gaps of understudied diverse nudging types and tailoring interventions to user characteristics. Thus, our core contributions (C) are (C1) to design and evaluate a diverse set of nudges in a controlled online experiment, and (C2) to demonstrate the fundamental potential of personalized nudging as an effective user-centred countermeasure against the spread of misinformation.

The paper begins by reviewing prior research on countering misinformation spread on social media nudges, and the relevance of personalization (Sect. 11.2). We then present our experimental methodology, (Sect. 11.3) and the results regarding the potential of personalized nudging to reduce misinformation (Sect. 11.4). Finally, we discuss the implications of the results, including the benefits of personalization and the need to consider ethical implications (Sect. 11.5).

11.2 Related Work

This work contributes to the design of personalized nudges as effective, lightweight digital interventions to combat misinformation on social media, addressing a significant socio-technological challenge in HCI research. The impact of misinformation on social media, extensively discussed across disciplines (Sect. 11.2.1), has prompted the design and evaluation of various digital interventions, including nudges as a low-cost interventions [214] (Sect. 11.2.2). In addition, personalization (Sect. 11.2.3) has been proposed as a promising way to increase efficacy of user interventions.

11.2.1 Misinformation on Social Media

Social media has proven to be a relevant information source in both disruptive situations and daily life [195, 460]. However, platforms like Twitter/X, Facebook, and TikTok have long been hubs for misinformation propagation [130, 533, 535]. While the existing literature on misinformation tends to focus heavily on the US (as most of the work described in Sects. 11.2.2 & 11.2.3) it is important to consider that misinformation and how people deal with it is always shaped by specific social and cultural contexts [473]. Lim and Perrault [336] provided insights from Singaporean society, which they characterise as a more collectivist culture, leading to less use of social media than in more individualistic cultures [256], and causing the need to *"design more effective interventions"* to combat misinformation specifically on instant messaging apps [336]. Further studies offer valuable perspectives from the Global South and emphasise that misinformation should be considered from this angle [545, 575, 599]. For example, considering the particular context of rural Bangladeshi villagers, Sultana and Fussell [545] highlighted the role of inaccessibility to modern fact-checking resources, while Wasserman and Madrid-Morales [599] highlighted low trust in social and national media in Kenya, Nigeria and South Africa as a crucial context for dealing with misinformation. Our study offers a vantage point for the study of transnational discourses of misinformation by focusing on German society as a rather individualistic society [121] with high use of social media and access to technical resources.

While misinformation plays a role on all social media platforms, a fairly large proportion of research on digital interventions focuses on Twitter/X [423, 584] or Facebook [318, 373]. A few studies focus on other content and platforms such as WhatsApp [412], Instagram [588], or TikTok [16]. The majority of misinformation research in HCI is more generic and potentially transferable to different (mostly text- and image-based) social media platforms and online content [214]. Research highlights that misinformation and other forms of online manipulation are a crucial concern from a perspective of CSCW [533]. HCI and CSCW research address misinformation and related phenomena from diverse angles. Besides the focus on the online content itself [92], many studies focus on the user perspective on social media to understand who shares misinformation, how and why, or how people generally perceive and handle misinformation online [97, 336, 462]. Findings reveal diverse perceptions and potentials for mitigation strategies that often depend on local contexts [336].

11.2.2 Digital Interventions to Combat Misinformation

On a solution-oriented level, studies in HCI explore how technology can aid in countering the effects of misinformation [299, 307, 560]—these investigations strongly guided our approach. Among strategies to combat misinformation, developing digital misinformation interventions is a complementary approach to initiatives like media literacy education in schools and strengthening professional journalism. In accordance with other studies [35, 479, 480] we understand 'digital misinformation interventions' as digital countermeasures that vary widely in their primary goal. On the one hand, this includes automatic detection approaches (e.g., using machine learning techniques) to distinguish between accurate information and misinformation. While they have been extensively studied in recent years [e.g., 518, 613], yielding promising results, they also reveal limitations in accuracy and explainability. On the other hand, this also includes approaches that take place after the successful detection of misinformation as a decision on what to do, e.g. with a misleading social media post. Digital misinformation interventions typically differ in their focus and intention to promote a learning effect. Some aim to foster critical reflection by presenting comprehensible indicators as cues for misinformation [31, 491, 620] (e.g., attention-grabbing layout, emotional language). While initial studies indicate promising results, others emphasise limited potential as the approach relies on rationality [62] and creates a false sense of confidence [75]. Conversely, interventions that focus on informing users about the credibility of content through warnings or corrections [29, 60, 328] show mixed effectiveness, with some studies highlighting limitations [391].

Such mixed findings emphasise the challenge of identifying effective digital misinformation interventions. Therefore, in addition to interventions that rely on rationality and may be limited by it [62], there is a need for interventions that also affect users at a more subconscious level. Nudging is a concept that aims to address this necessity. It has been widely applied in many contexts, such as health and cybersecurity [48, 218], and also as a lightweight digital intervention for the misinformation context [53, 84, 271]. Nudging offers a less paternalistic approach, allowing freedom of choice, as opposed to interventions that are more invasive and may provoke reactance. Like Lu et al. [344], some nudging interventions go beyond unconsciously steering users by providing additional information, and are sometimes referred to as 'hybrid nudges' [633] or 'enriched nudges' [454]. While in this work we mostly focus on classical nudges that have a subconscious effect on users and do not provide any additional information, we incorporated one hybrid nudge due to promising results observed in other domains like cybersecurity [218].

When designing nudges in the context of misinformation, various approaches can steer user behavior. First, nudges can encourage autonomous assessment of the credibility of content [53, 261] and fact-checking [557], to promote informed decision-making and learning. This might also be achieved by nudging users to consume credible information from trustful sources and steer them away from misinformation in the first place [138]. Another key objective in countering the impact of misinformation is to reduce its spread. For example, von der Weth, Vachery, and Kankanhalli [581] developed the browser extension "ShareAware" which provides users with an immediate feedback on a post's sources, including factuality labels based on linguistic features. Furthermore, studies use social norms [150, 182], for example, to nudge users to think twice before sharing misinformation, as *"most responsible people"* would do [22]—a relatively understudied type of nudge [467]. 'Accuracy nudges' encourage users to share less by making them think about accuracy and to act more thoughtfully in general. This concept has been applied in several studies [84, 271, 581] and reviewed in a meta-analysis [418], often but not always with promising results.

Like other misinformation interventions, the concept of nudging is controversially discussed as it can also be used in harmful ways, by representing a potential for subconscious manipulation or by unintentionally nudging users to make bad decisions. For instance, Lu et al. [344] revealed that while providing AI-based credibility indicators can nudge users to make informed decisions, this also nudges to blindly trust in the AI even if it is incorrect. Therefore, nudging can only be seen as one of several components that work together to combat misinformation.

11.2.3 Personalization to Improve Digital Interventions

Researchers in various fields, including cybersecurity and misinformation, face the challenge of developing lightweight yet effective digital strategies. Not only, but also in the field of nudging, both promising and limited evidence of effectiveness can be observed when planned interventions are not as effective as desired [25, 173]. Presently, most digital interventions, including nudges, target the average user, leaving certain user groups underserved. Studies from other contexts suggest that personalized interventions based on individual characteristics can enhance effectiveness [56, 146]. More specifically, showing personalized nudges in accordance with user characteristics instead of one-size-fits-all nudges have been investigated [118, 368] and have been found to be promising, e.g., in critical situations concerning privacy and security [218, 300]. While this has not yet been widely applied to the context of misinformation interventions, there are studies that argue for personalizing interventions to increase effectiveness and user acceptance. For example, Jahanbakhsh, Zhang, and Karger [260] identified solutions to provide more individualised user agency in curating social media feeds and assessing the exposure to misinformation by offering the ability to rate the accuracy of posts, evaluate trust in other users, and configure filters accordingly. In another study [262], they designed a personalised AI method, trained directly on each user's assessment of content to predict their assessment of other content, finding that user interactions with AI predictions can influence their judgement. Ali and Qazi [13] found that educational feedback, personalised based on the measuring of their previous assessment skills, has increased users' ability to identify misinformation in contrast to interventions with general educational feedback. Another approach is investigating personalisation that focuses on the strength of individual user interaction in order to increase interaction with verified posts [520]. Besides, Danry et al. [120] developed a wearable AI system that analyses the accuracy of arguments and suggest that future work should generate personalised explanations for individuals based on background knowledge and context rather than a one-size-fits-all approach. Similarly, Altoe and Pinto [17] contribute to personalization efforts in misinformation interventions by proposing a system that gives tailored creative, affective, personality-based or cognitive style-based explanations. Indeed, social media users and contexts of use differ significantly and not everyone is equally affected by misinformation [515]. On the one hand, vulnerability differs between demographic groups. For instance, older people [477], children, and teenagers [30] have shown to be more vulnerable to some extent. On the other hand, this may differ significantly in relation to specific platforms, as teenagers are more used to using video-based platforms such as TikTok, while older people are less familiar with the characteristics of this type of social media content. In addition to demographic characteristics, individual differences to approach decisions, perceptions, or assessments may influence the efficacy of misinformation interventions. In other contexts, segmenting users by their decision-making style or risk-taking attitudes, and providing them with tailored nudges has shown promising results [414, 496]. For example, people's decision-making styles were used to select personalized nudges for stronger passwords, which achieved better results when participants were given personalized nudges rather than one-size-fits-all nudges [414]. It has also been suggested that differences in cognitive style should be taken into account when designing effective personalized explanations for misinformation detection [17]. To our knowledge, there has been little research on the potential benefits of personalized digital interventions and more specific personalized nudges in the context of misinformation—a research gap that we aim to address in our work.

11.2.4 Research Gap

Our work contributes to the existing literature on user-centred misinformation intervention research by shedding light on personalized digital nudges and filling the following two main gaps:

First, in prior studies many nudging interventions to combat misinformation were designed as accuracy prompts [e.g., 84, 271, 418] or social norm nudges [150, 182, 441] with a clear emphasis on the former [467]. As systematically highlighted by Caraban et al. [85], there are many different ways to nudge, all coming with potentials and limitations. To offer a vantage point for research on diverse types of nudges on social media instead of focusing on a single specific nudging type (1st gap), we designed and evaluated multiple nudges in a controlled online experiment.

Second, while nudging is an established tool in misinformation research as well as in other contexts [e.g., 218, 414, 565], little research has been done on personalization to make nudges more effective in countering misinformation (2nd gap). We address this gap by investigating how psychometric tests can be used to segment user groups and accordingly provide distinct social media users with a suitable nudge instead of a one-size-fits-all solution. This is a relevant research problem, as the needs and preferences of social media users vary widely [17, 30, 477]. To address the identified research gaps, we aim to answer the following overarching research question:

RQ: Which potential do personalized nudges have for mitigating misinformation sharing on social media?
To answer this research question, we evaluated four aspects thereof. First, we evaluated the (1) results of the psychometric questionnaires used to assign users their personalized nudges and the corresponding nudge distribution. Second, we compared the degree of (2) misinformation sharing between the personalized, non-personalized and no nudge control groups on the group level. Third, we conducted a (3) detailed nudge analysis to identify specific effects on the level of nudges. Finally, we evaluated (4) user acceptance of the nudges from a post-experiment survey.

11.3 Methodology

To assess the potential of personalized nudging to mitigate the spread of misinformation on social media, we conducted an online experiment with participants from Germany. The general study design, the personalization approach, and the nudge design are presented in the following sections.

11.3.1 Study Approach: Procedure

Participants were initially assigned to one of three groups: (1) *without nudges*, (2) *non-personalized nudges* and (3) *personalized nudges*. Group (2) was randomly assigned to one of the five nudges, and those cases where the assigned nudge happened to be the personalized nudge were re-sorted into group (3) for statistical analysis. After providing socio-demographic information, participants completed three psychometric inventories that served as the basis for the personalized nudge assignment (see Sect. 11.3.4). Then, participants viewed 16 misinformation and eight truthful posts (with or without (personalized) nudges, depending on the assigned group) in a randomized order and answered questions about whether they would share such a post (see Sect. 11.3.2 for details on the stimuli and dependent variables).

Participants in the nudge groups were asked for additional feedback on whether they perceived the nudges, whether they found them annoying, and their thoughts on the potential usefulness of nudges in combating misinformation on social media. This last open-ended question was intended to provide qualitative insights to complement the core of our study, the quantitative findings, but was not systematically analysed. Instead, sample quotes were used to highlight perspectives and initial insights, supporting the statistical analysis. Afterward, participants were directed to the misleading content shown and provided with verified sources on the related topics.

11.3.2 Study Approach: Stimuli

We used a generic social media environment, similar to Facebook or Twitter/X, to create an experimental setup free from specific platform constraints in order to draw broader conclusions about nudges. We were particularly interested in findings applicable to text- and image-based platforms such as Facebook and Twitter/X. While this approach facilitated study design, it may not fully capture real-world social media behavior and its user implications, such as endless scrolling. We controlled the setting by limiting participants' exposure to a selected number of posts, allowing a focus on experimentally relevant variables and reducing distractions.

At the core of our experiment was the confrontation with misinformation and true information posts and the subsequent question, whether participants would share them. In the approach of selecting stimuli and measuring the dependent variables, we relied on recommendations on how to conduct behavioral research on misinformation [421] and on similar experiments in misinformation research [104, 299, 424]. We selected 16 debunked article headlines from various fact-checkers (correctiv.org, a German investigative journalism group; dpa / reuters factcheck, independent news agencies; mimikama, Austrian fact-checking site; and Snopes, a US fact-checking site). We also selected eight accurate headlines from esteemed mainstream media from a variety of sources, or headlines that had been verified by fact-checkers (see Tab. A.33 in Electronic Supplementary Material (ESM) for an overview of all topics). Two researchers have carefully curated a diverse set of headlines, covering both current and non-current topics, with the intention of avoiding a narrow focus on specific issues or crises. This involved choosing more unfamiliar topics that participants were unlikely to be aware of and ensuring that the content selected was thematically balanced.

To create the stimuli, we searched Twitter and Facebook for relevant social media posts arguing for these claims and transferred them to our generic social network, including any images used. We standardized all surrounding information (username, profile picture, number of comments, number of likes) across all posts to avoid introducing confounding variables. Participants were prompted with questions below each social media post, focusing primarily on the likelihood of sharing the post on a scale of 1 (extremely unlikely) to 6 (extremely likely). To mask the goal of our study, additional questions were asked about the posts, such as familiarity with the content, relevance of the topic, and perceived qualities (e.g., interesting, funny, exciting). Accuracy ratings were deliberately not prompted in order to avoid unintentionally influencing subsequent sharing behavior [421].

11.3.3 Study Approach: Participants

Based on an expected medium effect size of d = .5, an α-level of .05 and a desired statistical power of .8, we calculated the optimal sample size to detect differences between the three experimental groups to be be $N = 157$. In order to be able to make statements about differences at the level of individual nudges, we also aimed to fill all nudge groups with at least 30 participants (yielding a statistical power of 0.8 for medium-strong effect sizes of d = .65). This resulted in a total sample size of $N = 396$. As we ultimately found only a small effects, the power to detect differences between the three experimental groups was approximately .77. We recruited those participants via *Prolific*[1], a platform specifically designed to provide samples for scientific studies. Several studies have confirmed the reliability of Prolific and its ability to collect high quality data [4, 416]. Our study focused on individuals from Germany, selected based on residence, German language proficiency and social media use, with no additional strict criteria. While Prolific primarily draws from Western (European) states, it delivers diverse data in educational and income backgrounds [4, 416]. Within this German-speaking pool, we sought diverse perspectives by including participants of various ages and educational backgrounds. All participants had at least basic technical skills through their engagement with an online platform such as Prolific. This approach, albeit within the German sample, aimed to offer nuanced insights into how individuals with diverse experiences and technical abilities engage with online misinformation. The participants, aged 18 to 72 years, were 44% female, 54% male, and 2% diverse (see Tab. A.13 in ESM). The experiment was conducted online, and participants received €4 for their involvement.

[1] https://prolific.co

11.3.4 Personalization Approach

To personalise the nudge presentation, we used the three constructs and corresponding inventories *General Decision Making Style (GDMS)* [500], *Need for Cognition (NfC)* [80] and *Consideration of Future Consequences (CFC)* [542]. Our aim in combining several inventories was to cover several relevant dimensions and to improve the matching of nudges.

The GDMS scale measures how individuals make decisions and has previously successfully been used to personalise nudges [218, 414]. We hypothesized that rational decision-makers might benefit from alternative information sources, while intuitive decision-makers might need hints to scrutinize gathered information. Dependent decision-makers might follow others' behavior, and avoidant decision-makers might prefer nudges that relieve them of making choices.

The NFC scale is a measure of an individual's enjoyment of mental activities [80], and can divide individuals into two categories based on their engagement in mental activities: those with a low NFC score prefer simple tasks and tend to avoid effortful thinking, and those with a high NFC score enjoy effortful thinking and prefer to engage in difficult tasks that require reflective thinking [80]. We hypothesized that individuals with high NFC scores would likely assess information veracity by processing more details like alternative sources. Conversely, those with low NFC scores might prefer straightforward advice, avoiding the need for extra information consideration.

The CFC scale is a behavior-psychological self-report measure used to assess an individual's tendency to consider distant future consequences over immediate outcomes [275, 542]. While individuals who consider the consequences of sharing a misleading post would be more likely to refrain from doing so, those with lower CFC scores may only consider the immediate outcomes without thinking about the broader consequences and share such posts, even if only for amusement. Thus, we hypothesized that people with low CFC scores would be receptive to more present-oriented nudges, while those with high CFC scores would be more receptive to nudges that emphasized longer-term consequences.

11.3.5 Nudge Design

To evaluate the effectiveness of personalized nudging, we used a pool of five nudges to address the different requirements based on GDMS, NFC and CFC. In developing these nudges, we used guidelines for technology-mediated nudging in human-computer interaction [85] and previously successfully applied nudges [e.g., 284, 414]. We relied on the four overarching categories *facilitate*, *confront*, *social influence* and *reinforce*, and assigned each of our finalized nudges to the corresponding nudging mechanisms in [85] to facilitate their categorization. The aim of the selection was to cover a wide range of criteria relevant to our assignment process. The five nudges are described in Tab. 11.1 along with the rationale for their personalization. Screenshots of all nudges can be found in Fig. A.15, ESM. Generally, for the application of the nudges we assume a previous successful detection of misinformation (e.g., via automatic detection by algorithms or manual detection by professional fact checkers). Nudges were only shown on the misinformation posts, not the truthful ones.

Table 11.1 Nudge descriptions (see ESM A.6.3 for a more detailed description)

Nudge	Description	Personalized for ...
Hiding (Fig. A.13a, ESM)	Conceals post content behind a blurred panel, requiring user interaction to view the post.	... avoidant decision makers and low NFC users as this nudge provides no additional information. By initially hiding a post, avoidant users can avoid their fear of blame and anticipated regret if they were forced to make a clear decision about the content of the post [46].
Just-in-Time-Prompt (Fig. A.13b, ESM)	Triggers upon user interaction, confronting them about potential misinformation sharing and asking for confirmation	... intuitive decision makers and immediate CFC users, prompting spontaneous, non-conscious, holistic responses [119] and immediate consequence confrontation, and for low NFC users due to minimal additional information provided that could be consciously processed [80].
Social Comparison (Fig. A.14a, ESM)	Displays user contacts' avoidance of sharing due to potential for misinformation.	... dependent decision makers, utilizing the *herd instinct bias* [85]. As they seek advice and guidance from others [160], they should be sensible for information on how other users behaved.

(continued)

Table 11.1 (continued)

Nudge	Description	Personalized for ...
Raising Visibility of User Actions (Fig. A.14b, ESM)	Uses colored backgrounds to indicate previous account posting behavior and reliability (green for reliable poster, red for unreliable poster).	... intuitive and dependent decision makers due to instantly recognisable colour coding and relevance of poster which dependent decision makers should be sensible for [160], and for distant CFC and low NFC users because their own profile should remain green and thus a good reputation should be maintained, without further explanation of the exact calculation mechanisms.
Different Viewpoints (Fig. A.15a, ESM)	Presents fact-checked articles related to post to discourage unwanted action by instilling doubt.	... rational decision makers with high NFC as it offers deeper research material to engage in with cognitive effort [80] which serves their desire to reduce uncertainty [101].

11.3.6 Nudge Assignment

To assign the personalized nudges according to the logic described above, each participant's questionnaire scores for GDMS, NFC and CFC were normalized and weighted according to the presumed relevance for the different nudge types. The hypothesised relevance was based on the characteristics of the psychometric constructs and their potential role in the different nudges (see Table 11.1). Since, to our knowledge, there is currently no research linking psychometric scores to concrete, usable nudge designs, our work in this regard is hypothesis-driven. For example, a related study attempting to personalize nudges in the security domain also assigned nudges in an exploratory manner, based on a pre-study in which certain nudges proved more effective than others for certain user groups, without a strong theoretical foundation [414]. The scoring described below is an operationalization of the presumed relevance of particular manifestations in the psychometric dimensions for the suitability of particular nudges.

Specifically, the different manifestations of the questionnaire scores (immediate & distant for CFC; low & high for NFC) were given a weight of zero if they were not considered relevant to the effectiveness of a particular nudge (e.g., scores representing consideration of distant future consequences for the nudge *Hiding*) or a weight greater than zero if they were considered relevant to a particular nudge (e.g., scores representing high need for cognition for the nudge *Different Viewpoints*). In designing the nudges, we also aimed to address at least one of the GDMS subscales. Therefore, we rated the expected effectiveness of each nudge, taking into account the GDMS scores, on a scale from 0 (no increase in effectiveness) to 3 (most relevant increase in effectiveness). The highest expected relevance with a weight of 3 for the *Social Comparison* nudge was the *dependent* subscale, for the *Hiding* nudge the *avoidant* subscale, for the *Just-in-Time-Prompt* nudge the *intuitive* subscale, and for the *Different Viewpoints* nudge the *rational* subscale. For the *Raising Visibility of User Actions* nudge, the *intuitive* and *dependent* subscales were considered similarly relevant with a weight of 2 because the nudge represents both reliance on others and an intuitive decision based on the colored background. Finally, a smaller but still expected increase in nudge effectiveness was also represented with a weight of 1 in this approach, for example, by the *dependent* subscale for the *Different Viewpoints* nudge, because third parties are involved, or by the *intuitive* subscale for the *Hiding* nudge, because the lack of further information should elicit an intuitive response.

While the exact calculation differed between the different questionnaires due to their different designs, all scores were normalized so that all questionnaires determined the personalized nudge to the same degree. After normalizing and summing the scores for each nudge across the three questionnaires for each participant, the nudge with the highest score was selected as the personalized nudge. The conceptual assignment scheme is shown in Fig. 11.1, and a detailed mathematical description is available in ESM A.6.4.

11.3.7 Ethics

The study adhered to our university's ethics committee requirements, ensuring participant anonymity and minimizing stress or harm, among other things. We collected limited personal information (age, gender, education) and did not gather sensitive data (e.g., ethnicity, religion, health). Data was obtained via *SoSci Survey*, hosted on GDPR-compliant servers in Germany[2]. Participants provided

[2] https://www.soscisurvey.de/en/privacy/

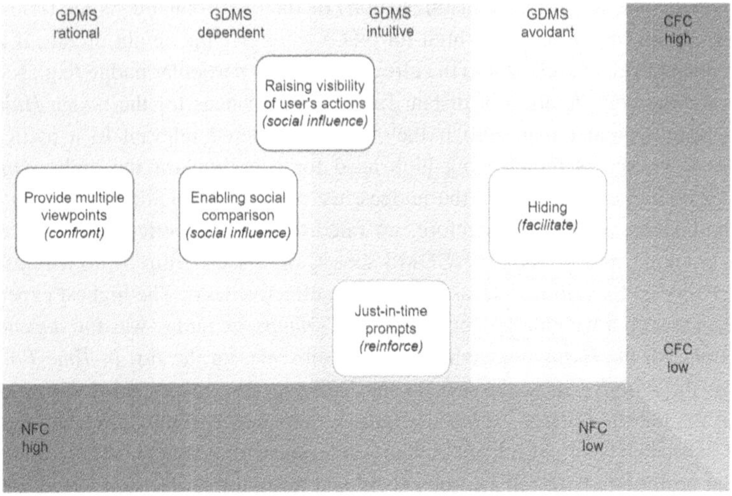

Fig. 11.1 Nudge assignment matrix (own illustration)

informed consent and could stop the study at any time. The data was secured and processed in accordance with the data protection provisions of the GDPR. After finishing the experiment, participants were debriefed and received transparent explanations regarding which posts contained misinformation, along with links to fact-checks detailing the level of misinformation.

11.3.8 Statistical Analysis

Overall, we compared three groups: (1) no nudges, (2) non-personalized nudges, and (3) personalized nudges. We hypothesized that the personalized nudges, in particular, would be associated with less misinformation sharing, as the tailored presentation of nudges would be more responsive to users' needs. As non-personalized nudges have previously been shown to reduce misinformation sharing, we hypothesized that the non-personalized nudge group would share less misinformation than the no nudge control group. To provide insights on aggregated responses and group level differences, we used the non-parametric Kruskal-Wallis test with subsequent Dunn's Post Hoc test to analyze the results initially. This test is a robust distribution-free method that uses a rank-based approach to determine the differences between the groups and is considered robust for unequal sample sizes [389, 396]—just like the Wilcoxon rank-sum test used to assess differences at the nudge level [346]. To avoid an overall increase in alpha errors due to multiple testing, the Benjamini-Hochberg correction for multiple comparisons [47] was applied. To substantiate the statistical analysis, we also created cumulative link mixed models (CLMM), an extension of cumulative link models (CLM) [332, 553]. This approach allowed for a more precise analysis by incorporating each individual response separately, in contrast to the Kruskal-Wallis analysis, where the summed intention to share, aggregated across all posts per participant, was analyzed. This method accounts for the non-independence of observations, as multiple participants responded to the same posts, and individual participants reacted to multiple posts. In this course, we modeled the posts and participants as random effects and the difference between the nudge and experimental groups as fixed effect using a logit link function. Subsequent pairwise comparisons were conducted using the 'emmeans' package[3], which provides estimated marginal means (EMMs), commonly known as least-squares means, enabling us to adjust for multiple comparisons and focus on the direct comparisons of interest within our model's framework. The CLMMs were estimated with the R 'ordinal' package[4], and all analyses were conducted in RStudio[5], version 4.2.2.

11.4 Results

The results are structured around four aspects that are considered relevant for answering the overarching research question about the potential of personalized nudging to mitigate misinformation sharing. First, we report on (1) descriptive results regarding the psychometric questionnaires used to assign personalized nudges and the corresponding nudge distribution (Sect. 11.4.1). Second, we compare the degree of (2) misinformation sharing between the experimental and control groups (Sect. 11.4.3). We then conduct a detailed nudge analysis to assess differences in the effectiveness of the pool of five nudges (Sect. 11.4.4). Finally,

[3] https://cran.r-project.org/web/packages/emmeans/index.html

[4] https://cran.r-project.org/web/packages/ordinal/index.html

[5] http://www.rstudio.com/

we report on a post-experiment survey to assess the (4) user acceptance of the nudges (Sect. 11.4.5).

11.4.1 Descriptive Results: Psychometric Questionnaires Scores and Nudge Assignment

Among the 396 participants, 104 were in the control group (no nudges), 146 in the non-personalized nudge group, and 146 in the personalized nudge group. We have allocated more participants to the nudge groups to accommodate the five nudge subgroups and applied statistical methods considered robust for unequal sample sizes [389, 396]. The distribution within the nudges was not even (see Tab. A.15, ESM), and thus also the distribution of participants types (i.e. which personalized nudge would presumably be suitable for which participant) was not even in each condition (see Tab. A.29, ESM). The main reason for this was the uneven distribution of the questionnaire scores, especially for the GDMS (see Fig. 11.2). While the rational thinking style was the most pronounced, followed by the intuitive thinking style, the scores for the dependent and especially the avoidant thinking styles were much lower. Both NFC and CFC scores were fairly balanced, with rashes in both the higher and lower regions.

These characteristics in terms of decision making and cognitive styles went hand in hand with an over-representation of the nudge *Different Viewpoints*, which was assigned mainly on the basis of a high level of rational thinking. Additionally, the nudge *Social Comparison* was assigned in the personalized nudge group only three times. The main reason for this under-representation was the similarity of assignment to the nudge *Raising Visibility of User Actions*. The latter nudge was also more likely to be assigned with higher scores on the *dependent* GDMS subscale, although to a lesser extent than the *Social Comparison* nudge. However, these attributional differences may have been too small, and the *Raising Visibility of User Actions* nudge also benefited more from the common *intuitive* GDMS subscale, and was more likely to be assigned when participants had distant CFC scores—whereas CFC was not considered to be relevant for the personalized assignment of the *Social Comparison* nudge. In order to obtain similar sample sizes for the different nudges and also to fill in the less frequent nudge personalizations, we screened out n = 493 whose personalized nudge samples had already been filled in and who therefore did not complete the experiment.

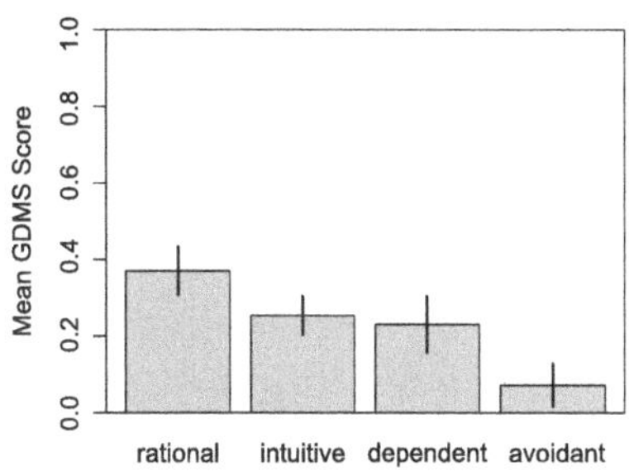

(a) Mean GDMS scores.

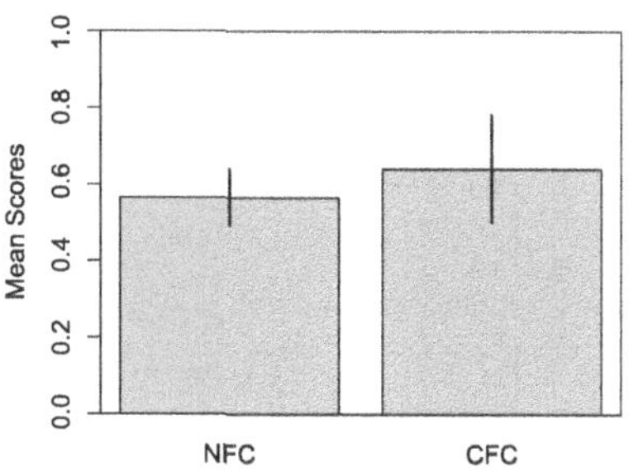

(b) Mean NFC and CFC scores.

Fig. 11.2 Mean values and standard deviation for all psychometric questionnaires

11.4.2 Descriptive Results: Social Media Post Evaluation

To assess potential confounding factors, we asked participants to rate whether they were familiar with the content of the post (yes/unsure/no), whether the post dealt with a topic relevant to the participants, and how they rated the content of the post (interesting, amusing, unappealing, exciting, upsetting, other) (see Fig. A.10, ESM). A cumulative link mixed model with post sharing as the dependent variable showed, among other things, that participants shared posts less often when they were unfamiliar with the topic of the post ($estimate = -.88$, $p < .001$) and shared more posts when they found the topic interesting ($estimate = .2.68$, $p < .001$) and amusing ($estimate = 1.82$, $p < .001$) (see Tab. A.14, ESM for detailed results). In general, participants indicated that they were somewhat familiar with the topic of the posts in 10.7 % of cases. Excluding these familiar posts, however, did not change the main results for the statistical inferences that follow (see Tab. A.17, ESM). As we were ultimately interested in the potential of the nudges to reduce misinformation sharing in general and not just regarding unknown posts, the following results are calculated with all posts included.

11.4.3 Comparison of Misinformation Sharing between Experimental and Control Groups

The main question to be answered in this study was whether personalized nudge presentation has the potential to reduce the sharing of misinformation on social media. All experimental and control groups taken together, Fig. 11.3b thereby shows that true information tended to be shared more often than the posts containing misinformation.

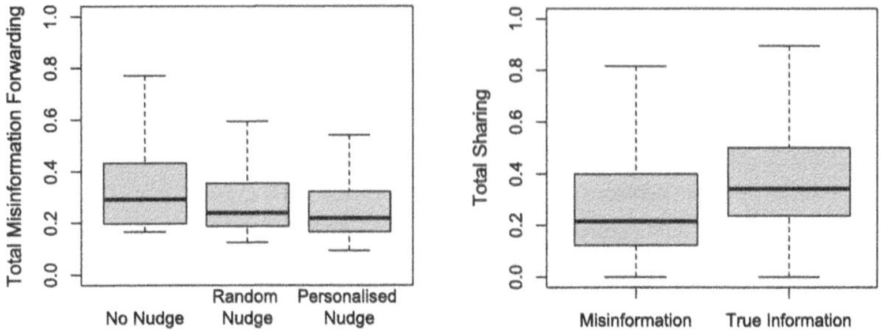

(a) Total misinformation sharing in the different experimental groups.

(b) Misinformation and true information sharing across all groups.

Fig. 11.3 Sharing between experimental groups and post content

A Kruskal-Wallis test revealed a significant difference in misinformation sharing between the no nudge, non-personalized nudge and personalized nudge group ($\chi^2(2) = 17.38, p < 0.001$, see also also Fig. 11.3). Post-hoc pairwise comparisons using Dunn's test showed significant differences between all three groups at the $\alpha = .05$ level (see Tab. A.16, ESM). Thus, both nudge groups reduced misinformation sharing compared to the no nudge control group, but the personalized nudge group did so with a larger—though still small—effect size ($r = .26$ vs. $r = .16$). In addition, the personalized nudge group reported significantly less misinformation sharing than the non-personalized nudge group ($Z = -2.22, p = .040, r = .13$). However, as the effect size is small, the impact of personalized nudges is limited to a certain extent. The results were supported by a cumulative link mixed model (see Table A.20, ESM) that included participants and social media posts as random effects to account for possible correlations within these categories. Due to the uneven distribution of participant types (i.e., which personalized nudge is likely to be appropriate for which participant) across the three experimental conditions, the general likelihood of sharing misinformation between these participant types was also evaluated in this model and included as fixed effect, just as the experimental group (no nudge, non-personalized nudge, personalized nudge). Thus, the model assessed both differences between participant types (including all participants across all experimental groups) and differences between experimental groups. Random effects analysis showed considerable variation, mainly due to individual differences between participants (SD = 1.75) and, to a lesser extent, variability in posts (SD = .44). The CLMM results (see Table A.20, ESM) and subsequent contrasts (see Table A.22, ESM) supported the results of the Kruskal-Wallis test and also revealed significant differences between the no nudge, non-personalized nudge and personalized nudge group. Post-hoc analyses evaluating contrasts between all subgroups further revealed that, apart from the group differences between all three experimental groups (see Table A.21, ESM), no differences were found for the different participant types (see Table A.22 and Fig. A.11, ESM). Thus, the differences in the experimental groups are unlikely to be caused by unequal group fillings and different general likelihoods of misinformation sharing for different participant types.

Table 11.2 Mean misinformation sharing in nudge groups (no nudge group mean: 1.96) and results of direct comparisons using Wilcoxon rank-sum test.

Nudge	Non-Pers. N.	Personalized N.	W	p	Effect size r
Hiding	1.47 (N = 33)	1.46 (N = 33)	537	.58	.01
Just-in-Time Prompt	1.70 (N = 32)	1.45 (N = 33)	448	.24	.13
Social Comparison	1.40 (N = 11)	2.27 (N = 3)	30	.98	.57
Raising Visibility of User Actions	2.10 (N = 37)	1.89 (N = 33)	510	.24	.14
Different Viewpoints	1.65 (N = 33)	1.38 (N = 44)	470	.02*	.30

11.4.4 Detailed Nudge Analysis

To better understand the dynamics of personalization, we conducted a post-hoc analysis at the nudge level. Here, we evaluated which nudges could generally reduce misinformation sharing even in their (1) non-personalized form and which nudges could reduce misinformation sharing in their (2) personalized form. Finally, (3) we evaluated direct differences between nudges in their non-personalized and personalized form.

Generally Effective Nudges. To evaluate nudge effectiveness regardless of personalization, we compared the no nudge and the non-personalized nudge group regarding the summed intention to share misinformation across all misinformation stimuli on the nudge level. A Kruskal-Wallis test revealed significant differences between the different nudge subgroups and the no nudge group

($\chi^2(5) = 17.99, p = 0.003$). Post-hoc pairwise comparisons using Dunn's test showed that only the *Hiding* nudge was effective in reducing misinformation sharing compared to the no nudge control group ($Z = 2.96, p = 0.023$) (see Tab. A.18 in ESM for all comparisons). This was supported by a CLMM calculated on post level that included participants and social media posts as random effects and nudge type as fixed effect. This model also showed only the *Hiding* nudge to significantly reduce misinformation sharing (estimate $= -1.10$, $Z = -3.04$, $p = 0.002$) compared to the no nudge control group (see Table A.24, ESM).

Effective personalized Nudges. In contrast, a detailed analysis of the personalized nudges (where a Kruskal-Wallis test also revealed significant differences between the different nudge subgroups and the no nudge group: ($\chi^2(5) = 28.23, p < 0.001$)) revealed that the *Different Viewpoints* ($Z = 4.34, p < 0.001$) and *Just-in-Time-Prompt* ($Z = 2.96, p = 0.015$) nudge in addition to the *Hiding* ($Z = 3.05, p = 0.017$) nudge were effective in reducing misinformation sharing compared to the no nudge control group (see Tab. A.23, ESM). This was supported by a CLMM calculated on post level that included participants and social media posts as random effects and nudge type as fixed effect. This model also showed showed the *Different Viewpoints* (estimate $= -1.73$, $Z = -4.87$, $p =< 0.001$), *Just-in-Time-Prompt* (estimate $= -1.13$, $Z = -2.97$, $p = 0.003$), and *Hiding* (estimate $= -1.17$, $Z = -3.06$, $p = 0.002$) nudge to significantly reduce sharing compared to the no nudge control group (see Tab. A.24, ESM).

Direct Nudge Comparison. A direct comparison of non-personalized and personalized nudges using Wilcoxon rank-sum tests showed a significant difference for the *Different Viewpoints* nudge, with the personalized version leading to a significantly lower sharing of misinformation than the non-personalized version, with a medium effect size of $r = .30$ (see Tab. 11.2 for all comparisons). The personalized *Different Viewpoints* nudge thereby led to the lowest overall sharing of misinformation (mean $= 1.38$). The *Just-in-Time Prompt* and *Raising Visibility of User Actions* nudges also showed a reduction in misinformation sharing at a descriptive level, while the sample size for the personalized *Social Comparison* nudge was too small to interpret meaningfully. Again, a CLMM calculated on post level with the participant and post modeled as random effect showed the *Different Viewpoints* nudge to differ between personalized and non-personalized nudge group significantly. To further ensure that this difference was not due to a lower tendency to share misinformation for the type of participant for whom the *Different Viewpoints* nudge was personalized, we additionally compared misinformation sharing for this type of participant between the no nudge, non-personalized nudge, and personalized nudge groups. The resulting CLMM (see Table A.27 in ESM) and posthoc contrasts (see Table A.28 in ESM) thereby still showed significant differences for the personalized nudge group both compared to the no nudge (estimate = 1.58, $p < .001$) and non-personalized nudge group (estimate = 1.26, $p < .001$) (see also Fig. A.12 in ESM). Thus, the reduced misinformation sharing in the personalized nudge group can be considered a specific effect of the personalization approach.

Taken together, the detailed nudge analysis suggests that only the *Hiding* nudge is generally effective in reducing the sharing intentions for misinformation, regardless of personalization. Personalization, on the other hand, is beneficial in that its application causes the *Just-in-Time-Prompt* and *Different Viewpoints* nudges to also significantly reduce the sharing of misinformation compared to the no-nudge group. Finally, clear evidence for the benefits of personalization exists for the *Different Viewpoints* nudge, where direct differences were found between the non-personalized and personalized forms.

11.4.5 User Perceptions and Opinions on Nudges

In order to obtain an assessment of the subjects' evaluation of the nudges, we asked final reflection questions (Fig. 11.4). Participants reported actively perceiving the nudges in 56% (non-personalized) and 60% (personalized) cases. While these figures are not particularly high, they are self-reported and not objectively measured. In addition, the processing of nudges is often subconscious. Participants generally rated the level of annoyance as relatively low, with a mean score of 1.60 on a scale of one to six. A Wilcoxon rank-sum test showed no difference between the personalized (mean = 1.64) and non-personalized (mean = 1.57) nudge groups ($W = 10450$, $p = .73$).

Similarly, in terms of opinions about the potential usefulness of nudges to reduce the spread of misinformation on social media, participants' opinions did not differ between the personalized (mean = 3.94) and non-personalized (mean = 3.88) nudge groups ($W = 10520$, $p = .85$). Overall, however, perceptions were generally positive, with an average score of 3.91 on a scale of one to six, although not significantly so. Open comments to the latter questions confirmed this potential ambiguity of nudges. Participants thereby explained the potential of nudges (across the different types of nudges) in terms of starting a thought process, for example: *"Yes, I think the nudges are useful, especially to assess the news correctly or to get out of one's own bubble." (P55)* or: *"It's really only very slightly annoying but can be very helpful in inspiring people to think more about the content of a post." (P117)* This might be especially useful for people with little social media experience, as one participant argued: *"It helps people with little experience in social media to become more critical of information sharing." (P129)* However, participants also emphasized that, in addition to the specific nudge, the rationale for the nudge and the wider context in which it is embedded are important in determining their agreement. For example, one participant noted that: *"It is important that the use of nudges is transparent, ethically responsible and democratically controlled." (P59)* Another participants also stressed that: *"If reputable sources of nudges are used, they can help distinguish fake news from real news." (P93)*

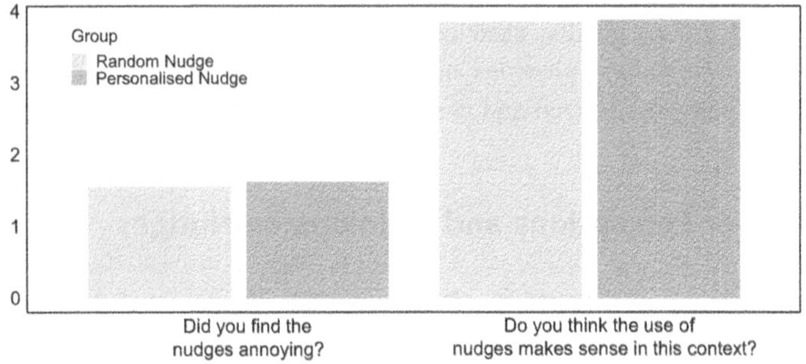

Fig. 11.4 Nudge reflection questions

Taken together, the low level of annoyance with nudges, combined with a rather positive view of the potential benefits, is promising—however, some potential caveats are raised by users and need to be considered when evaluating application scenarios.

11.5 Discussion

The spread of misinformation on social media continues to be a concern and the search for appropriate countermeasures remains highly relevant. This study evaluated the potential of personalized nudges to reduce this spread. The nudge presentation was personalized based on decision-making style, need for cognition and consideration of future consequences. An online experiment revealed that both non-personalized and personalized nudges resulted in decreased misinformation sharing compared to the no nudge control group. However, the personalized nudge group did so with a larger effect size and also showed significantly lower misinformation sharing compared to the non-personalized nudge group. The analysis showed that the *Hiding* nudge was generally effective, even in its non-personalized form. The personalization of the nudges resulted in two additional nudges, *Different Viewpoints* and *Just-in-Time Prompt*, significantly reducing the sharing of misinformation compared to the control group without nudges. A clear effect was found for the *Different Viewpoints* nudge, which also showed a significant reduction in misinformation when compared directly to its non-personalized variant. No differences were found between the two nudge groups in terms of their ratings of perceived annoyance and potential usefulness.

The research question we set out to answer was *"Which potential do personalized nudges have for mitigating misinformation sharing on social media"*. Our results suggest that by personalizing which nudge to present and aligning this process by certain participant characteristics such as decision making style, need for cognition and consideration of future consequences, the spread of misinformation can be mitigated effectively. Besides (C1) designing and evaluating a diverse set of nudges in a controlled online experiment, our main contribution is therefore to (C2) demonstrate the fundamental potential of personalized nudging as an effective user-centred countermeasure against the spread of misinformation. Specifically, we show that while nudges do not necessarily have a consistent overall effect, personalization may offer a way to better target user support and to more effectively reduce misinformation on social media. In addition, we shed light on the fact that while there are non-personalized effective nudges (*Hiding*), there are also nudges that prove effective only in their personalized form (*Different Viewpoints* and *Just-in-Time Prompt*) and increase effectiveness when personalized (*Different Viewpoints*). This also provides an explanation as to why nudging has not proven to be consistently effective and that it is worthwhile to consider specific user characteristics when developing user-centred support. We discuss this potential, ethical considerations and limitations due to the small effect size in the following sections.

11.5.1 The Benefits of Personalization

Many user interventions in the area of critical information handling are designed as one-size-fits-all approaches. This research demonstrates the potential of personalization to improve the management of misinformation and potentially reduce its spread. In demonstrating these results, this study is consistent with previous research highlighting the benefits of personalized support over one-size-fits-all solutions in the broader area of dealing with critical information regarding privacy and security [146, 218, 300, 414, 454]. In the specific context of misinformation, initial assessments of personalized approaches such as enabling more individualized user agency in curating social media feeds or providing personalized AI support proved promising as well [260, 262]. In terms of non-personalized nudging to reduce the spread of misinformation, it has shown promise in some studies [e.g., 52, 261, 449], but limited or even non-existent effects in others [e.g., 25, 173]. While some of this could be explained by the heterogeneity of the nudges used, the results of this study shed light on another explanation. Namely, individuals may respond differently to different types of nudges, depending on how well they match their cognitive characteristics. Our study results show that there is an effective non-personalized nudge (*Hiding*), but there are also nudges that only prove to be effective in their personalized form (*Different Viewpoints* and *Just-in-Time Prompt*) or increase effectiveness when personalized (*Different Viewpoints*). While the *Hiding* nudge has generally proved effective, its biggest caveat for application is its invasiveness. Because the content is initially hidden from users, it is not as lightweight as other nudges. Therefore, the findings that the nudges *Just-in-Time Prompt* and *Different Viewpoints* are only effective when personalized are particularly promising for developing lightweight interventions. Especially the *Different Viewpoints* nudge benefited from personalization. It was the only 'hybrid' nudge that provided additional information to evaluate the content of the post, and therefore has a high potential to help identify misleading information. However, it requires effort to process this rational information and cognitive alignment with these demands might make participants with high levels of need for cognition and rational thinking particularly receptive and more likely to engage with the additional information provided. The potential of appealing to the consideration of the veracity of the information to be shared has been researched previously [260–262, 422], but increasing the effectiveness of nudges by personalizing the presentation of a nudge that provides additional information adds here to the scientific literature. In contrast to previous studies, which have indicated differing misinformation sharing tendencies based on specific psychometric characteristics, such as analytic thinking [419] or reflexive open-mindedness [420], our study demonstrated no overall sharing differences between participant types (i.e. which personalized nudge is likely to be appropriate for which participant). This may be attributed to the aggregation of several psychometric variables, which may have resulted in the creation of less homogeneous circumscribed dimensions, such as analytical thinking or reflexive open-mindedness. The *Social Comparison* nudge was assigned too infrequently to be meaningfully interpreted, but showed potential at the descriptive level, in line with previous work [150, 182]. Finally, no effect could be found for the

Raising Visibility of User Actions nudge, possibly because the color coding was too obscure to be intuitively decoded.

11.5.2 Application Scenarios

The scope of this study is to examine nudging interventions to combat misinformation in the European context, with a focus on Germany. Research has highlighted the need to consider specific perspectives [336] as misinformation behavior is always shaped by its social and cultural environment [473]. Adding to perspectives from different regions of the world such as the Anglo-American sphere [53, 418], as well as regions of the Global South [545, 575, 599], we provide insights from a rather individualistic culture [121] which is linked to more extensive use of social media platforms for communication (e.g., in contrast to rather collectivist countries like Singapore [336]), making social media platforms an important application area for misinformation interventions in Germany. Despite concerted efforts to diversify within the constraints of the German-speaking population, we recognize that certain subgroups or underrepresented demographics may have been less represented in our sample. Nonetheless, the range of ages, educational backgrounds, and technical skills captured in our participant pool provided a robust basis for exploring how individuals from different backgrounds engage with and interpret online misinformation. Our findings should therefore be interpreted within the European, and more specifically the German, context and may not be universally transferable to other contexts.

In the European Union, platforms are obliged to take action against false information and (if possible) delete it [485]. In borderline cases, where a message is not completely false, but, for example, just taken out of context, nudges can help to limit the spread of the message and thus fulfill the platform's obligation without directly carrying out a complete deletion. The practical aspect of implementing personalized nudges revolves around utilizing available platform data. Platforms have access to necessary data, potentially enabling the personalization of nudges without individual creation. To increase the effectiveness of nudges, platforms could use the method presented to achieve the personalization of nudges. The nudges would not need to be created for each individual user, but personalization could be achieved with a selection of existing standard nudges. The social media platforms could technically perform the personalization, for example, by grouping users according to the four decision-making styles we used. Based on this, the user would then receive the appropriate nudge. Previous studies indicate that digital trace data can infer personality traits [32] or decision-making styles

[95] which could then be used directly to select a suitable nudge. In principle, the platforms have the necessary data available to determine the user's decision-making type. However, such an approach raises significant privacy concerns (see Sect. 11.5.3), which tend to be greater in Europe than in the US, for example [440, 514]. Due to the potentially sensitive relevant personalization dimensions, tailoring could also be done by third parties, subject to strict anonymization and secure data transmission principles. These third parties would need to be particularly trustworthy and could provide personalized nudge recommendations without revealing the details of why a particular nudge is particularly suitable for a particular individual. However, such a transfer would bring its own set of complex challenges, so a fundamentally privacy-preserving approach seems more feasible. The collection and classification of users' personality traits could be circumvented by having the platform initially randomize and test our proposed nudges for each user. Over time, the nudge that led to a lower amount of shared misinformation individually could then be identified and applied more frequently in the future. This method mirrors a commonly used A/B testing approach, ensuring user privacy while optimizing nudge effectiveness. Another possibility would be to simply allow users to choose the type of nudge they explicitly prefer, reducing the need to collect sensitive data—provided that users actually choose effective nudges. This user-driven approach empowers individuals to select interventions that resonate with them, potentially increasing the effectiveness of the nudge and user engagement. By integrating these alternative strategies, platforms could enable the delivery of personalized nudges while addressing privacy and ethical concerns. Future research should continue to explore these approaches to develop robust, scalable, and privacy-preserving personalization techniques.

11.5.3 Ethical Considerations

As personalized support for users to deal with critical information becomes increasingly common, it is crucial to consider the *personalization-privacy paradox* [450]. The dilemma arises from the potential conflict between privacy and personalization, where the latter often requires the collection of personal data, seemingly compromising privacy [450]. In addition to the use of questionnaires, personalization can also involve the collection of large amounts of behavioral data. However, this approach can have privacy implications, especially when based on large databases and algorithmic predictions of individuals' privacy preferences and behaviors. This issue becomes particularly relevant when personalization is used to potentially improve individuals' handling of critical information, because in the context of *securitization*, potentially detrimental effects may be more easily accepted [153]. Specifically, the desire to deliver personalized support creates an incentive structure to collect potentially sensitive personal information. Therefore, personalization should have limits and should only go so far as to include rather uncritical information such as decision making styles. More sensitive information, especially protected by the GDPR [571], such as health, religious or political information, should be avoided for personalizing nudges.

Beyond the personalization-privacy paradox, there are *ethical considerations about nudging* itself as they influence users, often on a subconscious basis. They can potentially restrict freedom of opinion and alter decisions that users would make rationally [337]. If the basis for providing nudges is flawed, they could also lead users to judge actually true information as false [344]. Accordingly, their use should be carefully considered and critically reflected upon [493]. This criticism was also raised by participants in this study where some stressed the need for transparency in the use of nudges and the importance of using reputable sources as a basis for presenting nudges. While in this work we have included classic nudges, which have a rather subconscious effect on users, we have also included the hybrid nudge *Different Viewpoints*, which has shown to be effective in its personalized form. Together with another effective personalized nudge (*Just-in-Time Prompt*), it may be more ethical to present the latter two nudges, as they are not as invasive as the *Hiding* nudge, which was also effective non-personalized, but effectively hides posts from users initially.

These findings shed light on the design trade-offs that need to be made when implementing interventions such as personalized nudging. In particular, it is crucial to balance the effectiveness of such interventions with ethical considerations such as privacy and potential perceived paternalism. Notably, while the *Hiding* nudge proved effective without the need to collect personal data, the implementation of other, less paternalistic, personalized nudges has the potential to engage a broader user base and potentially increase user acceptance [67, p. 37]. By covering these alternative types of nudges, we aim to explore a spectrum of approaches, each with its own trade-offs between personalization, effectiveness, and ethical considerations. Future research could further explore these dimensions, particularly how to optimize the balance between efficiency, personalization, and user privacy.

Overall, while a significant effect was observed for the personalized nudges, the smaller effect size underlines that they are not a panacea and that, in addition to powerful technical detection, fully transparent user interventions that explicitly encourage thought processes remain crucial—of which there is a wide range to choose from [214]. On the other hand, nudges are lightweight and offer more freedom of choice. In order for users to be sovereign, they should be sufficiently informed and be made aware that such personalized nudges may influence their decision-making process, and they should be given the opportunity to reject them or adjust their preferences—so that the general acceptance of personalized nudges is made in an informed way.

11.5.4 Limitations and Future Work

Some limitations of the current study need to be considered and potentially addressed in future work. First, it is important to consider the method used to assign participants to their personalized nudge. There is no standard protocol, and assigning based on presumed performance improvement (e.g., that the *Different Viewpoints* nudge works particularly well for people with a high degree of rational decision-making) remains somewhat exploratory and hypothesis-driven. While the assumptions about the relationship between psychometric characteristics and nudges have been derived to some extent from the literature, the exact scoring is an ad hoc operationalization of the presumed performance increase of nudges when a given psychometric characteristic is present. Again, given the lack of established methods for doing this, there is room for improvement in refining the accuracy of the mapping. In particular, given the uneven distribution of nudges, weaknesses in the assignment process have the potential to affect the extraction of the effects of personalized nudges. Therefore, future studies could explore alternative assignment methods like factor analysis or machine learning algorithms to refine the assignment precision. Alternatively, evaluating other ways to assign personalized nudges could be addressed by conducting additional studies that attempt to vary the assignment of nudges, such as the one conducted by Pe'er et al. [414]. Nevertheless, the increase in effectiveness in the personalized nudge groups suggests that the assignment to personalized nudges was indeed somewhat successful and did not result in arbitrary assignment. Overall, our aim was not to provide perfectly personalized nudges, but rather to improve the fit relative to generic, non-personalized nudges by relying on sound psychometric properties.

Second, our analysis and proposed method served as a conceptual examination of the potential of personalized nudges to mitigate misinformation sharing without explicitly evaluating their impact on truthful posts or falsely detected misinformation—a potential caveat, as detecting misinformation with 100% accuracy is unlikely. While this is an interesting additional angle, our goal was to examine the effects within the different nudge subgroups in as much detail as possible in order to answer the overarching research question about the potential of personalized nudges to mitigate misinformation sharing. In addition, the frequent and effective personalized nudge *Different Viewpoints* could only display trusted sources to reduce further the risk of misinformation being misidentified. However, future work could extend the investigation beyond reducing misinformation sharing to include the impact of personalized nudges on both truthful posts and falsely detected misinformation to provide a more comprehensive understanding of their real-world impact.

Third, the types of participants, and therefore the five personalized nudges, were not evenly distributed. Instead, the *Different Viewpoints* nudge was assigned particularly often. However, our aim was not necessarily to have a balanced distribution of nudges, but to provide individuals with a nudge that was more appropriate than an average nudge. The strong emphasis on a rational thinking style, which was the cause of the overrepresentation of the *Different Viewpoints* nudge, is thus also seen in other studies [e.g. 44, 500]. However, as participants were not randomly assigned within the personalized nudge group, but based on certain characteristics, this could potentially bias the interpretation of group differences. We mitigated this problem by (1) providing analyses showing no overall difference in misinformation sharing between participant types. Furthermore, for the most common participant type for whom the *Different Viewpoint* nudge is personalized, (2) we show that misinformation sharing is significantly lower in the personalized nudge group compared to the non-personalized and no nudge groups. Thus, the reduction in misinformation sharing can be specifically attributed to personalization rather than to a similar, particularly low tendency to share misinformation in all groups for this participant type.

11.6 Conclusion

The presence of misinformation hinders factual discourse on social media—a situation that can be improved by reducing the spread of posts containing misinformation. The results of this study shed light on the potential of providing personalised nudges to reduce the spread of misinformation on social media. The potential benefits of personalisation should thereby be carefully weighed against the potentially detrimental incentive structure for collecting personal data. However, given the complexity of the problem, personalised nudges can make a valuable contribution to a much-needed diverse and complex set of tools to counter misinformation on social media.

12 Nudge or Restraint: How do People Assess Nudging in Cybersecurity—A Representative Study in Germany

Abstract While nudging is a long-established instrument in many contexts, it has more recently emerged to be relevant in cybersecurity as well. For instance, existing research suggests nudges for stronger passwords or safe WiFi connections. However, those nudges are often not as effective as desired. To improve their effectiveness, it is crucial to understand how people assess nudges in cybersecurity, to address potential fears and resulting reactance, and to facilitate voluntary compliance. In other contexts, such as the health sector, studies have already thoroughly explored the attitude towards nudging. To address that matter in cybersecurity, we conducted a representative study in Germany ($N = 1,012$), asking people about their attitude towards nudging in that specific context. Our findings reveal that 64% rated nudging in cybersecurity as helpful, however several participants expected risks such as intentional misguidance, manipulation, and data exposure as well.

Original Publication *Katrin Hartwig and Christian Reuter. "Nudge or Restraint: How Do People Assess Nudging in Cybersecurity—A Representative Study in Germany". In:* Proceedings of the 2021 European Symposium on Usable Security (EuroUSEC). *Association for Computing Machinery, 2021, pp. 141–150. DOI:* 10.1145/3481357.3481514 (Minor language edits have been made to the original publication for the readability of this dissertation.)

Supplementary Information The online version contains supplementary material available at https://doi.org/10.1007/978-3-658-47885-8_12.

K. Hartwig, *Navigating Misinformation*, Technology, Peace and Security | Technologie, Frieden und Sicherheit,
https://doi.org/10.1007/978-3-658-47885-8_12

12.1 Introduction

The concept of nudging has evolved to be widely applied in many contexts and for many years to change people's behavior for the better. A nudge is *"any aspect of the choice architecture that alters people's behavior in a predictable way without forbidding any options or significantly changing their economic incentives"* [554]. While forcing people's decisions towards the desired outcome tends to create reactance, nudging aims to be perceived as less paternalistic, maintaining the freedom of choice. There are many ways in which nudges can be utilized, both harmfully and favorably. While the advertising industry nudges consumers to buy their products, nudges are also widely applied in the health sector to encourage healthy eating habits, or to prevent smoking [e.g., 209, 445]. In the context of privacy and security, nudging has recently become a growing research field [e.g., 3, 541]. For usable security, it is crucial to address user-centered approaches during the design process [55, 317, 484]. Recent works have started to investigate how bad security and privacy decisions of users can be nudged towards more beneficial decisions [3]. However, these studies are mostly based on one-size-fits-all nudges, while not being as effective as desired [146, 284]. Several studies therefore suggest personalized nudges, e.g., considering character traits [e.g., 146, 219, 300, 414, 565]. To facilitate personalization, it is crucial to understand how specific people assess nudges in specific contexts. To make nudges effective, people need to willingly endorse them and have positive attitudes towards specific nudges [413].

Until now, there are several studies on the attitude towards nudging, which compare nudges in different contexts [e.g., 204, 277, 453, 546] in one or several countries. While the health context has already been addressed on a larger scale and there are initial investigations for the context of privacy [134, 278], studies investigating the attitude towards nudging in cybersecurity are largely missing. To address that gap, this article presents an initial investigation on how people assess nudging in the context of cybersecurity. Hence, our research questions are:

RQ1: *How do people assess nudging in cybersecurity?*
In order to answer this question, we want to address the following sub-questions:
RQ1.1 Do they see risks of nudging in cybersecurity?
RQ1.2 Does the attitude differ concerning different scenarios?
RQ1.3 Do user characteristics or priming conditions affect the attitude?

RQ2: *What are design implications for nudging in cybersecurity?*

To answer our research questions, we conducted a representative survey in Germany ($N = 1,012$ after data cleansing), including priming with nudges for password strength and two standardized psychometric scales. The article is organized as follows: Section 12.2 presents related work on the attitude towards nudging in multiple-context studies and context-specific studies, highlighting the research gap. In Sect. 12.3 we present our survey design, the characteristics of our survey participants, and the method of analysis. In Sect. 12.4, we present the empirical results before discussing (Sect. 12.5) and concluding (Sect. 12.6) our contributions and limitations.

12.2 Related Work and Research Gap

Within the research area of attitudes towards nudging, two distinct groups can be identified: First, some studies investigate the attitude concerning nudges in multiple contexts. They often present descriptions of several nudges for different contexts so that the participants can assess them in comparison or give a more general statement [e.g., 204, 277]. Second, some studies focus on one context to evaluate how the participants perceive nudging concerning that specific matter [e.g., 278].

12.2.1 Attitude Towards Nudging in Multiple-Context Studies

The majority of studies do not focus on a specific context but on general attitudes. Hagman et al. [204] explored how American and Swedish citizens perceived different nudge policies regarding organ donation and climate compensation, among others. To explore individual differences, they conducted the Rational-Experiential Inventory scale (REI) and the Cultural Cognition Worldview Group scale. In an online survey, they found that while the overall support for nudges was high in both countries, the general acceptance was higher in Sweden compared to the United States. Concerning influencing factors, they found that preferences in analytical thinking were not significant [204], while participants who scored high on the REI-rational scale perceived nudges as more intrusive. In a multi-country study, Reisch and Sunstein [453] explored the attitude of Europeans towards nudges from different contexts on a larger scale. They focused on nudges in the contexts of healthy eating and organ donation, among others. Analyzing cross-country differences, they found that nudges that have been adopted in democratic nations were supported in all listed countries. More recently, Pe'er et al. [413] investigated the attitude towards nudging of two distinct minority groups in Israel (Israeli Arabs and Ultra-Orthodox Jews). Similar to other studies, they focused on the contexts of healthy eating, organ donation, privacy settings on social networks, and smoking, among others. They stress the importance of attitudes of minorities or other groups within a country towards nudging, which is largely overlooked by national and cross-national studies. They found that nudges were less supported when they were not consistent with a minority group's social norms [413].

12.2.2 Attitude Towards Nudging in Context-Specific Studies

Several studies have explored people's attitudes towards nudges for a healthier lifestyle. For instance, conducting in-depth semi-structured interviews in the UK, Junghans, Cheung, and De Ridder [278], investigated how consumers approve of specific nudges, whether they believed in their effectiveness, and whether who designed those nudges mattered. The authors found that in the context of health, most consumers approved of nudging, when they were beneficial to individuals and society and the targeted behavior was transparent [278]. Diepeveen et al. [132] conducted a systematic literature review on studies exploring the attitude in Europe, North America, Australia, and New Zealand towards interventions to change tobacco and alcohol use, diet, and physical activity. They found that the public acceptability of interventions in the health sector is greatest for the least intrusive nudges [132]. Evers et al. [152] investigated whether citizens from different European countries approved of nudges for healthy eating. Despite cross-country differences, they further explored if characteristics such as gender, age, or Body Mass Index and the level of intrusiveness of the nudges were influencing factors on the attitude. They found that women were more likely than men to approve of nudges in eating behavior. Again, less intrusive nudges had generally higher approval rates than intrusive nudges.

12.2.3 Research Gap

While there is an upcoming trend of nudges in privacy and security, little has been explored about the public attitude. Dogruel [134] compared the attitude towards policy interventions in information privacy between the United States and Germany. They found that privacy nudges were generally supported, and nudges for education and information were preferred [134]. Although nudges in the context of cybersecurity are a growing research field [3], to the best of our knowledge, there is no study that explicitly explores the attitude towards nudging within that context. Several studies have designed nudges for privacy and evaluated their effectiveness [e.g., 2, 36, 100, 255, 311, 339, 364]. Other studies have explored the effectiveness of nudges in cybersecurity [e.g., 69, 223, 345, 385, 414, 448, 569, 570]. For instance, Zimmermann and Renaud [634] evaluated nudges and hybrid nudges, including information provision for different contexts in cybersecurity (e.g., password creation and choice of public WiFi). They found that for some contexts, hybrid nudges were even more effective than a simple nudge, complementing a previous study on hybrid nudges [456]. However, in other cases, nudges in cybersecurity are often not as effective as desired because they are mostly designed for the average user, showing one-size-fits-all nudges [146, 284] and neglecting public attitude towards nudging in the design process, even though nudging *"strongly relies on voluntary compliance [and] public attitude towards specific nudges"* [413].

Personalization is an upcoming trend [e.g., 69, 146, 300, 414, 457, 565] which can be facilitated by exploring the attitude of people towards nudging in specific contexts. While people's attitudes have already been addressed in other contexts on a wider scale [e.g., 134, 152, 204, 277, 278, 453, 546], to the best of our knowledge, studies investigating the attitude towards nudging in cybersecurity are largely missing. We suggest that it remains important to expand knowledge on this matter, identifying both a general trend on people's attitudes for that specific context and differences between people's characteristics. We focused on the attitude of a representative sample from Germany to provide an opportunity for cross-country comparability in future studies.

12.3 Approach

To give an overview of the attitudes of the German population towards nudging in cybersecurity, we conducted a representative survey in Germany and investigated a general tendency as well as differences between demographic subgroups and the effect of different priming scenarios.

12.3.1 Survey Design

We conducted a representative online survey with 1,012 participants (after data cleansing) from the German adult population (18 to 74 years) in July 2019, using the panel provider *Respondi*. Each participant was paid a small allowance (€1). To ensure good quality answers, we included multiple attention check questions (e.g., *"Please select answer option number three."*) and excluded 92 participants who did not answer them correctly. In this work, we particularly investigate four survey questions. The first question aims at gaining insights into who had already consciously noticed nudges online (*yes, no, no answer*). In the second question, we asked five items: how helpful, dangerous, necessary, patronizing, or superfluous our participants assessed nudging in cybersecurity on a Likert scale from one to five. To get more specific information, we further asked in question three on a five-step Likert scale how useful nudging in cybersecurity was assessed in eight concrete contexts (see Sect. 12.4.2). Since some participants might not have enough knowledge of specific contexts to answer the question reasonably, answering those items was optional. To gain deeper inductive information, with question four, we included an optional free-text question, asking more openly about the attitude towards nudging in cybersecurity and potential risks.

To investigate potential influencing factors, we included two psychometric scales (see Sect. A.7.2, Electronic Supplementary Material (ESM)): the *General Decision Making Style scale* (GDMS) with two subscales: *rational* (making decisions logically and systematically) and *dependent* (rarely making important decisions without consulting other people) and the *Rational-Experiential Inventory* (REI) with two subscales: *REI-RA: rational ability* (ability to think logically and analytically) and REI-RE: *rational engagement* (reliance on and enjoyment of thinking in an analytical, logical manner) [402, 500]. Both psychometric scales are measured on a scale from 1 to 5. While the GDMS scale measures how individuals make decisions, the REI measures preferences for information-processing. Particularly the REI was already successfully used by Pe'er et al. [413] to personalize nudges and by Hagman et al. [204] to identify influencing factors on the attitudes towards nudges. We intend to investigate if they also influence attitudes towards nudging in the specific cybersecurity context.

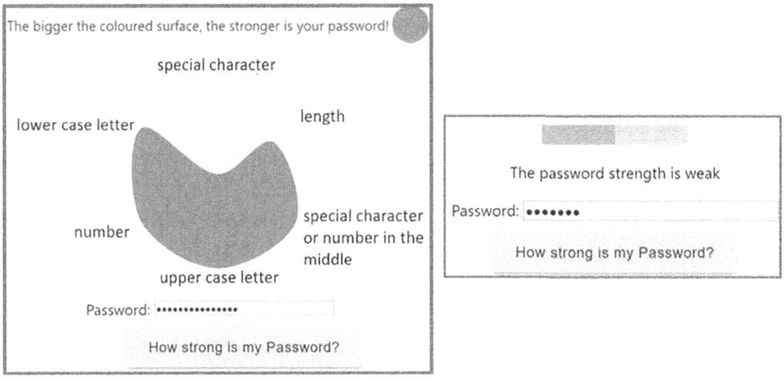

Fig. 12.1 Radar chart (left) and password meter (right) as priming conditions. (Own figure)

While all participants were presented with identical definitions and descriptive examples of nudging before asking our survey questions (see A.7.3, ESM), there were three distinct settings of additional priming (see Fig. 12.1). One-third of our participants had interacted with a common password meter as a nudge for password strength before answering the survey questions; another third had interacted with a radar chart visualization of password strength as a nudge [219], and the last third was not presented with any interactive nudge beforehand. Hence, in our evaluation, we further investigated whether those varied primings had an effect on the attitude towards nudging in cybersecurity. The sample sizes between our three groups slightly differ as participants were excluded that did not pass the quality check questions.

While we used password nudges that encourage complexity, it is important to acknowledge that recommending complex passwords has been ruled out due to usability conflicts [189]. However, state-of-the-art password nudges still use disaggregated information on what to do to strengthen a password [569] using complexity recommendations and some studies argue that focusing on password length alone does not always result in better results as many users still tend to prefer shorter passwords [219]. Hence, for our evaluation, we decided to include a nudge that encourages complexity as well.

12.3.2 Characteristics of Survey Participants

After data cleansing, 1,012 participants out of 1,104 were included for the analysis of the single and multiple-choice questions. For the free-text question, we performed an additional data cleansing and included a total of 489 participants. Our survey on the attitude of German citizens towards nudging in cybersecurity is representative of the German population in all groups regarding gender, age from 18 to 74, education, and income. To analyze potential group differences, we formed clusters for age (18–29; 30–39; 40–49; 50–59; 60–74 years) and income (< € 2,000; € 2,000 – € 4,000; > € 4,000). We also asked participants to select their highest level of education from three groups (without a school diploma; certificate of secondary education ("Hauptschulabschluss"), general certificate of secondary education ("mittlere Reife"), qualification for university entrance ("Abitur"); university degree). Regarding geographical spread, our sample is widely and approximately proportionally distributed across all 16 German federal states.

12.3.3 Analysis

We used *SPSS* and *R* for data preparation and statistical analysis and *Tableau* to create data figures. In a first step, we calculated the basic frequencies, including all demographic information (see Sect. A.7.1, ESM). To verify the representativeness of our sample, we determined significant differences in gender, age, education, income, and federal states applying the χ^2-test of independence. For our survey questions that were answered on a Likert-scale, we determined main effects of group differences performing a multivariate analysis of variance (*MANOVA*). As dependent variables, we included the five items on general attitudes as well as the eight items on context-specific attitudes.

We were interested in the following independent variables: priming condition (interaction with nudge), age, gender, education, as well as decision-making and information-processing styles. Before performing the *MANOVA*, we tested for homogeneity of variance as necessary condition, using the *Levene's test*. As the test was significant for some dependent variables, we adjusted the respective data with a *Box-Cox Power-transformation*. To identify which mean values were responsible for significant *MANOVA* results, we subsequently performed post-hoc tests. Therefore, we used the *Tukey's HSD*-test. We set the significance level to $\alpha = 0.05$. To counteract the problem of multiple comparisons, we applied a Bonferroni correction of the p-values.

While there were 1,012 valid answers to the Likert scale questions, we excluded 523 participants from the optional free-text question who decided not to answer it or explicitly stated they were too undecided to give specific input. The remaining 489 textual answers to our open question were evaluated in detail, performing a qualitative content analysis. Hence, we deductively assigned each answer to a cluster (rating nudges as useful, undecided, and thoroughly evaluated answers, rating nudges as inappropriate, comments on potential risks) by thoroughly analyzing the answers while looking for keywords (e.g., "dangerous"). Clusters were assigned in a two-step approach where answers were first grouped roughly into unambiguously positive or negative contents as well as contents that needed a more intensive reflection before decision. Afterward, the answers were assigned to one of the four final clusters while especially reflecting on the initially ambiguous contents. Clustering was conducted independently by two people with *RQDA* and the inter-coder reliability was calculated, resulting in a substantial level (Cohen's kappa coefficient of $\kappa = 0.76$). Based on the assigned clusters, we later determined significant differences regarding specific participant characteristics such as demographics.

12.4 Empirical Results

We present the findings of our representative online survey, by starting with the general attitude towards nudging in cybersecurity (see Sect. 12.4.1). Afterward, we suggest our results regarding people's attitudes in more specific scenarios within the context of cybersecurity (see Sect. 12.4.2). Further, we will give an overview of effects of priming and user characteristics on the attitude in Sect. 12.4.3.

12.4.1 General Attitude Towards Nudging in Cybersecurity

Aiming to contextualize our findings, we first asked our participants if they ever noticed nudges online. We found that the majority (72%) stated to not have noticed them, while 21% noticed nudges online and 7% did not make a statement.

We further asked if it was important to our participants to realize when someone tried to nudge them online. 69% agreed it was important to them. To get a better understanding of the general attitude, we first asked our participants on a Likert scale from 1 to 5, how helpful, dangerous, necessary, patronizing, or superfluous they perceived nudging in cybersecurity. You can see an overview of the results in Fig. 12.2. Further, we need to consider that two-thirds of the participants were assigned to a priming condition before answering the survey questions, which slightly influenced the general attitude towards nudging for the better, unconditionally whether the participants interacted with the password meter or the radar chart. For instance, they were slightly less likely to agree with nudging in cybersecurity being dangerous ($M_{nudge} = 2.24, SD_{nudge} = 1.0, M_{none} = 2.73, SD_{none} = 1.1; F(1, 769) = 48.41, p < .0005; d = 0.25$).

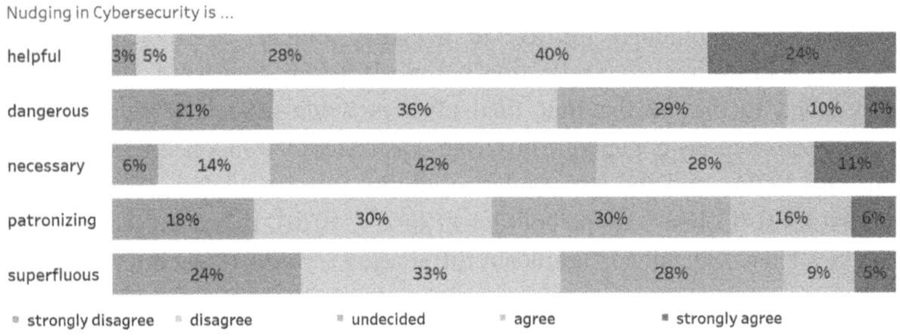

Fig. 12.2 Attitudes towards nudging in cybersecurity without priming

We found that many participants assessed nudging in cybersecurity in a positive way. For instance, 64% (priming: 68%) agreed that it was a helpful instrument. Further, 38% (priming: 42%) stated to agree that nudging in cybersecurity was necessary. However, many (42%) were undecided concerning this matter, and 20% disagreed (priming: 41%, 17%). Correspondingly, 57% (priming: 62%) disagreed with nudging in cybersecurity being superfluous. To gain a general insight into the perception of possible risks, we asked our participants how dangerous they assessed nudging in cybersecurity. 57% stated to not rate it as a dangerous instrument while 14% assessed it as dangerous (priming: 64%, 10%). As avoiding reactance is an important aim of nudging, we asked how patronizing our participants rated nudging in cybersecurity. 48% stated to disagree with it being patronizing while 30% were undecided and 21% agreed with nudging in cybersecurity being patronizing (priming: 52%, 31%, 17%).

While our Likert scale questions aimed at gaining a quantitative insight into the general attitude towards nudging in cybersecurity, we decided to include an optional question in free-text format for deeper inductive information. Hence, we more openly asked our participants if they considered nudges to be a sensible way to steer online behavior in a secure direction or if they saw any risks. In a deductive approach, two researchers independently assigned each answer to one of four clusters using RQDA. As the reliability was sufficient, we report the mean values.

Rating Nudges as Useful. We found that 56% explicitly made a positive statement on nudging in cybersecurity. Among the most frequently mentioned benefits were paying more attention to online security, support for unaware users, protection from certain risks and preservation of freedom of choice. For instance, one participant stated: *"I see it as a useful method. The freedom of choice remains, as the hint can be ignored"* (P253). Another participant wrote: *"Yes, a bit of Internet education does not hurt as long as you can decide for yourself"* (P309). Further, P301 stated: *"Yes because many are very careless with their data and information. A hint in the right direction would be just right"*. Others only stated they found nudging in cybersecurity a sensible instrument without giving any reasons (e.g., P293: *"Yes, it is useful"*).

Undecided and Thoroughly Evaluated Answers. 15% intensively evaluated benefits and risks but were undecided whether nudging in cybersecurity was a sensible instrument or not. Several participants highlighted that they assessed nudging in cybersecurity as a useful instrument as long as nudges were not followed blindly (e.g., P113: *"Yes, I see risks, namely that the users rely blindly on the programs written by humans [...] without having to think about it themselves"*). Moreover, our participants highlighted the importance of nudges being transparent (e.g., P180: *"[...] The way the nudges work and make their decisions must be completely transparent"*). Further, several participants stated it was important that nudges came from a trustworthy source (e.g., P437: *"As long as the creator of the nudges is morally sound, I see it as a useful method [...]"*). Some participants stated that nudges should appear only occasionally to prevent feeling annoyed (e.g., P439: *"As long as it happens occasionally, it's okay and helpful. But I believe that the benefits quickly turn into the opposite and people feel that things are just annoying and [nudges are] overlooked"*). Hence, when looking deeper into the answers, we identified four main aspects that were mentioned as crucial for making nudges in cybersecurity a sensible instrument: Not following blindly, transparency of benefits and intentions, trustworthy sources, and only occasional appearance of nudges.

Rating Nudges as Inappropriate. Some participants (11%) explicitly disagreed with nudging being useful in that specific context. For instance, one participant stated: *"Actually, I do not find it useful. I find it patronizing"* (P15). Another participant stated: *"No, not really. I find it annoying at a certain point"* (P72). Similarly, P122 stated: *"No real risk, but annoying. I do not like, for example, to be told how to choose my password. I know how to generate an (allegedly) strong password but I think that it is weaker than a simple one that I can remember* *without writing it down. That's why I find nudging more harmful"*. More critically, P11 wrote: *"In the end, it only serves to spy on personal data. No normally, rationally thinking person needs that"*. Furthermore, P20 wrote: *"It is a deliberate and intended manipulation of the Internet user, which I consider very dangerous. A subconscious control of the user takes place"*. Others simply claimed they did not find nudging in cybersecurity useful without specifying a reason (e.g., P70: *"No, I do not think it is useful"*). When analyzing all answers in cluster C, we found that the most frequently given reasons for rating nudges in cybersecurity as inappropriate were: Being superfluous, being annoying and being manipulative. We discuss explicitly mentioned risks more precisely in the following cluster.

Comments on Potential Risks. Others (17%) only named potential risks while not answering if they assessed nudging in cybersecurity as a useful instrument or not. When looking deeper into the given answers, we found that the most frequently named risks were *intentional misguidance / manipulation*, followed by *data exposure / data collection*. One participant wrote: *"I am not sure how far such nudges can be used to intercept data"* (P463). Four participants brought up the term "fake nudges" as a potential risk, which was used as a matching part to the commonly discussed term "fake news" (e.g., P217: *"I only see a risk in fake nudges"*). Further mentioned risks were hackers, paternalism, people are unlearning thinking, lack of independent control, and fear of censorship. For instance, P113 stated: *"Yes, I see risks. Namely that the users rely blindly on the programs written by humans. They only serve the purpose of increasing their security without having to think about it themselves"*.

12.4.2 Context-Specific Attitude Towards Nudging in Cybersecurity

After gaining general insight into the attitude, we asked more specifically on a five-step Likert scale about how useful our participants assessed nudging in eight concrete contexts of cybersecurity that are not exhaustive but exemplary. Scenarios were chosen by looking for commonly implemented nudges in cybersecurity in scientific literature [e.g., 569, forpasswordcreation] and adding other, partly more generic scenarios for user-related cyber incidents (loss of data). Although we set the question as optional, for all items, a minimum of 835 participants felt capable of answering. We included the following contexts and scenarios: (1) password creation for important accounts, (2) password creation for unimportant accounts, (3) management of cryptocurrencies, (4) reminders of backups regarding important data, (5) protection from loss of money, (6) protection from loss of data, (7) prevention from sharing private data with strangers and (8) prevention from risky behavior on the Internet.

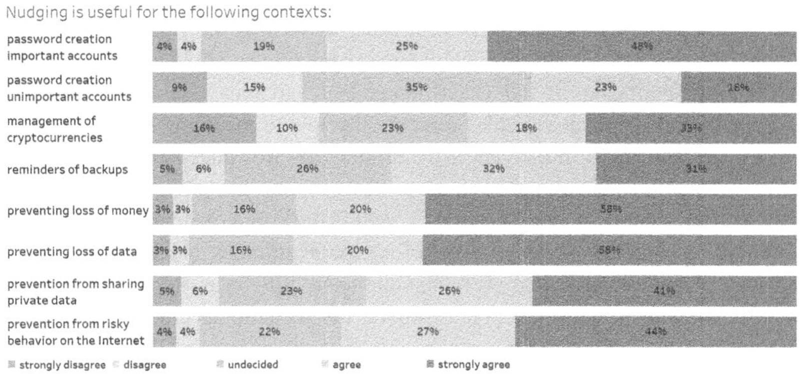

Fig. 12.3 Context-specific attitudes towards nudging in cybersecurity

For all investigated contexts, many participants assessed nudging as a useful instrument. The interaction with one of the nudges had a priming effect only for password creation ($M_{nudge} = 4.21, SD_{nudge} = 1.0, M_{none} = 3.85, SD_{none} = 1.2$; $F(1, 769) = 22.83, p < .0005; d = 0.17$). In the context of *password creation*, 73% agreed with nudging being useful regarding important accounts, while 19% were undecided and 8% disagreed (N = 976; priming: 76%, 16%, 5%). For unimportant accounts, still, 41% (priming: 44%) rated nudges as useful (N = 978). Furthermore, in the context of *cryptocurrencies*, 51% agreed with nudging being useful (N = 835). For *reminders of backups* regarding important data, a total of 63% agreed with nudges being useful (N = 959). Moreover, we asked our participants if they assessed nudges as useful when they were protecting against loss of money or loss of data. For *loss of money*, 78% agreed with nudges being a useful instrument (N = 950). Similarly, concerning *loss of data*, 78% agreed (N = 949). Regarding the *prevention from sharing private data with strangers*, a total of 67% agreed with nudges being a useful instrument, while 23% were undecided and 10% disagreed (N = 960). More generally, we asked about nudges to *prevent from risky behavior on the Internet*. Here, 71% rated nudges as useful, 22% were undecided, and 8% rated nudges as not useful (N = 971). A visualization of the results is presented in Fig. 12.3.

12.4.3 Effects of User Characteristics on Attitudes

We investigated if the attitude towards nudging was influenced by specific user characteristics. Hence, in the following, we will suggest our findings regarding effects of demographics and decision-making as well as information-processing styles. We tested for statistical significance applying a *MANOVA*. The overall model was significant for all tested dependent variables. Main effects of the investigated independent variables were mostly low to medium.

Demographics. To investigate if the attitude towards nudges in cybersecurity differed regarding demographic characteristics of our participants, we included demographic groups in the *MANOVA* as well. Concerning gender and education, we found no significant difference both for general attitudes and scenario-specific attitudes. Regarding age, there were no significant differences in general attitudes towards nudging in cybersecurity. However, the scenario-specific assessment of usefulness slightly differed for management of cryptocurrencies ($F(4,769) = 3.24$, $p = .012$; $d = 0.13$) between the age groups. More specifically, the post-hoc tests reveal that younger participants tend to rate nudging in those contexts as more helpful than older participants.

Decision-Making and Information-Processing Styles. Similar to the work of Hagman et al. [204], we further investigated if individual differences in rational thinking and decision-making styles were an influencing factor on the attitude towards nudging in cybersecurity by conducting two subscales each of the *Rational-Experiential Inventory scale* (REI) and the *General Decision Making Style scale* (GDMS) [204, 402, 500]. We found no significant differences regarding general and scenario-specific attitudes towards nudges in cybersecurity between participants with a high score (from 3.0 to 5) and participants with a low score (from 1 to 2.9) on the rational ability subscale (REI-RA: ability to think logically and analytically). However, when comparing the results of the rational engagement subscale (REI-RE: reliance on and enjoyment of thinking in an analytical, logical manner), we found that participants with a low score rated nudging in cybersecurity slightly but significantly less necessary ($M_{high} = 3.42, SD_{high} = 1.1, M_{low} = 3.17, SD_{low} = 1.0; F(1, 769) = 9.20, p = .003; d = 0.11$) than participants with a high score. When examining the results of the GDMS-subscales, we found that for both the general and the scenario-specific attitudes it made no significant difference if participants were scoring low (from 1 to 2.4) or high (from 2.5 to 5) on the rational-subscale (making decisions logically and systematically). For the dependent-subscale (rarely making important decisions without consulting other people) however, we found slight differences regarding the attitudes. Participants with a high score rated nudging in cybersecurity as slightly more necessary than those with a low score ($M_{high} = 23.30, SD_{high} = 1.0, M_{low} = 3.02, SD_{low} = 1.1; F(1, 769) = 12.21, p = .001; d = 0.13$).

12.5 Discussion

Nudging has emerged to be a relevant research field in the context of cybersecurity to steer people's behavior in a more secure direction. While several works have already developed and evaluated different types of nudges for cybersecurity, they are often not as effective as desired. To gain deeper insights on requirements for successful nudging in cybersecurity, Pe'er et al. [413] suggest including knowledge about the public attitude in the design process for the specific context. As personalization is an emerging trend to make nudges more effective, exploring people's attitudes for a specific context and identifying differences between people's characteristics can facilitate that idea. In other contexts, such as health and privacy, attitudes towards nudging have already been investigated. Furthermore, multiple-context studies have explored general attitudes towards nudging for several scenarios. However, the context of cybersecurity has largely been excluded from those investigations. Hence, our scientific contribution is an initial exploration of general and scenario-specific attitudes towards nudging in cybersecurity through a representative survey in Germany while identifying differences for user characteristics and priming conditions.

To answer our first research question (*RQ1: How do people assess nudging in cybersecurity?*), our findings suggest a partially positive attitude of the German population. Similar to studies on the health and privacy contexts, *(1)* many agreed that nudging was a helpful instrument in cybersecurity. However *(2)*, many were also undecided if it was a necessary tool. When asking free-text questions, we had the opportunity to gain deeper insights into the participants' assessments. For instance, 52% made an explicitly positive statement on nudging in cybersecurity, while *(3)* frequently revealing benefits such as *paying more attention to online security, support for unaware users, protection from certain risks and preservation of freedom of choice. (4)* Interestingly, 69% agreed that it was important to them to realize when someone tried to nudge them online. We gained insights into requirements, both concerning end-users and designers, that make nudges a useful instrument in cybersecurity: *(5) not following blindly, transparency of benefits and intentions, trustworthy sources, and only occasional appearance of nudges.* The highlighted relevance of transparency corresponds to the ethical guidelines for nudging by Renaud and Zimmermann [455] who suggest that nudges should be transparent to the nudgees and *"should only be deployed when the benefit is clear"*. **Thus, we suggest focusing on transparent nudges for the context of cybersecurity where benefits are evident.**

Diving deeper into peoples' assessments, we evaluated sub-question *RQ1.1 (Do people see risks of nudging in cybersecurity?)*. We found that *(6)* generally around half of the participants did not assess nudging in cybersecurity as dangerous, while however, 21% agreed with nudging in cybersecurity being patronizing. Again, we gained deeper inductive insights by evaluating free-text formats. Our qualitative content analysis revealed that *(7)* 12% explicitly rated nudging in cybersecurity as inappropriate while most frequently giving reasons such as *being superfluous, being annoying, and being manipulative.* When asking more specific about potential risks *(8), intentional misguidance / manipulation, data exposure / data collection, hackers, paternalism, unlearning thinking, lack of independent control, and fear of censorship* were further mentioned. Interestingly, four participants brought up the term "fake nudges" in that context as a potential risk. **We suggest to exhaustively evaluate the potential risks when designing nudges in cybersecurity, aiming to address the concerns of end-users.**

We further investigated sub-question *RQ1.2. (Does the attitude differ concerning different scenarios?)*. Here we found a widely consistent attitude across most addressed scenarios. For all investigated scenarios *(9)*, many assessed nudging as a useful instrument while revealing the largest approval for nudges to protect against loss of data. **We suggest to further extend the design of nudges to other scenarios in cybersecurity, such as protection against loss of data and money.**

For further insights we evaluated sub-question *RQ1.3. (Do user characteristics or priming conditions affect the attitude?)*. We investigated if interacting with an exemplary nudge for stronger passwords affected the attitudes. We found that indeed *(10)* participants who beforehand interacted with an exemplary nudge were less likely to agree with nudging in cybersecurity being dangerous, patronizing, or superfluous. While demographic characteristics had almost no effect on the attitudes, we found *(11)* that decision-making and information-processing styles were slightly affecting attitudes, too. People who do not rely on and do not enjoy thinking in an analytical way assessed nudging in cybersecurity as slightly less necessary. Hence, consistently with the work of Hagman et al. [204], differences in the REI score did slightly affect attitudes towards nudging in cybersecurity. Furthermore, participants who tend to make important decisions consulting other people rated nudging in cybersecurity as slightly more necessary than participants. Hence, our findings suggest that decision-making as well as information processing styles make small but significant differences for attitudes towards nudging in cybersecurity. **To facilitate personalization, we suggest considering user characteristics for the design of nudges in cybersecurity where sensible. We, however, suggest identifying other psychometric scales that have a greater impact on attitudes.**

Our findings on *RQ1* reveal a partially positive attitude towards nudging in cybersecurity. However, we further found that people perceive a number of potential risks as well. Although the majority stated to have never noticed nudges for cybersecurity, it was striking that many were able to thoroughly think about potential benefits and risks which can partly be addressed by recommendations retrieved from our findings. As Pe'er et al. [414] have stated, the effectiveness of nudges *"strongly relies on voluntary compliance"* [and] *public attitude towards specific nudges can play an important role"'* [414].

Summarizing our previously demonstrated main findings on *RQ1*, we make several suggestions to address challenges and make nudges in that specific context more effective and to facilitate the emerging trend of personalization (see Table 12.1), answering our second research question *(RQ2: What are design implications for nudging in cybersecurity?)*. Given the concern that users might rely blindly on the feedback of nudges when moving through the digital space (challenge 1), we consider it crucial to design nudges for cybersecurity that encourage individual users to keep thinking for themselves while being assisted by nudges. Similarly, we found that making transparent that final decisions are up to the end-user may be an important step to avoid reactance and feelings of paternalism (challenge 2). Further, to design effective nudges, our findings reveal that considering users' requirements and concerns is essential. We found that this applies to making benefits of nudges salient (challenge 3), taking fear of potential risks seriously (challenge 4), and enabling users to understand if a nudge comes from a trustworthy source (challenge 5). While we have identified several design implications for nudges in cybersecurity, it is still important to not overuse nudges, even when designed adequately, and to implement nudges that do not significantly delay the respective process. Our findings reveal that users tend to feel annoyed when nudges are used when not necessary (challenge 6).

Table 12.1 Design implications for nudging in cybersecurity

Challenges	Design Implication
Challenge 1: Unlearning thinking	Enabling and encouraging the individual user to keep thinking for himself while being assisted by the nudge instead of relying blindly.
Challenge 2: Reactance	Make transparent that the final decisions are up to the end-user to avoid feeling patronized.
Challenge 3: Benefits are not clear	Enabling the user to understand the benefits that result from a specific nudge.
Challenge 4: Fear of potential risks	Make transparent to the nudgee how a nudge works at any time. Address concerns in the nudge design by transparently showing what data is being used. Evaluate potential risks iteratively during nudge design.
Challenge 5: Trust	Enable users to understand if a nudge comes from a trustworthy source.
Challenge 6: Annoyance	Nudge only when necessary and without delaying the respective process to prevent users from feeling annoyed.

12.6 Conclusion and Limitations

Nudging in cybersecurity is perceived as helpful by many German citizens. However, they see potential risks as well. We suggest to address the perceived concerns and to extend research on how far those fears pose a realistic threat. The majority states to have never consciously noticed nudges for cybersecurity. Interestingly, many participants were still able to evaluate possible benefits and disadvantages. However, our survey questions were based on a priming description of nudges that highlighted specific application scenarios (especially password creation) more than others. Hence, we suggest gaining richer insights into the attitudes based on different scenarios to enhance a vivid imagination of benefits and risks, preferably in realistic field study. We propose that a comprehensive knowledge on people's attitudes is a first step to enhance effectiveness of nudges in cybersecurity as it facilitates personalization and helps to address concerns adequately.

While our work is an initial contribution, it also has its limitations: (1) The results were acquired using a survey, which is a method prone to social desirability biases and relies on self-report. While it is a challenge to measure attitudes indirectly, we suggest utilizing complementing techniques of data collection (e.g., thinking-aloud studies) to find more reliable results. (2) Also, it is important to consider that we used a selection of priming conditions as exemplary password nudges for comparison with a control group without a nudge, other than an exhausting selection of nudges of all inquired contexts. Of course, different types of nudges (e.g., fear appeals) and priming nudges for other contexts than password creation may result in different context-related attitudes. Also, our experiment was conducted using a web browser on a PC and may differ when displaying the selected nudges on a smartphone or with different size, timing, and frequency. Hence, we suggest taking a thorough look at a more exhaustive selection of priming conditions in future studies. (3) Although the sample was representative concerning several characteristics, the method implies the risk of participants being more technophile than the average internet user. (4) Our survey questions were not based on a standardized test; hence, other methods need to confirm that we measured attitudes in an adequate way. (5) Furthermore, many differences cannot be considered strong. Thus, they reveal existing tendencies other than game-changing influences. Our work provides an opportunity for cross-country comparability in future studies which may contribute to a more accurate understanding of the perception of nudging as an instrument in cybersecurity.

13 Finding Secret Treasure? Improving Memorized Secrets Through Gamification

Abstract Users tend to bypass systems that are designed to increase their personal security and privacy while limiting their perceived freedom. Nudges present a possible solution to this problem, offering security benefits without taking away perceived freedom. We have identified a lack of research comparing concrete implementations of nudging concepts in an emulated real-world scenario to assess their relative value as a nudge. Comparing multiple nudging implementations in an emulated real-world scenario, including a novel avatar nudge with gamification elements, this publication discusses the advantages of nudging for stronger user-created passwords regarding efficacy, usability, and memorability. We investigated the effect of gamification in nudges, performing two studies ($N_1 = 16, N_2 = 1,000$) to refine and evaluate implementations of current and novel nudging concepts. Our research found that a gamified nudge, which integrates a personalizable avatar guide into the registration process, performs less effectively than state-of-the-art nudges independently of participants' gaming frequency.

Original Publication *Katrin Hartwig, Atlas Englisch, Jan Pelle Thomson, and Christian Reuter. "Finding Secret Treasure? Improving Memorized Secrets Through Gamification". In:* Proceedings of the 2021 European Symposium on Usable Security (EuroUSEC). *ACM, 2021, pp. 105–117. DOI:* 10.1145/3481357.3481509 (Minor language edits have been made to the original publication for the readability of this dissertation.)

Supplementary Information The online version contains supplementary material available at https://doi.org/10.1007/978-3-658-47885-8_13.

K. Hartwig, *Navigating Misinformation*, Technology, Peace and Security | Technologie, Frieden und Sicherheit,
https://doi.org/10.1007/978-3-658-47885-8_13

13.1 Introduction

Despite the existence of various strong systems for user authentication like hardware tokens, one-time passwords, or password management tools, users still prefer using passwords as their main authentication factor [415, 632]. Unlike physical authentication devices, passwords—as plain text or in derived formats such as hashes—may be stolen in great quantities [109], for instance, to be used in illegal data or identity theft schemes. Even if stolen, login data is protected by hashing, although even a hash will always be guessed ("cracked") given infinite time or computational resources. Therefore we must increase demands on either time or resources required to break a hash.

One option to increase the time needed to crack a password is by increasing its length and complexity so that it is not feasible to brute-force the password, and it is less susceptible to dictionary-based attacks. Unfortunately, users tend to choose passwords of weaker complexity or misjudge short but complex and seemingly random character compositions as suitable and secure passwords [19, 170]. To mitigate these risks, current research is evaluating *nudging* concepts to steer users towards stronger password choices. A nudge *"alters people's behavior predictably without significantly changing their economic incentives"* [554, p. 6]. After being successfully established in contexts such as the health sector, it has more recently become a research topic in cybersecurity as well [3].

This paper heavily focuses on the user as the decisive factor in the information security chain. As Garfinkel and Lipford [175, p. 8] state, *"only by simultaneously addressing both usability and security concerns will we be able to build systems that are truly secure."* In other words, users will always find ways to ignore or disable security measures forced on them—e.g., for login processes, this could be achieved by slightly altering their universally used password to fit the system's complexity requirements, with diminishing returns. We must instead strive to include users in the process of securing their data in a way that does not overwhelm or hinder them. Nudging is a promising concept of assisting users in adopting a strong password policy without forcing it on them.

Current guidelines [189] advise service providers not to impose restrictions on the composition of passwords, except for limiting the password choices to those of eight characters or longer and possibly checking against a ban list of common passwords. This puts the responsibility of choosing a strong password solely on the user. Thus, providing users with nudges to increase password strength without altering the requirements presented by the service provider could lead to greater security. At the present time, research is ongoing regarding the efficacy of such nudges and ideal nudge selection [219]. To contribute to existing research, we compared current nudging approaches with novel concepts to increase password security and provide transparent help cues to nudgees. Some approaches include elements of gamification to make nudges more effective and approachable by increasing the user's motivation to change their learned behavior [302, 308]. According to Karimi and Nickpayam [287, p. 34], gamification describes *"a strategy that employs game mechanics, techniques, and theory in areas that traditionally do not function like a game."* Silic and Lowry [521], for instance, applied this concept to the context of cybersecurity.

We expected gamified nudges to resonate more strongly with audiences used to playing digital games than nudges that do not provide gamification elements and aimed for users to accept these as a less authoritative nudging option compared to strict policies and forced restrictions. Previous research [414] has highlighted individualized nudges as more effective than a one-size-fits-all approach. Hence, contributing to the trend of personalization, we evaluated if gamified nudges were more suitable for audiences that play digital games while providing different nudging options, which may then be selected based on the target audience of the provided service. We avoided approaches that significantly lengthen the registration process, as well as nudges that are severely intrusive as not to frustrate users [426, 612].

In conclusion, we address the following research questions:

RQ1: *How do gamified nudges compare to a state-of-the-art password meter?*
RQ2: *Does a user's gaming frequency increase the acceptance of a gamified nudge?*

The paper is structured as follows: First, we present the body of related literature resulting in a research gap (see Sect. 13.2). Subsequently, we describe our research design (see Sect. 13.3). In Sect. 13.4, we present the study design and results of our preliminary think-aloud study, including a detailed description of all investigated nudges, followed by Sect. 13.5 where we present method and results of our main study. We complement our work by a discussion of results (see Sect. 13.6), entailing a presentation of limitations of our work and relevant implications for the future, followed by the conclusion in Sect. 13.7.

13.2 Related Work and Research Gap

We can identify several current trends within the research field of nudging in cybersecurity. In the following we will focus especially on two aspects: gamification and personalization. Some works have highlighted the potential of enhancing motivation to engage in a more secure behavior through gamification. Other studies have brought up the idea of using personalized nudges instead of one-size-fits-all nudges to be more effective.

13.2.1 Gamified Nudges in Cybersecurity

While there are several studies dealing with nudging in cybersecurity, researchers have pointed out that nudges are often not as effective as desired in this case. For example, Kankane, Dirusso, and Buckley [284] evaluated the effectiveness of different nudges for password management. They found that none of the nudges was effective enough to significantly change the individuals' behavior regarding password creation. Against this background, the concept of gamification has more recently emerged as a potentially promising trend in cybersecurity to encourage engagement and intrinsic motivation and, thus, might make nudges more effective. When looking at the approaches in detail, we can differentiate between user interventions that use gamification as preventive training and gamified user interventions that take place at the point of a critical decision, e.g., as a nudge when about to create a password.

On the training interventions, Silic and Lowry [521, p. 5] conducted a long-term field study with 420 employees within an organization. As the gamified user intervention, they developed and evaluated a gamified website including an avatar and different game mechanisms such as the possibility to earn points after completing quizzes and reading tips about security education topics. Hence, the applied user intervention focuses strongly on educational elements that take place as preventive training. They found that gamified educational user interventions in the form of security training can indeed have a significantly positive effect on security behavior, fulfilling users' motivations and coping needs [521]. The idea of educational games in cybersecurity is not completely new, as Sheng et al. [512] designed and evaluated ANTI- PHISHING PHIL in 2007, an educational online game that teaches users not to fall for phishing attacks. The authors compared several training conditions (e.g., reading anti-phishing tutorials, reading online training material, or playing the Anti-Phishing Phil game) and found that participants performed better in identifying fraudulent websites when being assigned to the game condition beforehand. Canova et al. [83] followed up on that idea by developing NoPhish, an anti-phishing education app to teach users to access, parse, and check URLs regarding potential phishing attacks, raising security awareness. Focusing on the context of password security awareness, Scholefield and Shepherd [495] developed a mobile-based application with gamification features. The basis of this application is a multiple-choice quiz, educating users on topics such as choosing a strong password and avoiding common passwords. Overall, the application received positive feedback from participants of a small pilot study ($N = 17$).

Less attention has been paid to approaches that integrate gamification elements as a user intervention at the critical point of decision. For instance, Takada and Hattori [548] investigated how users can be motivated to voluntarily use a secure pattern-based authentication when provided with a role-playing game function to the authentication process. The authors state that while there are methods such as password policies and password meters to assist users in using secure credentials, those measures often come with drawbacks in usability and user experience. In a preliminary online experiment with eight participants, they found that a role-playing game function within the authentication process does indeed have the potential to motivate users to use stronger credentials [548]. Also, Ur et al. [570] evaluated if a dancing bunny as a gamified password meter led to longer passwords, resulting in better efficacy but mixed qualitative feedback. Further, Micallef and Arachchilage [362] designed and evaluated a gamified nudge to improve users' memorability of security questions. They adapted the 4 Pics 1 Word mobile game and found that a gamified nudge has the potential to improve the memorability of answers to security questions, constituting a promising approach to reduce the trade-off between usability and security in fall-back authentication [362]. Ophoff and Dietz [394] performed an online experiment with 232 participants to compare the effectiveness of password strength feedback with and without gamification elements. Therefore, the participants interacted either with a common password meter or a gamified feedback method where they could earn points while completing the authentication process. Different from our experiment, they chose a simple bar indicating strength by color (e.g., green to red), length of the bar, and text (e.g., "strong") instead of the state-of-the-art password meter by Ur et al. [569]. The gamified feedback dynamically displays a score of password strength for each character that was added or removed, showing the difference of strength to the previous password. When comparing the results of the test conditions in the online experiment, password cracking time did significantly differ between participants that were assigned to the password meter and those that interacted with the gamified nudge, showing potential for the gamified feedback [394].

13.2.2 Personalized Nudges

Currently, nudges are mostly implemented as a one-size-fits-all solution. However, to include users in the process of securing their data in a way that does not overwhelm or hinder them, personalization of nudges seems to be a promising approach. It focuses on adapting the design to individual users' requirements to make nudges in cybersecurity more effective and beneficial to a diverse set of end users instead of the average user alone [146]. Various studies have already investigated this idea or suggested potential benefits in their works [146, 270, 300, 414, 457, 565]. Some argue that such an approach might improve compliance significantly [146]. For instance, Knijnenburg [300] points out that showing tailored nudges can support users in making better privacy decisions.

To provide end users with the subjectively most effective nudge, it is fundamentally necessary to identify distinct user groups. Hartwig and Reuter [217] explored people's attitude towards nudging in cybersecurity for different contexts to gain an understanding of how to address certain users. More practically, various approaches use psychometric scales to segment users. For instance, Egelman and Peer [146] use decision-making styles and risk-taking attitudes to predict privacy and security behavior and segment user groups accordingly. This follows the assumption that understanding users' attitudes towards computer security helps to contextualize their observed behaviors as well as predict their future behaviors. Further, Dupree et al. [139, p. 5228] were able to identify distinct categories of end users by analyzing the participants' attitudes and behaviors towards security practices. Based on these findings, they suggest utilizing these clusters in the design of new privacy and security tools.

Even though the personalization of nudges is considered promising by several studies, only a few have implemented the concept within the cybersecurity context. Pe'er et al. [414] tested people's decision styles to personalize nudges for stronger passwords in two online experiments ($N = 2,047$) and argue that choosing a nudge from a pool of multiple existing nudges could be more effective than showing the same nudge to everyone. They achieved stronger passwords with personalization than with one-size-fits-all nudges and showed that decision-making styles can indeed be used to personalize nudges [414]. Research from Jeske, Coventry, and Briggs [270] also suggests that user differences play a role in security decision-making. They point out that the effectiveness of nudges depends on user characteristics, such as their impulse control, when selecting a public wireless network. Hence, the personalization of nudges in cybersecurity seems to be a more promising approach than one-size-fits-all nudging.

To allow many individuals to benefit, it is necessary to create a pool of different nudges. For instance, gamification was identified as a promising concept to motivate secure behavior and to encourage engagement and intrinsic motivation. Hence, we consider it worthwhile to investigate the effect of integrating elements of gamification in nudging and to shed light on the question if users with different gaming preferences favor different types of nudges.

13.2.3 Research Gap

Nudging in cybersecurity has recently emerged into a growing research field, resulting in the design and evaluation of different nudges for that specific context [e.g., 3, 219, 270, 300, 362, 414, 457, 570]. Yet, researchers have found that nudges in cybersecurity are often not as effective as desired [284]. Some argue that personalization instead of one-size-fits-all nudges is a promising trend to enhance effectiveness [e.g., 300]. Also, using elements of gamification in nudges can be considered a potentially effective measure to increase motivation of end users to behave in a more secure way. The benefits of gamification have already been shown in the context of cybersecurity. Micallef and Arachchilage [362] have investigated the potential of gamified nudges in the context of fall-back authentication, potentially reducing the trade-off between usability and security. Further, Ophoff and Dietz [394] have made initial investigations to evaluate gamified password feedback in comparison to a simple password meter. However, most current user interventions using gamified elements in cybersecurity focus on preventive training instead of nudging at a point of critical decision-making. The context of manual password creation is still highly relevant in usable security, as passwords are still the number one choice of user authentication despite various strong alternatives, and end users tend to struggle following password requirements while creating usable and strong passwords. To our knowledge, password nudges with gamification elements considering gaming preferences as an indicator for personalization have not been investigated on a larger scale. We hypothesize that especially an avatar as a customizable assistant has the potential to combine benefits of gamification and personalization as motivating factors.

Therefore, we suggest taking a closer look at password nudges with gamification elements and comparing their effectiveness among users with different gaming preferences. By doing so, the pool of effective nudges in cybersecurity may be extended, facilitating personalization. Therefore, we performed a two-fold evaluation of different nudges for stronger passwords, focusing on users within the German population to provide an opportunity for comparison with other countries in the future.

13.3 Research Design

In a two-fold evaluation, we first conducted a qualitative think-aloud study with seven nudge prototypes. The initial set of nudges contained both state-of-the-art nudges as well as novel nudges incorporating password feedback with gamification elements. For our subsequent main study, we reduced the sample of nudges to a novel gamified nudge that received positive feedback, a state-of-the-art nudge for password strength, and a state-of-the-art nudge for memorability. We refined these three nudges, considering feedback we obtained from our participants during the preliminary study. In our main study, we conducted an online experiment with 1,000 participants, testing the effect of the three different and promising nudges on password strength, short-term memorability, and usability in comparison to a control group without a nudge. Both studies were conducted in accordance with the requirements of the local ethics committee of the TU Darmstadt.

In the following, we present important background information on how we calculated password strength and which password criteria were considered for our nudges. An important resource for password creation rule sets is the 2016 NIST report [189], which states that the only rules to be forced on users may be a minimum password length of eight characters and that the password may not be included in an optional ban list, e.g., containing previously leaked passwords. To assist users in creating stronger passwords, we consider it sensible to give optional suggestions for possible password compositions (e.g., including digits). In accordance with the NIST report, we decided to give those suggestions not as a forced password criterion but as a suggestion to encourage variety. Slightly modifying the idea of Komanduri et al. [304], we have decided on a uniform standard for password criteria and suggestions in our nudges. As password criteria, we requested at least eight characters as suggested by NIST. We further checked against a ban list of 100,000 common passwords included in the service library NBP. For optional suggestions, we proposed including two or more characters per character class (upper case letters, lower case letters, digits, special characters). Additionally, we have included an informative text regarding the creation and criteria of strong passwords, featuring an example of "mnemonics", a memorable password strategy following work by Kiesel, Stein, and Lucks [293]. As Kaleta, Lee, and Yoo [282] suggest, we have included a section explaining why secure password choices are important and have an impact on information security (see Figure A.16 in Electronic Supplementary Material (ESM)).

13.4 Preliminary Think-aloud Study

We conducted a preliminary study ($N = 16$) in individual think-aloud sessions to qualitatively investigate the comprehensibility and usability of several different password nudges. The participants were acquired through Twitter and Discord channels (e.g., the university channel) and were between 20 and 30-year-old students, mostly of a technical degree program. The sample strategy has to be seen in the context of our two-fold study design. While the participants in the preliminary study were selected from the investigators' direct environment without aiming at a systematic sampling, the subsequent large-scaled online experiment of our main study is based on a broader demography and validates and expands preliminary findings of the think-aloud study quantitatively. Comprehensibility and usability are important factors for password nudges to be effective. Therefore, conducting a qualitative study as a first step helps us to identify potentially promising directions for novel nudges in comparison to established state-of-the-art nudges. The think-aloud study further helped us to refine promising nudges and to choose a smaller set of nudges for our large-scale main study. In the following, we present how the initial set of seven nudges was chosen and how the evaluation was conducted. We follow up with the presentation of the results of our preliminary study, motivating the research design of our subsequent main study.

13.4.1 Nudge Selection

For the preliminary study, we have created seven prototypes (see Fig. 13.1), of which three have been refined for use in the quantitative study (marked with an * asterisk in Table 13.1).

Table 13.1 Initial set of nudges. Nudges with the * asterisk were refined for the main study.

Key	Description	Source
N1*	Dynamic meter with checklist	[569]
N2*	Generated default password	[18, 85]
N3	Radar chart	[219]
N4	Expectation and Reflection	[85, 456]
N5	Reminding of consequences	[85, 213]
N6	Interactive chat	*novel*
N7*	Avatar creation	*novel*

Based on existing literature on password nudges, for the initial set of nudges we made sure to include state-of-the-art nudges for password strength and memorability to have a valid basis for comparison. Therefore, the dynamic meter with an integrated checklist by Ur et al. [569] was included as N1, representing the state-of-the-art for strength enhancement. We further included a generator of default passwords (N2) as a state-of-the-art nudge for memorability, based on the ideas of Caraban et al. [85] and Al-Ameen, Wright, and Scielzo [18]. As we aim to compare gamified with non-gamified nudges, we additionally included promising nudges that take advantage of different psychological effects, such as reminding of the consequences and reflection without using gamification as N4 and N5, following the suggestions of Caraban et al. [85]. As gamified nudges, we included a radar chart with potential gamification elements for password feedback (N3) that has already been investigated by Hartwig and Reuter [219] to some extent, missing information on memorability. Finally, we included two novel nudges with gamification elements, namely an interactive chat and the creation of an avatar as a customizable assistant (N6 and N7). In the following, we describe each nudge in more detail.

N1*: Dynamic Password Meter. The dynamic password meter with a checklist is currently one of the most successful approaches to password nudging, albeit competing with a more simple password meter which is commonly used in registration forms. In our prototype, both are featured: a data-driven password meter making use of a traffic light color scheme, and a checklist detailing which required and suggested password criteria are met (see Fig. 13.1 at the top left), similar to the work of Ur et al. [569]. The nudge combines the "at a glance" nature of a meter with the transparent approach of a checklist. Data-driven password meters in the style of Ur et al. [569] are regarded as the state-of-the-art concept in nudging and, modified to fit the password criteria specified, have been our baseline for testing the efficacy of novel nudges.

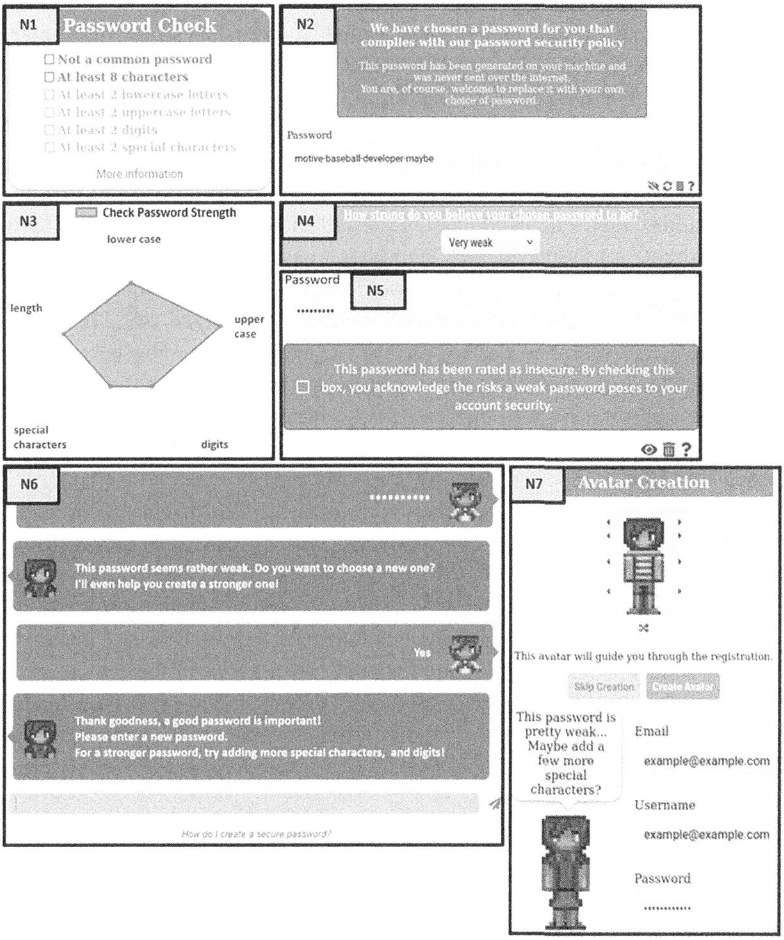

Fig. 13.1 Representation of all seven investigated nudges N1 to N7. (Own figure)

N2*: Generated default password. Caraban et al. [85] discussed the concept of providing a default option for users, which we have implemented based on the generation method suggested by a XKCD webcomic [379]. As examined in related work [18, 180], this method should provide significant memorability benefits compared to random string generation, making it our baseline for testing the short-term memorability of our nudges. The user is presented with a default password generated on the client, which intends to provide a secure base choice, suggesting what a secure password may look like and encouraging the use of a service-unique, random password (see Fig. 13.1 at the top right). We examined if users tend to choose a generated or a custom password. The user may repeatedly choose to generate a new default password to aid in finding an easy-to-memorize secret.

N3: Radar chart. The radar chart, as described by Hartwig and Reuter [219], aims to provide a transparent view into how a password's strength is measured and how a user may improve their password. Along every axis radiating from the center, a criterion of the total password strength is quantified (see Fig. 13.1 at the middle left). As this is a chart, reading off and understanding how well the criteria are satisfied should feel more intuitive to the nudgee than an aggregated score. This approach features elements of gamification, which could allow us to compare our novel gamification-inspired nudges to one already discussed in the present literature.

N4: Expectation and Reflection. As human decision-making tends to fall into two categories, driven by either fast and instinctive or slow and deliberate thinking [281], this nudge aims to make use of the more deliberate system for password selection. By providing the user with the option to rate their password's strength on a subjective scale before submitting the registration, our intent was to invoke reflection and re-evaluation regarding the security of the chosen secret. As this nudge does not provide active feedback about the password strength, we took interest in comparing it to the novel nudges, which provide explicit suggestions and feedback. Similar work to this nudge has been done by Renaud and Zimmermann [456].

N5: Reminding of consequences. Very similar in concept to the previous nudge N4, but different in how reflection is achieved, N5 checks the ZXCVBN score of the password and, if rated weak, forces the user to check a box confirming they understand some of the risks and consequences of choosing a weak password as detailed in the info text (see Figure A.16, ESM). Our implementation is inspired by the eponymous section in Caraban et al. [85]. We examined which of these two nudges is more effective in strengthening a password so that one may be chosen for comparison in the main study if deemed promising.

N6: Interactive chat. Some services have adopted gamification elements into their sign-up process by implementing a chat-based registration. The user chats with an automated system (a *bot*), which gives feedback on password strength. It queries the user's information by asking questions in natural language and processing the reply (see Fig. 13.1 at the bottom left), although it usually does not understand responses in natural language (e.g., "My name is Bob."), but rather concise information (e.g., "Bob"). Chatbots have already been discussed in academic research [e.g., 45]. However, to our knowledge, the efficacy of a chat-based approach as a password nudge was not previously evaluated.

N7*: Avatar creation. Following the trend of gamification, we have introduced a digital avatar to the registration process (see Fig. 13.1 at the bottom right). In accordance with related research [e.g., 591], the appearance of the avatar is customizable. It intends to create a bond between the user and the avatar, with the targeted outcome of increasing the value of its suggestions and encouraging secure passwords by offering context-aware positive feedback. Following our goal of not significantly lengthening the process, the avatar runs parallel to the registration. It offers information without requiring interaction and may be skipped entirely (however, not in our experiment). The hypothesized success of this concept is based on multiple psychological effects, predominantly the "IKEA effect" [388], giving weight to the avatar's cues by including a "do-it-yourself"-styled personalization section in addition to the perceived "sunk cost" of spending time customizing, which adds further incentive to satisfy the avatar's suggestions.

13.4.2 Methodology

We implemented high-fidelity and fully functional prototype versions of the nudges for our preliminary study. Those provided the opportunity for user interaction within a role-playing registration form. A mock registration form was implemented to provide a realistic registration process, allowing a role-playing approach for preliminary evaluation of the nudges. Previous research [154] has shown this strategy to increase the probability of study participants acting similarly to real-life situations. In individual think-aloud sessions, the acceptance of various nudges was assessed, and explicit feedback on implementation and concept was gained. Hence, we focused on qualitative feedback. With this data, we gained qualitative insights, filtered out unpromising nudges, and refined the remaining approaches for use in the second large-scale study. The participants of the preliminary study each reviewed a subset of three or four of the initial nudges in random order, sharing their reasoning and impressions while completing the mock registration processes. Each nudge was tested by eight people. We used online-conferencing tools that allowed both screen sharing as well as audio communication to create audiovisual recordings for later evaluation.

The participants were briefed on the think-aloud method [163] and agreed to the recording of their session. For each nudge, the participants were then asked to role-play creating an account for a fictional service provider. If participants fell silent for a few seconds, they were reminded to keep thinking aloud. After they successfully created an account, the participants were asked several open questions (see Table A.34, ESM), which were tailored to the specific insights we wanted to gain for each nudge. While we had a prepared set of questions, we gave enough space for topics beyond. Afterward, the participants were asked to perform a memorability test while thinking aloud, where they were redirected to log in with the password used during registration. The user was informed if an attempt was incorrect. We transcribed the audiovisual feedback we received and systematically collected our findings regarding user assessments for each nudge in qualitative content analysis.

13.4.3 Results

The goal of the preliminary study was to gain first qualitative insights about novel and state-of-the-art nudges with and without gamification elements before conducting a subsequent large-scale online experiment with the nudges that got largely positive feedback. We further used this as an opportunity to collect feedback on possible improvements to the nudges' presentation.

N1*: Dynamic Meter. The state-of-the-art dynamic meter nudge is one of the most common approaches to password nudging. The input of one of our participants reflects this, who described the nudge as a "known environment". Because of the common use of this nudge, the acceptance of this nudge seemed quite high. Even though one participant criticized that the checklist seems to "apply pressure", most of the participants felt like the nudge improved their password and described the tool as "intuitive". Therefore, it served as our baseline for testing the efficacy of novel nudges in the subsequent online experiment. For the main study, we improved the distinction between mandatory and optional requirements as well as the distinction between password meter and checklist.

N2*: Generated default password. Compared to random string generation, this method should offer significant memorability benefits. Therefore, this nudge serves as our baseline for testing the short-term memorability for our nudges. We received some negative feedback on that nudge. For instance, participants were concerned about dictionary attacks when using default passwords. We chose to examine this nudge in our main study, although noting some participants expressed concerns that a recommended password feels insecure and a password consisting of four words might be too long to remember.

N3: Radar chart. This nudges' aim is a comprehensible view of how password strength is measured as well as how a user may improve their password strength using a chart-based visualization. Therefore, the user is quite involved in the process. One could argue that this nudge features gamification elements. However, we received rather negative feedback on this nudge in our preliminary study. Participants described this nudge as too distracting from the actual password creation, unnecessarily complicated, non-intuitive, and overly long. Additionally, they presumed this type of chart to be rather unknown to most users. Many participants noted that they "felt urged to fill out all axes" (P #140) of the chart, disregarding the creation of a complex and memorable password, instead appending numbers and special characters until the chart was filled. One participant suggested color codes like a traffic light color scheme within the chart might be helpful, another participant stated they would like the idea of this nudge as a plug-in or as default on lots of websites. Due to the mainly negative feedback, however, we did not consider N3 for our main study.

N4: Expectation and Reflection. This nudge makes use of the more deliberate system of password selection by providing the user with the option to rate their password's strength on a subjective scale before submitting the registration. This type of nudge does not provide feedback about the password strength. A lot of our participants strongly criticized this feature as being confusing because they did not know why they were asked to reflect on their password. They described this lack of suggestion as "unhelpful". As a consequence, it was quite hard for the participants to evaluate how to strengthen their passwords. In addition, this feature made the nudge feel unintuitive due to the lack of consequences of the selection the participant executed. We did not further evaluate it in our main study.

N5: Reminding of consequences. Similar to N4, this nudge does not provide feedback and also relies on self-reflection. However, the nudge tests the ZXCVBN-score of the password the user chose and if it is rated weak, the user has to check a box confirming they understand some risks and consequences of choosing a weak password to continue the process. Participants noted that this warning is quite interesting and helpful as well as reasonable and not being intrusive. At the same time, participants describe this warning as not being transparent. The user is left in the dark as to why their password is weak. Therefore, some participants are left irritated, others even mistrust the tool because of this lack of transparency. Hence, we chose not to further evaluate this nudge in our main study.

N6: Interactive chat. This nudge is a chat-based registration that uses gamification elements. Some of our participants described the avatar as "cute" and "entertaining". The feedback using natural language via the bot attracted positive attention from some. It was noted that it felt like undergoing the registration process with somebody by one's side. Thus, this nudge was described as a user-friendly tool that could be quite helpful for users who are not very computer-oriented. At the same time, participants raised awareness to some security concerns: "it felt like you have to reveal your password to someone". Additionally, it stood out negatively that the nudge did not provide enough feedback about one's password strength. Aside from that, it did not feel transparent, because it was not clear what information was needed and how long the process would take. Moreover, participants criticized that the nudge is not appropriate for users who are technically oriented, and it takes away the known workflow of creating a password. Furthermore, the participants noted different faults of the design and criticized the "unnecessarily complex design" (P #861). Even though this nudge received mixed feedback, the negative responses predominated. Hence, we chose not to further evaluate it in our main study.

N7*: Avatar creation. Similar to N6, this nudge uses gamification elements in the form of a digital avatar for the registration process. This nudge received much better feedback from the participants. Some described the avatar as "cute" and "friendly", while others described it as "childish" and suggested it might, therefore, not be fitting for all services. Even though the phrasing of the hints was sometimes seen as "unprofessional", most of the participants rated the hints as useful and easy to understand. After improving the dialogue of this nudge as well as the hints for the avatar creation, we used it as our gamification nudge in our main study. Furthermore, to get uniform results in the main study, we deactivated the button to skip the avatar creation.

Our preliminary study aimed to gain first insights, which will be validated and quantified in our subsequent online experiment for a reduced sample of nudges. Feedback for N3, N4, N5 and N6 was rather critical. By comparison, N1 and N7 scored better in the preliminary study. N2 received mixed feedback, but serves as our baseline for testing the short-term memorability. Therefore, we implemented the technical feedback we received for N1, N2, and N7 and used them in the subsequent online experiment.

13.5 Main Study: Representative Evaluation

We conducted a large-scale ($N = 1,000$) quantitative study to determine the efficacy, acceptance, and short-term memorability of our final selection of nudges, compared to a control group that did not interact with a nudge. Our goal was to compare our approaches regarding these categories and draw conclusions about the performance of our novel avatar nudge N7 as a user intervention with gamification elements. As the final set of nudges for our main study, we used N1, N2, and N7. N1 implements a state-of-the-art nudging concept by Ur et al. [569], which includes a checklist for password criteria and a one-dimensional password meter displaying the password complexity [604], providing a baseline for efficacy in nudging for stronger passwords. It was randomly assigned to 239 participants. We further included N2 "Default Password Generator", as our memorability-heavy baseline nudge - an additional nudge without gamification elements. It presents the user with a locally generated suggested password, which should provide a secure default option as well as good password memorability for those who picked the suggested password [180] and was randomly assigned to 256 participants. Further, we included N7, the avatar as our novel gamified nudging concept, which provides natural language cues to the user. We hypothesized that when letting the user customize their own avatar, the sunk cost would give weight to the avatar's cues and suggestions. It was randomly assigned to 258 participants. We further included a control group ($N = 247$) that did not interact with a nudge while creating a password. The main study was based on a mock-up registration for a fictional streaming platform.

13.5.1 Method of Data Collection and Analysis

Using the German panel provider RESPONDI, we conducted an online experiment and integrated survey with 1,000 participants approximately representative of the German demography regarding age from 18 to 74 years, gender, and education. The survey was implemented using the software LIMESURVEY, and did not introduce the participants to the topic of nudging so as not to bias them, instead presenting a focus on optimizing security in online registration processes. After performing demographic checks regarding age, gender, and education, the participants were invited to complete a mock-up online registration for a fictional streaming platform, which included either a randomly assigned nudge or the control group. Afterwards, a SYSTEM USABILITY SCALE (SUS) [70] questionnaire regarding the presented nudge was filled out by the participants. As our participants were German, we used the German SUS adaption by Ruegenhagen and Rummel [471]. It provides a "quick and dirty" method for measuring usability and acceptance of a system. The control group was not exposed to any explicit nudging concept and was therefore also not assigned to the SUS. Additionally, all participants were asked to provide further context to the experiment's results, e.g., by stating how worthy of protection they assess login data for a streaming platform. Finally, the participants were asked to perform a short-term memorability test identical to the one used in our prestudy: The participants were prompted to log in with the password used during registration and were informed if an attempt was incorrect. After a maximum of five attempts to recall the correct password, the test was closed. We tracked the number of attempts required to input the correct password.

For our main study, we collected the ZXCVBN- SCORE which approximates the magnitude of guesses ($\log(guesses)$) an adversary would need to crack each respective password, and grouped them by individual nudge or control group (see Table A.35 in ESM). Additionally, we compared the capped zxcvbn-score from 1 to 4 to gain more simplified and intuitive insights. For interpretation of the results, it is, however, crucial to note that those capped results do not consider higher scores than 4. For our *suggested password* nudge N2, we also introduced the subgroups N2.d (only participants that picked the offered *default* option) and N2.c (participants that chose to use their own, *custom* password) in the test. Further, we collected the number of attempts to recall a password in our short-term memorability test (capped at five tries) and the SUS score for usability testing. We also included demographic data and information about gaming habits, asking about the frequency with which the participants played games (mobile, desktop or any other digital format). That was crucial to identify potentials for personalization.

For pre-processing and analysis we used PYTHON and R. To counter the problem of multiple comparisons, the determined p-values were adjusted with Bonferroni correction. Plots and tables were then derived and created using R. We first performed analyses of variance (ANOVA) to investigate if a significant statistical relevance was present. In a post-hoc TUKEY HSD test, we then tested if individual data pairs significantly differed from each other. As the data for cracking time ($\log(guesses)$) does not meet the requirements of an ANOVA due to outliers, in that case, we performed the non-parametric KRUSKALL- WALLIS RANK SUM TEST and a post-hoc WILCOXON TEST.

13.5.2 Results

In the following, we will present the results of our main study, including user characteristics and the comparison of nudging efficacy regarding password strength, usability, and memorability. We thereby especially focus on potential effects of gaming habits.

13.5.2.1 User Characteristics

Even though the study does not absolutely represent the German demography, it comes very close concerning age, gender, and education. There were 47.8% (German average: 49.6%) female participants whereas 52.3% (50.4%) were male. Unfortunately, our panel provider does not yet support diverse genders—this limited the amount of self-reported diverse-gendered participants to 1. Age 60–74 years constitute the most represented group with 25.4% (26.3%), followed by the age of 18–29 with 22.6% (18.8%). Meanwhile, the age group of 50-59 years constituted 19.9% (20.7%), the age group of 40–49 years 18% (18.2%), and lastly, the age group of 30-39 years was represented by 14.6% (16.1%) of our participants. We gathered information on the highest level of education in groups (without a school diploma/certificate of secondary education ("Hauptschulabschluss"), a general certificate of secondary education ("mittlere Reife"), qualification for university entrance ("Abitur"), or university degree). We observed that our pool of participants skewed towards users of "high" education. 42.8% held a university degree or the qualification for university entrance (33.7%). 34.6% (32.5%) held a general certificate of secondary education, 23.1% (33.9%) have a certificate of secondary education or were without a school diploma. To gain context on the experiment's results, we asked our participants on a Likert scale from 1 to 5 how worthy of protection they assess login data for a streaming platform (such as in the experiment). The mean answer was $M = 3.44$ with $SD = 1.07$. For the usability score of our nudges, it made no difference, however, participants that rated the login data as very worthy of protection created significantly stronger passwords (e.g., for the capped zxcvbn-score $M = 3.08, SD = 1.01$ when the answer was "5" versus $M = 2.58, SD = 1.12$ when the answer was "1").

13.5.2.2 Comparison of Password Strength

The Kruskall-Wallis test for log(*guesses*) showed a significant difference between the test conditions (N1, N2, N7, control group): $H(4) = 518.59, p < .001$. The Post-hoc Wilcoxon test revealed that, with the exception of N1 and N2.c ($p = .57$), all group pairings showed significant statistical differences ($p = .04$ for N7 with N2.c and $p < .001$ for all other combinations) regarding password strength. We hypothesize that N2.c performed comparably to N1 regarding password strength because the *suggested password* nudge increased the participants' motivation to create a password that they assumed to be of similar strength to the suggested option. We have found that, compared to the control group in particular, all nudge groups show a statistically significant improvement in password strength, the least difference yielded by N7 and the largest improvement yielded by N2.d (see Fig. 13.2).

The ANOVA test for the capped zxcvbn-score from 1 to 4 showed significant differences as well ($F(3, 996) = 129.7, p < .001$). The post-hoc Tukey HSD test revealed significant differences between all groups. The descending order of mean scores is as followed: N2.d ($M = 4, SD = 0$), N1 ($M = 3.07, SD = 0.92$), N2.c ($M = 2.91, SD = 1.07$), N7 ($M = 2.65, SD = 0.96$) and N0 ($M = 2.27, SD = 1$). Hence, while all evaluated nudges were effective regarding password strength compared to the control group, the generated default passwords were the strongest. Also, the dynamic password meter (N1) following the idea of Ur et al. [569] yields good results. The novel gamified avatar nudge (N7) is still an effective nudge; it is not as strong as the other evaluated nudges. Interestingly, participants who were assigned to the generated default nudge (N2) and decided not to use the suggested password but created their own (N2.c) yielded surprisingly strong scores as well. When comparing the efficacy of our nudges separately among the user characteristics age and education for insights into potential personalization, we found no significant differences (e.g., regarding age and N7: $F(4, 253) = 0.8, p = .5$).

13.5.2.3 Comparison of Usability Rating

To measure the relative acceptance and usability of our test conditions, we performed a similar analysis over the SYSTEM USABILITY SCALE scores, although not adding the subgroups for N2 and leaving out the control group without a nudge, which did not perform the usability test (see Fig. 13.3). Our ANOVA test showed a significant difference in the set with $F(2, 750) = 9.48, p < 0.001$). The post-hoc Tukey HSD tests revealed a significant statistical difference between N1 ($M = 76.36, SD = 16.51$) and N2 ($M = 70.66, SD = 16.30$) as well as N1 and N7 ($M = 70.77, SD = 16.87$). N2 and N7 showed no significant difference between each other. The difference in SUS score puts the default password generator (N2) and the novel gamified avatar nudge (N7) in the adjective rating bracket "OK", not quite reaching the dynamic password meter's (N1) "Good" rating. All nudges fulfill the highest acceptability rating of "Acceptable", as rated by Brooke [70, p. 20].

13.5.2.4 Effect of Gaming Frequency

In our work we are especially interested in the effect of gaming frequency on the acceptance and effectiveness of our nudges. Regarding the SUS scores in general, we found no significant differences when factoring in the levels of gaming frequency ($F(4, 748) = 1.90, p = 0.11$). For password strength, however, we did detect a significant effect on the capped zxcvbn-score from 1 to 4 ($F(4, 995) = 2.72, p = .029$). When looking at the post-hoc Tukey HSD test, we can see that the significant difference appears between "weekly" and "never", where the zxcvbn-score for "weekly" is $M = 3.08, SD = 1.08$ and for "never" it is $M = 1.84, SD = 1.04$. When comparing the strength regarding $\log(guesses)$, there is no significant difference between the gaming groups ($F(4, 253) = 0.94, p = .441$).

While the general SUS scores and strength measurements give insights about the general assessment of all nudges, we intend to evaluate differences in acceptance and effectiveness of gamified nudges versus non-gamified nudges, comparing groups with different gaming habits. Thus, we looked at how specific nudges performed for those groups. Regarding the novel gamified avatar nudge (N7), all participants assigned to that nudge rated its usability similarly, independently of their gaming frequency ($F(4, 253) = 0.51, p = .732$). For all

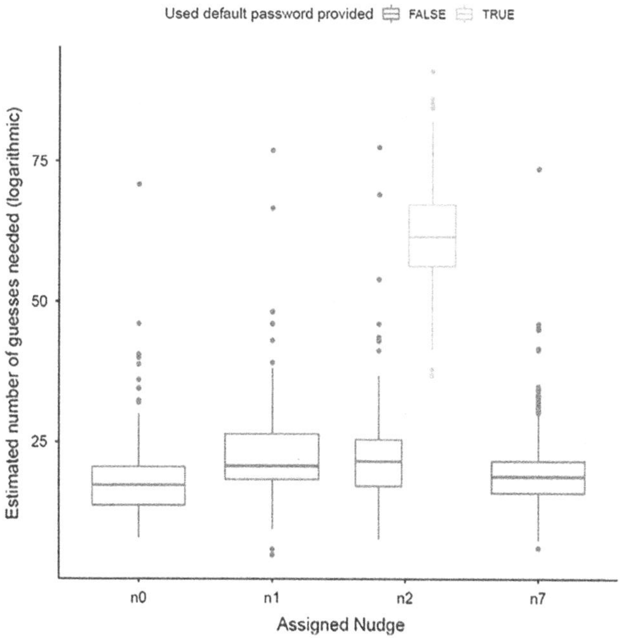

Fig. 13.2 Strength of passwords per condition in estimated number of guesses needed (logarithmic)

other nudges, we did not detect significant differences either. We performed the same analysis concerning password strength. For none of the test conditions (N1, N2, N7, control group) the password strength differed between groups of gaming frequencies (e.g., for N2: $F(4, 251) = 0.16, p = .960$). Hence, for nudge usability or effectiveness it made no difference if participants with a specific gaming frequency were assigned to a gamified or non-gamified nudge.

13.5.2.5 Comparison of Short-term Memorability

We tracked the amount of failed tries to recall the password, limited at 4. An amount of 0 failed tries declares that the user recalled their password immediately. We performed ANOVA tests over all test conditions, including the subgroups N2.d and N2.c, and found a significant difference: $F(4, 995) = 74.59, p < 0.001$. A post-hoc Tukey HSD test showed that only N2.d differs significantly from all other groups, with its memorability being much worse than that of all other test groups where users created their own passwords (e.g., comparing N2.d with N1 the difference is highly significant with $p < .001$ and a mean difference of −

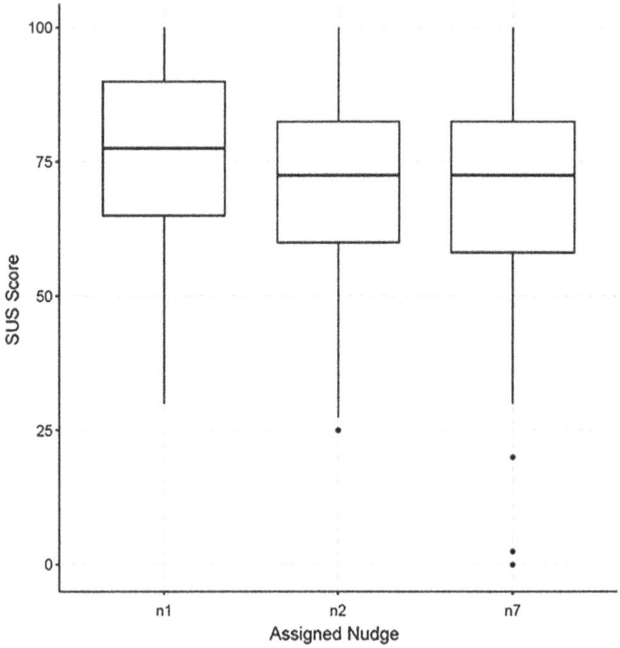

Fig. 13.3 SUS score per nudge

0.63). To sum it up, all evaluated nudges except for the default password generator when using the default option, performed similarly regarding memorability. None of the other nudges explicitly facilitated or hindered memorability in comparison to the control group without a nudge.

13.6 Discussion of Results and Limitations

The goal of this work is to evaluate the potential of gamified nudges in comparison to non-gamified nudges and state-of-the-art concepts to contribute to the pool of effective nudges and facilitate personalization in cybersecurity. We have identified a lack of research regarding large-scale evaluations of concrete nudge implementations, comparing novel nudges against state-of-the-art nudges such as the dynamic password meter of Ur et al. [569], as well as regarding gamification in nudges for cybersecurity. To make initial efforts towards filling this gap, we iteratively designed novel gamified nudges for password strength, in particular an interactive, customizable avatar nudge guiding the user through the registration process using natural language cues, and compared their usability, effectiveness and short-term memorability to other (established) nudges without gamification elements. Our evaluation took place in a two-fold study design, consisting of a preliminary think-aloud study for qualitative insights, followed up by a large-scale online experiment with 1,000 participants to quantitatively compare usability, effectiveness, and memorability. Notably, we asked about the gaming frequency of our participants, as we hypothesized that gaming frequency affects the acceptance of our gamified nudge. We gained some interesting insights, contributing to the human-centered research in password nudges:

We found that *(1)* while most evaluated nudges were having rather small but significant effects on password strength, our gamified avatar nudge N7 was less effective than the other nudges. Contrary to our expectations, that finding holds true *(2)* also when considering only participants that frequently play online games and, thus, may feel more appealed to gamification. While Jeske, Coventry, and Briggs [270] among others suggest that user differences play a role in security decision-making, we could not confirm that for gaming preferences in our setting. Among the evaluated samples, the generated default passwords under condition N2.d achieved the best results regarding password strength, followed by the dynamic password meter in the style of Ur et al. [569]. Interestingly *(3)*, the default generator nudge was comparably effective regarding password strength also when participants chose to not use the suggested password but created their own under condition N2.c.

Memorability is a crucial factor for password nudges as well. Due to the limitations of our online experiment, we decided to evaluate short-term memorability to gain first comparable insights. While the default-generated passwords in condition N2.d yielded great results concerning password strength, the nudge comes with an essential shortcoming in short-term memorability. Indeed *(4)*, it performed significantly worse than all other evaluated nudges, suggesting that our short-term memorability study can not reproduce the high memorability results of Ghazvininejad and Knight [180]. This finding is not surprising, as the participants did not have to interact with the password at all while executing the registration process. In all other test conditions, the participants were forced to invest a minimum of thoughts into the password creation which yielded better short-term memorability. While Furnell, Alotaibi, and Esmael [169] suggest gamification as an enhancing factor for memorability, our avatar nudge did not appear to be effective concerning that matter.

Our study revealed interesting insights into how participants assessed different nudges with and without gamification elements. During our large-scale online experiment, we found *(5)* that the state-of-the-art dynamic password meter nudge N1 performed significantly better than our gamified nudge N7 and the default password generator nudge N2 while N2 and N7 did not significantly differ from each other regarding usability. We assume this to be caused by N1 being a well-established, state-of-the-art nudge, as multiple participants of the preliminary study praised it for being a *"known environment"* (e.g., P #970, P #140, P #493). However *(6)*, all evaluated nudges achieved satisfying SUS scores.

To sum up our findings, we propose the following answers to our research questions:

RQ1: How do gamified nudges compare to a state-of-the-art password meter?
The gamified avatar as a novel nudge did not encourage password strength comparably to the dynamic password meter by Ur et al. [569]. Still, the avatar nudge can be considered an effective nudge as it indeed enhanced password strength compared to the control group without a nudge. Regarding usability, the avatar nudge yielded sufficient results, however, it was outperformed by the usability of the state-of-the-art dynamic password meter.

RQ2: Does a user's gaming frequency increase the acceptance of a gamified nudge?
Contrarily to our expectations and the results of other studies [e.g., 219], the users' gaming frequency had no effect on neither the acceptance nor the effectiveness of gamified versus non-gamified nudges.

Limitations and Future Work: While the study we report is a first step towards extending the pool of effective nudges in cybersecurity for personalization, it has bears limitations. The evaluation of a nudge in an online experiment does not lead to information about efficacy as realistic as in a real-world scenario. Future studies may complement evaluations on the efficacy and memorability of the gamified nudge by using a long-term study design, also including more participants of lower education. While our work is a first step to assess the potential of gamified versus non-gamified nudges for cybersecurity on a short-term basis, we cannot make a final conclusion about their overall efficacy before testing them under more realistic circumstances. Hence, we suggest utilizing other data collection techniques in the future. Additionally, we propose that individualizing nudges with regard to the intended audience of a service (e.g., readers of an information security news feed possibly being more tech-savvy) could be an easier and less costly implementation of individualizing nudges than identifying individual user traits by questionnaire. This could lead to greater acceptance and efficacy of the shown nudges, as discussed by Knijnenburg [300].

13.7 Conclusion

In this work, we examined a sample of nudges for stronger passwords, focusing on a novel gamified nudge in comparison to state-of-the-art concepts and other nudges without gamification elements. We evaluated the nudges in a two-fold study design, conducting a qualitative think-aloud preliminary study and a large-scale online experiment with 1,000 participants. Considering password strength, usability score, and short-term memorability, we found that users still tend to prefer well-known concepts for password creation assistance. While our gamified nudge showed significant effects on password strength compared to a control group without a nudge, the effects are rather small. Interestingly, we further found that the gaming frequency of our participants made no difference in the effect of our gamified nudge.

Bibliography

1. Ahmed Abbasi and Hsinchun Chen. "Writeprints: A Stylometric Approach to Identity-Level Identification and Similarity Detection in Cyberspace". In: *ACM Transactions on Information Systems* 26.2 (2008), 7:1–7:29. https://doi.org/10.1145/1344411.1344413.
2. Alessandro Acquisti. "Nudging Privacy: The Behavioral Economics of Personal Information". In: *IEEE security & privacy* 7.6 (2009), pp. 82–85.
3. Alessandro Acquisti et al. "Nudges for Privacy and Security: Understanding and Assisting Users' Choices Online". In: *ACM Computing Surveys (CSUR)* 50.3 (2017), pp. 1–41.
4. Troy Adams, Yuanxia Li, and Hao Liu. "A Replication of Beyond the Turk Alternative Platforms for Crowdsourcing Behavioral Research—Sometimes Preferable to Student Groups". In: *Transactions on Replication Research* 6 (2020).
5. Sisira Adikari, Craig McDonald, and Penny Collings. "A Design Science Approach to an HCI Research Project". In: *Proceedings of the 18th Australia Conference on Computer-Human Interaction: Design: Activities, Artefacts and Environments*. OZCHI '06. Association for Computing Machinery, 2006, pp. 429–432. https://doi.org/10.1145/1228175.1228265.
6. Sadia Afroz, Michael Brennan, and Rachel Greenstadt. "Detecting Hoaxes, Frauds, and Deception in Writing Style Online". In: *2012 IEEE Symposium on Security and Privacy*. IEEE, 2012, pp. 461–475. https://doi.org/10.1109/SP.2012.34.
7. Zhila Aghajari, Eric Baumer, and Dominic Difranzo. "Reviewing Interventions to Address Misinformation: The Need to Expand Our Vision Beyond an Individualistic Focus". In: *Proceedings of the ACM on Human-Computer Interaction* 7.CSCW1 (2023), pp. 1–34. https://doi.org/10.1145/3579520.
8. Jon Agley et al. "Intervening on Trust in Science to Reduce Belief in COVID-19 Misinformation and Increase COVID-19 Preventive Behavioral Intentions: Randomized Controlled Trial". In: *Journal of Medical Internet Research* 23.10 (2021), e32425. https://doi.org/10.2196/32425.
9. Araz Ramazan Ahmad and Hersh Rasool Murad. "The Impact of Social Media on Panic During the COVID-19 Pandemic in Iraqi Kurdistan: Online Questionnaire Study". In: *Journal of Medical Internet Research* 22.5 (2020). https://doi.org/10.2196/19556.

K. Hartwig, *Navigating Misinformation*, Technology, Peace and Security I Technologie, Frieden und Sicherheit, https://doi.org/10.1007/978-3-658-47885-8

10. Wasim Ahmed et al. "COVID-19 and the 5G Conspiracy Theory: Social Network Analysis of Twitter Data". In: *Journal of Medical Internet Research* 22.5 (2020). https://doi.org/10.2196/19458.
11. Michael Aird et al. "Does Truth Matter to Voters? The Effects of Correcting Political Misinformation in an Australian Sample". In: *Royal Society Open Science* 5.12 (2018), p. 180593. https://doi.org/10.1098/rsos.180593.
12. Syeda Zainab Akbar et al. "Misinformation as a Window into Prejudice: COVID-19 and the Information Environment in India". In: *Proceedings of the ACM on Human-Computer Interaction* 4.CSCW3 (2021), pp. 1–28. https://doi.org/10.1145/3432948. URL: https://doi.org/10.1145/3432948.
13. Ayesha Ali and Ihsan Ayyub Qazi. "Countering Misinformation on Social Media through Educational Interventions: Evidence from a Randomized Experiment in Pakistan". In: *Journal of Development Economics* 163 (2023), p. 103108. https://doi.org/10.1016/j.jdeveco.2023.103108. URL: https://www.sciencedirect.com/science/article/pii/S0304387823000639.
14. Malik Almaliki. "Misinformation-Aware Social Media: A Software Engineering Perspective". In: *IEEE access : practical innovations, opensolutions* 7 (2019), pp. 182451–182458. https://doi.org/10.1109/ACCESS.2019.2960270.
15. Malik Almaliki. "Online Misinformation Spread: A Systematic Literature Map". In: *Proceedings of the 2019 3rd International Conferenceon Information System and Data Mining*. ICISDM 2019. Association for Computing Machinery, 2019, pp. 171–178. https://doi.org/10.1145/3325917.3325938.
16. Nadia Alonso-López. "Beyond Challenges and Viral Dance Moves: TikTok as a Vehicle for Disinformation and Fact-Checking in Spain, Portugal, Brazil, and the USA". In: *Analisi: Quaderns de Comunicacio i Cultura* 64.1 (2021), pp. 65–84.
17. Filipe Altoe and H. Sofia Pinto. "Towards a Personalized Online Fake News Taxonomy". In: *Proceedings of the 31st ACM Conference on User Modeling, Adaptation and Personalization*. ACM, 2023, pp. 96–105. https://doi.org/10.1145/3565472.3592963.
18. Mahdi Nasrullah Al-Ameen, Matthew Wright, and Shannon Scielzo. "Towards Making Random Passwords Memorable: Leveraging Users' Cognitive Ability through Multiple Cues". In: *Proceedings of the 33rd Annual ACM Conference on Human Factors in Computing Systems*. ACM, 2015, pp. 2315–2324.
19. Matteo Dell' Amico, Pietro Michiardi, and Yves Roudier. "Password Strength: An Empirical Analysis". In: *2010 Proceedings IEEE INFOCOM* (2010), pp. 1–9. https://doi.org/10.1109/infcom.2010.5461951.
20. Zaid Amin, Nazlena Mohamad Ali, and Alan F. Smeaton. "Visual Selective Attention System to Intervene User Attention in Sharing COVID-19 Misinformation". In: *International Journal of Advanced Computer Science and Applications* 12.10 (2021), pp. 36–41. https://doi.org/10.14569/IJACSA.2021.0121005.
21. Jonathan Anderson. "Lix and Rix: Variations on a Little-known Readability Index". In: *Journal of Reading* 26.6 (1983), pp. 490–496.
22. Simge Andi and Jesper Akesson. "Nudging Away False News: Evidence from a Social Norms Experiment". In: *Digital Journalism* 9.1 (2020), pp. 106–125. https://doi.org/10.1080/21670811.2020.1847674.

23. Theo Araujo et al. “In AI We Trust? Perceptions about Automated Decision-Making by Artificial Intelligence”. In: *AI & SOCIETY* 35.3 (2020), pp. 611–623. https://doi.org/10.1007/s00146-019-00931-w.
24. Alberto Ardevol-Abreu, Patricia Delponti, and Carmen Rodriguez-Wanguemert. “Intentional or Inadvertent Fake News Sharing? Fact-checking Warnings and Users’ Interaction with Social Media Content”. In: *Profesional de la Informacion* 29.5 (2020), pp. 1–13. https://doi.org/10.3145/epi.2020.sep.07.
25. Kevin Aslett et al. “News Credibility Labels Have Limited Average Effects on News Diet Quality and Fail to Reduce Misperceptions”. In: *Science Advances* 8.18 (2022), eabl3844. https://doi.org/10.1126/sciadv.abl3844.
26. Reema Aswani, Arpan Kumar Kar, and P. Vigneswara Ilavarasan. “Experience: Managing Misinformation in Social Media—Insights for Policymakers from Twitter Analytics”. In: *J. Data and Information Quality* 12.1 (2019). https://doi.org/10.1145/3341107.
27. Shubham Atreja, Libby Hemphill, and Paul Resnick. “Remove, Reduce, Inform: What Actions Do People Want Social Media Platforms to Take on Potentially Misleading Content?” In: *Proc. ACM Hum.-Comput. Interact.* 7.CSCW2 (2023), pp. 1–33. https://doi.org/10.1145/3610082.
28. Patricia Aufderheide. *A Report of the National Leadership Conference on Media Literacy*. Tech. rep. The Aspen Institute Wye Center, 1992, pp. 9–16.
29. Kevin Autry and Shea Duarte. “Correcting the Unknown: Negated Corrections May Increase Belief in Misinformation”. In: *Applied Cognitive Psychology* 35.4 (2021), pp. 960–975. https://doi.org/10.1002/acp.3823.
30. Carl-Anton Werner Axelsson, Mona Guath, and Thomas Nygren. “Learning How to Separate Fake from Real News: Scalable Digital Tutorials Promoting Students’ Civic Online Reasoning”. In: *Future Internet* 13.3 (2021), p. 60. https://doi.org/10.3390/fi13030060.
31. Jackie Ayoub, X. Jessie Yang, and Feng Zhou. “Combat COVID-19 Infodemic Using Explainable Natural Language Processing Models”. In: *Information Processing & Management* 58.4 (2021), p. 102569. https://doi.org/10.1016/j.ipm.2021.102569.
32. Danny Azucar, Davide Marengo, and Michele Settanni. “Predicting the Big 5 Personality Traits from Digital Footprints on Social Media: A Meta-Analysis”. In: *Personality and Individual Differences* 124 (2018), pp. 150–159. https://doi.org/10.1016/j.paid.2017.12.018. URL: https://www.sciencedirect.com/science/article/pii/S0191886917307328.
33. Ingrid Bachmann and Sebastian Valenzuela. “Studying the Downstream Effects of Fact-Checking on Social Media: Experiments on Correction Formats, Belief Accuracy, and Media Trust”. In: *Social Media + Society* 9.20563051231179694 (2023), pp. 1–13. https://doi.org/10.1177/20563051231179694.
34. Bence Bago, David G. Rand, and Gordon Pennycook. “Fake News, Fast and Slow: Deliberation Reduces Belief in False (but Not True) News Headlines.” In: *Journal of Experimental Psychology: General* 149.8 (2020), pp. 1608–1613. https://doi.org/10.1037/xge0000729.
35. Joseph B. Bak-Coleman et al. “Combining Interventions to Reduce the Spread of Viral Misinformation”. In: *Nature Human Behaviour* 6.10 (2022), pp. 1372–+. https://doi.org/10.1038/s41562-022-01388-6.

36. Rebecca Balebako and Lorrie Cranor. "Improving App Privacy: Nudging App Developers to Protect User Privacy". In: *IEEE Security & Privacy* 12.4 (2014), pp. 55–58.
37. Jack M. Balkin. "Free Speech in the Algorithmic Society: Big Data, Private Governance, and New School Speech Regulation". In: *SSRN Electronic Journal* 615 (2017), pp. 1–68. https://doi.org/10.2139/ssrn.3038939.
38. Dipto Barman and Owen Colan. "Does Explanation Matter? An Exploratory Study on the Effects of Covid–19 Misinformation Warning Flags on Social Media". In: *2023 10th International Conference on Behavioural and Social Computing (BESC)*. IEEE, 2023, pp. 1–7. https://doi.org/10.1109/BESC59560.2023.10386371.
39. Ava Bartolome and Shuo Niu. "A Literature Review of Video-Sharing Platform Research in HCI". In: *Proceedings of the 2023 CHI Conference on Human Factors in Computing Systems*. CHI '23. Association for Computing Machinery, 2023, pp. 1–20. https://doi.org/10.1145/3544548.3581107.
40. Ranojoy Barua et al. "F-NAD: An Application for Fake News Article Detection Using Machine Learning Techniques". In: *2019 IEEE Bombay Section Signature Conference (IBSSC)*. Institute of Electrical and Electronics Engineers, 2019, pp. 1–6. https://doi.org/10.1109/IBSSC47189.2019.8973059.
41. Zapan Barua et al. "Effects of Misinformation on COVID-19 Individual Responses and Recommendations for Resilience of Disastrous Consequences of Misinformation". In: *Progress in Disaster Science* 8 (2020), p. 100119.
42. Corey H. Basch, Grace C. Hillyer, and Christie Jaime. "COVID-19 on TikTok: Harnessing an Emerging Social Media Platform to Convey Important Public Health Messages". In: *International Journal of Adolescent Medicine and Health* 34.5 (2022), pp. 367–369. https://doi.org/10.1515/ijamh-2020-0111.
43. Federico Battaglia et al. *Cats, War and Creators: TikTok as Place for (Dis)Information for Teenagers (Translated from German)*. Tech. rep. Amadeu Antonio Stiftung, 2023.
44. Jozef Bavol̆ár and Ol̆ga Orosová. "Decision-Making Styles and Their Associations with Decision-Making Competencies and Mental Health". In: *Judgment and Decision Making* 10.1 (2015), pp. 115–122. https://doi.org/10.1017/S1930297500003223. URL: https://www.cambridge.org/core/product/identifier/S1930297500003223/type/journal_article.
45. Anshul Bawa et al. "Do Multilingual Users Prefer Chat-Bots That Code-Mix? Let's Nudge and Find Out!" In: *Proc. ACM Hum.-Comput. Interact.* 4.CSCW1 (2020). https://doi.org/10.1145/3392846.
46. Jane Beattie et al. "Psychological Determinants of Decision Attitude". In: *Journal of Behavioral Decision Making* 7.2 (1994), pp. 129–144. URL: https://doi.org/10.1002/bdm.3960070206.
47. Yoav Benjamini and Yosef Hochberg. "Controlling the False Discovery Rate: A Practical and Powerful Approach to Multiple Testing". In: *Journal of the Royal Statistical Society Series B: Statistical Methodology* 57.1 (1995), pp. 289–300. https://doi.org/10.1111/j.2517-6161.1995.tb02031.x. URL: https://academic.oup.com/jrsssb/article/57/1/289/7035855.
48. Kristoffer Bergram et al. "The Digital Landscape of Nudging: A Systematic Literature Review of Empirical Research on Digital Nudges". In: *CHI Conference on Human Factors in Computing Systems*. ACM, 2022, pp. 1–16. https://doi.org/10.1145/3491102.3517638. URL: https://doi.org/10.1145/3491102.3517638.

49. Puneet Bhargava et al. "How Effective Are TikTok Misinformation Debunking Videos?" In: *Harvard Kennedy School Misinformation Review* 4.2 (2023), pp. 1–17. https://doi.org/10.37016/mr-2020-114.
50. Prasanta Bhattacharya, Raj Kumar Gupta, and Yinping Yang. "Exploring the Contextual Factors Affecting Multimodal Emotion Recognition in Videos". In: *IEEE Transactions on Affective Computing* 14.2 (2023), pp. 1547–1557. https://doi.org/10.1109/TAFFC.2021.3071503.
51. MD Momen Bhuiyan et al. "Designing Transparency Cues in Online News Platforms to Promote Trust: Journalists' and Consumers' Perspectives". In: *Proceedings of the ACM: Human Computer Interaction(PACM): Computer-Supported Cooperative Work and Social Computing* 5.CSCW2 (2021), 395:1–395:31. https://doi.org/10.1145/3479539.
52. Md Momen Bhuiyan et al. "FeedReflect: A Tool for Nudging Users to Assess News Credibility on Twitter". In: *Companion of the 2018 ACM Conference on Computer Supported Cooperative Work and Social Computing*. CSCW '18. Association for Computing Machinery, 2018, pp. 205–208. https://doi.org/10.1145/3272973.3274056.
53. Md Momen Bhuiyan et al. "NudgeCred: Supporting News Credibility Assessment on Social Media Through Nudges". In: *Proceedings of the ACM on Human-Computer Interaction* 5.CSCW2 (2021), 427:1–427:30. https://doi.org/10.1145/3479571.
54. Tom Biselli, Katrin Hartwig, and Christian Reuter. "Mitigating Misinformation Sharing on Social Media through Personalised Nudging". In: *Proceedings of the ACM: Human-Computer Interaction(PACM): Computer-Supported Cooperative Work and Social Computing(CSCW)*. ACM, 2025. https://doi.org/10.1145/3711034.
55. Tom Biselli and Christian Reuter. "On the Relationship between IT Privacy and Security Behavior : A Survey among German Private Users". In: *16th International Conference on Wirtschaftsinformatik* March (2021), pp. 1–17.
56. Tom Biselli, Laura Utz, and Christian Reuter. "Supporting Informed Choices about Browser Cookies: The Impact of Personalised Cookie Banners". In: *Proceedings on Privacy Enhancing Technologies* 2024.1 (2024), pp. 171–191. https://doi.org/10.56553/popets-2024-0011. URL: https://petsymposium.org/popets/2024/popets-2024-0011.php.
57. Carl-Hugo Björnsson. *Læsbarhed*. Paedagogisk Orientering 9. Gad, 1971.
58. Robert A. Blair et al. "Interventions to Counter Misinformation: Lessons from the Global North and Applications to the Global South". In: *Current Opinion in Psychology* 55.101732 (2024), pp. 1–10. https://doi.org/10.1016/j.copsyc.2023.101732.
59. Leticia Bode and Emily K. Vraga. "In Related News, That Was Wrong: The Correction of Misinformation Through Related Stories Functionality in Social Media". In: *Journal of Communication* 65.4 (2015), pp. 619–638. https://doi.org/10.1111/jcom.12166.
60. Leticia Bode and Emily K. Vraga. "See Something, Say Something: Correction of Global Health Misinformation on Social Media". In: *Health Communication* 33.9 (2018), pp. 1131–1140. https://doi.org/10.1080/10410236.2017.1331312.
61. Mike Bostock. *D3.Js*. 2020. URL: https://d3js.org/.
62. Danah Boyd. "Did Media Literacy Backfire?" In: *Journal of Applied Youth Studies* 1.4 (2017), pp. 83–89.

63. Lia Bozarth et al. "Wisdom of Two Crowds: Misinformation Moderation on Reddit and How to Improve This Process—A Case Study of COVID-19". In: *Proc. ACM Hum.-Comput. Interact.* 7.CSCW1 (2023), pp. 1–33. https://doi.org/10.1145/3579631.
64. Nadia M. Brashier et al. "Timing Matters When Correcting Fake News". In: *Proceedings of the National Academy of Science of the United States of America* 118.5 (2021), e2020043118. https://doi.org/10.1073/pnas.2020043118.
65. Adrian M. P. Braşoveanu and Răzvan Andonie. "Semantic Fake News Detection: A Machine Learning Perspective". In: *Advances in Computational Intelligence*. Ed. by Ignacio Rojas, Gonzalo Joya, and Andreu Catala. Vol. 11506. Springer International Publishing, 2019, pp. 656–667. https://doi.org/10.1007/978-3-030-20521-8_54.
66. Virginia Braun et al. "Thematic Analysis". In: *Handbook of Research Methods in Health Social Sciences*. Ed. by Pranee Liamputtong. Springer, 2019, pp. 843–860. https://doi.org/10.1007/978-981-10-5251-4_103.
67. Sharon S. Brehm and Jack W. Brehm. *Psychological Reactance: A Theory of Freedom and Control*. Academic Press, 2013.
68. Michael Brennan and Rachel Greenstadt. "Practical Attacks Against Authorship Recognition Techniques". In: *Proceedings of the Twenty-First Conference on Innovative Applications of Artificial Intelligence*. AAAI, 2009, pp. 60–65.
69. Debora Briggs et al. "Nudging Whom How: IT Proficiency, Impulse Control and Secure Behaviour". In: *Networks. An International Journal* 49 (2014), p. 18.
70. John Brooke. "SUS—A Quick and Dirty Usability Scale". In: *Usability evaluation in industry* 189.194 (1996), pp. 4–7.
71. Yana Brovdiy. *Disinformation in Times of COVID-19: Reinforcing the Responses of the European Union and the United States*. Tech. rep. College of Europe, 2020.
72. Étienne Brunet. *Le Vocabulaire de Jean Giraudoux: Structure et Évolution: Statistique et Informatique Appliquées à l'étude Des Textes à Partir Des Données Du Trésor de La Langue Française*. Slatkine, 1978.
73. Sonja Buchegger and Jean-Yves Le Boudec. "The Effect of Rumor Spreading in Reputation Systems for Mobile Ad-hoc Networks". In: *WiOpt'03: Modeling and Optimization in Mobile, Ad Hoc and Wireless Networks*. IEEE, 2003, pp. 1–10.
74. Klara Austeja Buczel et al. "How Do Forewarnings and Post-Warnings Affect Misinformation Reliance? The Impact of Warnings on the Continued Influence Effect and Belief Regression". In: *Memory & Cognition* 52.2 (2024), pp. 1–17. https://doi.org/10.3758/s13421-024-01520-z.
75. Monica Bulger and Patrick Davison. "The Promises, Challenges, and Futures of Media Literacy". In: *Journal of Media Literacy Education* 10.1 (2018), pp. 1–21. https://doi.org/10.23860/JMLE-2018-10-1-1.
76. Cody Buntain et al. "YouTube Recommendations and Effects on Sharing Across Online Social Platforms". In: *Proceedings of the ACM on Human-Computer Interaction* 5.CSCW1 (2021), 11:1–11:26. https://doi.org/10.1145/3449085.
77. Grégoire Burel et al. "Co-Spread of Misinformation and Fact-Checking Content During the Covid-19 Pandemic". In: *Social Informatics*. Ed. by Samin Aref et al. Vol. 12467. Springer International Publishing, 2020, pp. 28–42. https://doi.org/10.1007/978-3-030-60975-7_3. URL: https://doi.org/10.1007/978-3-030-60975-7_3.
78. Judee K. Burgoon et al. "Detecting Deception through Linguistic Analysis". In: *Proceedings of Intelligence and Security Informatics*. Ed. by Hsinchun Chen et al. Lecture

Notes in Computer Science. Springer, 2003, pp. 91–101. https://doi.org/10.1007/3-540-44853-5_7.

79. Pete Burnap et al. "Tweeting the Terror: Modelling the Social Media Reaction to the Woolwich Terrorist Attack". In: *Social Network Analysis and Mining* 4.1 (2014), p. 206. https://doi.org/10.1007/s13278-014-0206-4. URL: https://doi.org/10.1007/s13278-014-0206-4.
80. John T. Cacioppo and Richard E. Petty. "The Need for Cognition". In: *Journal of Personality and Social Psychology* 42.1 (1982), pp. 116–131. https://doi.org/10.1037/0022-3514.42.1.116.
81. Kelly Caine. "Local Standards for Sample Size at CHI". In: *Proceedings of the 2016 CHI Conference on Human Factors in Computing Systems*. ACM, 2016, pp. 981–992. https://doi.org/10.1145/2858036.2858498.
82. Danielle Caled and Mário J. Silva. "Digital Media and Misinformation: An Outlook on Multidisciplinary Strategies against Manipulation". In: *Journal of Computational Social Science* 5.1 (2021), pp. 123–159. https://doi.org/10.1007/s42001-021-00118-8.
83. Gamze Canova et al. "NoPhish: An Anti-Phishing Education App". In: *Lecture Notes in Computer Science* 8743 (2014), pp. 88–192. https://doi.org/10.1007/978-3-319-11851-2_14.
84. Valerio Capraro and Tatiana Celadin. ""I Think This News Is Accurate": Endorsing Accuracy Decreases the Sharing of Fake News and Increases the Sharing of Real News". In: *Personality and Social Psychology Bulletin* 0.0 (2022), pp. 1–11. https://doi.org/10.1177/01461672221117691.
85. Ana Caraban et al. "23 Ways to Nudge: A Review of Technology-Mediated Nudging in Human-Computer Interaction". In: (2019), pp. 1–15. https://doi.org/10.1145/3290605.3300733.
86. Kevin Matthe Caramancion. "Quantification of Infographic Intervention Effect on Mis/Disinformation Vulnerability". In: *2022 IEEE 17th International Conference on Computer Sciences and Information Technologies (CSIT)*. IEEE, 2022, pp. 297–300. https://doi.org/10.1109/CSIT56902.2022.10000489.
87. Bruno Castanho Silva, Federico Vegetti, and Levente Littvay. "The Elite Is Up to Something: Exploring the Relation Between Populism and Belief in Conspiracy Theories". In: *Swiss Political Science Review* 23.4 (2017), pp. 423–443. https://doi.org/10.1111/spsr.12270. URL: https://onlinelibrary.wiley.com/doi/abs/10.1111/spsr.12270.
88. Carlos Castillo, Marcelo Mendoza, and Barbara Poblete. "Information Credibility on Twitter". In: *Proceedings of the 20th International Conference on World Wide Web*. 2011, pp. 675–684. https://doi.org/10.1145/1963405.1963500.
89. Abhijnan Chakraborty et al. *Stop Clickbait: Detecting and Preventing Clickbaits in Online News Media*. 2016.
90. Aimee Challenger, Petroc Sumner, and Lewis Bott. "COVID-19 Myth-Busting: An Experimental Study". In: *Bmc Public Health* 22.1 (2022), p. 131. https://doi.org/10.1186/s12889-021-12464-3.
91. Man-pui Sally Chan and Dolores Albarracin. "A Meta-Analysis of Correction Effects in Science-Relevant Misinformation". In: *Nature Human Behaviour* 7.9 (2023), pp. 1514–1525. https://doi.org/10.1038/s41562-023-01623-8.

92. Xunru Che, Danaë Metaxa-Kakavouli, and Jeffrey T. Hancock. “Fake News in the News: An Analysis of Partisan Coverage of the Fake News Phenomenon”. In: *Companion of the 2018 ACM Conference on Computer Supported Cooperative Work and Social Computing*. ACM, 2018, pp. 289–292. https://doi.org/10.1145/3272973.3274079. URL: https://dl.acm.org/doi/10.1145/3272973.3274079.
93. Guangyu Chen, Paolo Ciuccarelli, and Sara Colombo. “VisualBubble: Exploring How Reflection-Oriented User Experiences Affect Users’ Awareness of Their Exposure to Misinformation on Social Media”. In: *Extended Abstracts of the 2022 CHI Conference on Human Factors in Computing Systems*. CHI EA ’22. Association for Computing Machinery, 2022, pp. 1–7. https://doi.org/10.1145/3491101.3519615.
94. Jiangjie Chen et al. “LOREN: Logic-Regularized Reasoning for Interpretable Fact Verification”. In: *Proceedings of the AAAI Conference on Artificial Intelligence* 36.10 (2022), pp. 10482–10491. https://doi.org/10.1609/aaai.v36i10.21291.
95. Juh-Jen Chen et al. “Predicting Consumers’ Decision-Making Styles by Analyzing Digital Footprints on Facebook | International Journal of Information Technology & Decision Making”. In: *International Journal of Information Technology & Decision Making* 18.2 (2019), pp. 601–627. URL: https://www.worldscientific.com/doi/abs/10.1142/S0219622019500019.
96. Liang Chen and Hongjie Tang. “Examining the Persuasion Process of Narrative Fear Appeals on Health Misinformation Correction”. In: *Information, Communication & Society* 26.15 (2022), pp. 1–19. https://doi.org/10.1080/1369118X.2022.2128849.
97. Sijing Chen, Lu Xiao, and Akit Kumar. “Spread of Misinformation on Social Media: What Contributes to It and How to Combat It”. In: *Computers in Human Behavior* 141.2023 (2022), p. 107643. https://doi.org/10.1016/j.chb.2022.107643.
98. Yimin Chen, Niall J. Conroy, and Victoria L. Rubin. “Misleading Online Content: Recognizing Clickbait as “False News””. In: *Proceedings of the 2015 ACM on Workshop on Multimodal Deception Detection*. WMDD ’15. Association for Computing Machinery, 2015, pp. 15–19. https://doi.org/10.1145/2823465.2823467.
99. Tosti H. C. Chiang, Chih-Shan Liao, and Wei-Ching Wang. “Impact of Artificial Intelligence News Source Credibility Identification System on Effectiveness of Media Literacy Education”. In: *Sustainability* 14.8 (2022), p. 4830. https://doi.org/10.3390/su14084830.
100. Eun Kyoung Choe et al. “Nudging People Away from Privacy-Invasive Mobile Apps through Visual Framing”. In: *IFIP Conference on Human-Computer Interaction*. Springer, 2013, pp. 74–91.
101. Charles L. Citroen. “The Role of Information in Strategic Decision-Making”. In: *International Journal of Information Management* 31.6 (2011), pp. 493–501. URL: https://www.sciencedirect.com/science/article/pii/S0268401211000235.
102. Nathan Edward Clark et al. “Exploring the Impacts of Social Media and Crowdsourcing on Disaster Resilience”. In: *Open Research Europe* (2024). https://doi.org/10.12688/openreseurope.13721.3.
103. Aaron Clauset, M. E. J. Newman, and Cristopher Moore. “Finding Community Structure in Very Large Networks”. In: *Physical Review E* 70.6 (2004). https://doi.org/10.1103/physreve.70.066111.

104. Katherine Clayton et al. "Real Solutions for Fake News? Measuring the Effectiveness of General Warnings and Fact-Check Tags in Reducing Belief in False Stories on Social Media". In: *Political Behavior* 42.4 (2020), pp. 1073–1095. https://doi.org/10.1007/s11109-019-09533-0.
105. Meri Coleman and T. L. Liau. "A Computer Readability Formula Designed for Machine Scoring". In: *Journal of Applied Psychology* 60 (1975), pp. 283–284. https://doi.org/10.1037/h0076540.
106. Coleman, Keith. *Introducing Birdwatch, a Community-Based Approach to Misinformation.* 2020. URL: https://blog.twitter.com/en_us/topics/product/2021/introducing-birdwatch-a-community-based-approach-to-misinformation.
107. T. Comes, Kenny Meesters, and Stina Torjesen. "Making Sense of Crises: The Implications of Information Asymmetries for Resilience and Social Justice in Disaster-Ridden Communities". In: *Sustainable and Resilient Infrastructure* 4 (2017), pp. 1–13. https://doi.org/10.1080/23789689.2017.1405653.
108. Jennifer M. Connolly et al. "Communicating to the Public in the Era of Conspiracy Theory". In: *Public Integrity* 21.5 (2019), pp. 469–476. https://doi.org/10.1080/10999922.2019.1603045.
109. Lucian Constantin. *Sony Stresses That PSN Passwords Were Hashed.* 2011. URL: https://news.softpedia.com/news/Sony-Stresses-PSN-Passwords-Were-Hashed-198218.shtml.
110. Correctiv.org. *Wie erkenne ich Falschmeldungen?* 2023. URL: https://correctiv.org/faktencheck/faktencheck-tipps/2023/05/01/wie-erkenne-ich-falschmeldungen-2/.
111. Laura Courchesne, Julia Ilhardt, and Jacob Shapiro. "Review of Social Science Research on the Impact of Countermeasures against Influence Operations". In: *Harvard Kennedy School Misinformation Review* (2021). https://doi.org/10.37016/mr-2020-79.
112. Michael Craig and Santosh Vijaykumar. "One Dose Is Not Enough: The Beneficial Effect of Corrective COVID-19 Information Is Diminished If Followed by Misinformation". In: *Social Media + Society* 9.20563051231161298 (2023), pp. 1–14. https://doi.org/10.1177/20563051231161298.
113. Limeng Cui et al. "DEFEND: A System for Explainable Fake News Detection". In: *Proceedings of the 28th ACM International Conference on Information and Knowledge Management.* CIKM '19. Association for Computing Machinery, 2019, pp. 2961–2964. https://doi.org/10.1145/3357384.3357862.
114. Shih-Chieh Dai et al. "Ask to Know More: Generating Counterfactual Explanations for Fake Claims". In: *Proceedings of the 28th ACMSIGKDD Conference on Knowledge Discovery and Data Mining.* KDD '22. Association for Computing Machinery, 2022, pp. 2800–2810. https://doi.org/10.1145/3534678.3539205.
115. Yue Dai. "The Effects of Message Order and Debiasing Information in Misinformation Correction". In: *International Journal of Communication* 15.2021 (2021), pp. 1039–1059.
116. Dharma Dailey and Kate Starbird. " "It's Raining Dispersants": Collective Sensemaking of Complex Information in Crisis Contexts". In: *Proceedings of the 18th ACM Conference Companion on Computer Supported Cooperative Work & Social Computing.* CSCW' 15 Companion. Association for Computing Machinery, 2015, pp. 155–158. https://doi.org/10.1145/2685553.2698995.

117. Dharma Dailey and Kate Starbird. "Journalists as Crowdsourcerers: Responding to Crisis by Reporting with a Crowd". In: *ComputerSupported Cooperative Work (CSCW)* 23 (2014), pp. 445–481.
118. Sandor Dalecke and Randi Karlsen. "Designing Dynamic and Personalized Nudges". In: *Proceedings of the 10th International Conference on Web Intelligence, Mining and Semantics*. Ed. by Richard Chbeir. ACM Digital Library. Association for Computing Machinery, 2020, pp. 139–148. https://doi.org/10.1145/3405962.3405975.
119. Erik Dane and Michael G. Pratt. "Exploring Intuition and Its Role in Managerial Decision Making". In: *Academy of Management Review* 32.1 (2007), pp. 33–54. URL: https://journals.aom.org/doi/10.5465/amr.2007.23463682.
120. Valdemar Danry et al. "Wearable Reasoner: Towards Enhanced Human Rationality Through A Wearable Device With An Explainable AI Assistant". In: *Proceedings of the Augmented Humans International Conference*. AHs '20. Association for Computing Machinery, 2020, pp. 1–12. https://doi.org/10.1145/3384657.3384799.
121. Abdel-Fattah E. Darwish and Günter L. Huber. "Individualism vs Collectivism in Different Cultures: A Cross-Cultural Study". In: *Intercultural Education* 14.1 (2003), pp. 47–56. https://doi.org/10.1080/1467598032000044647. URL: http://www.tandfonline.com/doi/abs/10.1080/1467598032000044647.
122. William Davies. *What's Wrong with WhatsApp*. 2020. URL: https://williamdavies.blog/2020/07/02/whats-wrong-with-whatsapp/.
123. Madeleine de Cock Buning. *A Multi-Dimensional Approach to Disinformation. Report of the Independent High Level Group on Fake News and Online Disinformation*. Tech. rep. 2018. https://doi.org/10.2759/739290.
124. Jonas De Keersmaecker and Arne Roets. " 'Fake News': Incorrect, but Hard to Correct. The Role of Cognitive Ability on the Impact of False Information on Social Impressions". In: *Intelligence* 65 (2017), pp. 107–110. https://doi.org/10.1016/j.intell.2017.10.005.
125. Nora Denner, Benno Viererbl, and Thomas Koch. "Effects of Repeated Corrections of Misinformation on Organizational Trust: More Is Not Always Better". In: *International Journal of Strategy Communication* 17.1 (2023), pp. 39–53. https://doi.org/10.1080/1553118X.2022.2135098.
126. Saoirse Connor Desai and Stian Reimers. "Does Explaining the Origins of Misinformation Improve the Effectiveness of a given Correction?" In: *Memory & Cognition* 51.2 (2023), pp. 422–436. https://doi.org/10.3758/s13421-022-01354-7.
127. Amelie Desvars-Larrive et al. "A Structured Open Dataset of Government Interventions in Response to COVID-19". In: *Scientific Data* 7.1 (2020), pp. 1–9.
128. Jacob Devlin et al. "BERT: Pre-training of Deep Bidirectional Transformers for Language Understanding". In: *Proceedings of the2019 Conference of the North American Chapter of the Association for Computational Linguistics: Human Language Technologies, Volume 1 (Long and Short Papers)*. Association for Computational Linguistics, 2019, pp. 4171–4186. https://doi.org/10.18653/v1/N19-1423.
129. Giandomenico Di Domenico, Daniel Nunan, and Valentina Pitardi. "Marketplaces of Misinformation: A Study of How Vaccine Misinformation Is Legitimized on Social Media". In: *Journal of Public Policy & Marketing* 41.4 (2022), pp. 319–335. https://doi.org/10.1177/07439156221103860.

130. Lindsay Levkoff Diamond et al. "The Polyvocality of Online COVID-19 Vaccine Narratives That Invoke Medical Racism". In: *CHI Conference on Human Factors in Computing Systems*. ACM, 2022, pp. 1–21. https://doi.org/10.1145/3491102.3501892.
131. Carlos Diaz Ruiz and Tomas Nilsson. "Disinformation and Echo Chambers: How Disinformation Circulates on Social Media Through Identity-Driven Controversies". In: *Journal of Public Policy & Marketing* 42.1 (2023), pp. 18–35. https://doi.org/10.1177/07439156221103852.
132. Stephanie Diepeveen et al. "Public Acceptability of Government Intervention to Change Health-Related Behaviours: A Systematic Review and Narrative Synthesis". In: *BMC public health* 13.1 (2013), p. 756.
133. Tom Dobber et al. "The Effect of Traffic Light Veracity Labels on Perceptions of Political Advertising Source and Message Credibility on Social Media". In: *Journal of Information Technology and Politics* (2023), pp. 1–16. https://doi.org/10.1080/19331681.2023.2224316.
134. Leyla Dogruel. "Privacy Nudges as Policy Interventions: Comparing US and German Media Users' Evaluation of Information Privacy Nudges". In: *Information, Communication & Society* 22.8 (2019), pp. 1080–1095.
135. Shawn Domgaard and Mina Park. "Combating Misinformation: The Effects of Infographics in Verifying False Vaccine News". In: *Health Education Journal* 80.8 (2021), pp. 974–986. https://doi.org/10.1177/00178969211038750.
136. Chiara Patricia Drolsbach and Nicolas Pröllochs. "Diffusion of Community Fact-Checked Misinformation on Twitter". In: *Proc. ACM Hum.-Comput. Interact.* 7.CSCW2 (2023), pp. 1–22. https://doi.org/10.1145/3610058.
137. Elena-Alexandra Dumitru. "Testing Children and Adolescents' Ability to Identify Fake News: A Combined Design of Quasi-Experiment and Group Discussions". In: *Societies* 10.3 (2020), p. 71. https://doi.org/10.3390/soc10030071.
138. Megan Duncan. "What's in a Label? Negative Credibility Labels in Partisan News". In: *Journalism & Mass Communication Quarterly* 99.2 (2020), pp. 390–413. https://doi.org/10.1177/1077699020961856.
139. Janna Lynn Dupree et al. "Privacy Personas: Clustering Users via Attitudes and Behaviors toward Security Practices". In: *Proceedings of the 2016 CHI Conference on Human Factors in Computing Systems. ACM* (2016), pp. 5228–5239. https://doi.org/10.1145/2858036.2858214.
140. Jared Duval et al. "Chasing Play on TikTok from Populations with Disabilities to Inspire Playful and Inclusive Technology Design". In: *Proceedings of the 2021 CHI Conference on Human Factors in Computing Systems*. CHI '21. Association for Computing Machinery, 2021, pp. 1–15. https://doi.org/10.1145/3411764.3445303.
141. Ullrich K. H. Ecker, Joshua L. Hogan, and Stephan Lewandowsky. "Reminders and Repetition of Misinformation: Helping or Hindering Its Retraction?: Journal of Applied Research in Memory and Cognition". In: *Journal of Applied Research in Memory and Cognition* 6.2 (2017), pp. 185–192. https://doi.org/10.1037/h0101809.
142. Ullrich K. H. Ecker, Stephan Lewandowsky, and Matthew Chadwick. "Can Corrections Spread Misinformation to New Audiences? Testing for the Elusive Familiarity Backfire Effect". In: *Cognitive Research: Principles and Implications* 5.1 (2020), pp. 1–25. https://doi.org/10.1186/s41235-020-00241-6.

143. Ullrich K. H. Ecker et al. "Correcting False Information in Memory: Manipulating the Strength of Misinformation Encoding and Its Retraction". In: *Psychonomic Bulletin & Review* 18.3 (2011), pp. 570–578. https://doi.org/10.3758/s13423-011-0065-1.
144. Ulrich K. H. Ecker, Lucy H. Butler, and Anne Hamby. "You Don't Have to Tell a Story! A Registered Report Testing the Effectiveness of Narrative versus Non-Narrative Misinformation Corrections". In: *Cognitive Research: Principles and Implications* 5.1 (2020), pp. 1–26. https://doi.org/10.1186/s41235-020-00266-x.
145. Economic and Social Research Council. *Choosing What Social Media You Use.* 2020. URL: https://www.ukri.org/councils/esrc/impact-toolkit-for-economic-and-social-sciences/how-to-use-social-media/choosing-what-social-media-you-use/.
146. Serge Egelman and Eyal Peer. "The Myth of the Average User: Improving Privacy and Security Systems through Individualization". In: *Proceedings of the 2015 New Security Paradigms Workshop.* ACM, 2015, pp. 16–28.
147. Jennifer Elias. *YouTube to Add Labels to Some Health Videos amid Misinformation Backlash.* 2021. URL: https://www.cnbc.com/2021/07/19/youtube-labeling-some-health-videos-amid-misinformation-backlash.html.
148. Rob Ennals, Beth Trushkowsky, and John Mark Agosta. "Highlighting Disputed Claims on the Web". In: *Proceedings of the 19th International Conference on World Wide Web.* ACM, 2010, pp. 341–350. https://doi.org/10.1145/1772690.1772726.
149. Ziv Epstein, Gordon Pennycook, and David Gertler Rand. "Will the Crowd Game the Algorithm? Using Layperson Judgments to Combat Misinformation on Social Media by Downranking Distrusted Sources". In: *Proceedings of the 2020 CHI Conference on Human Factors in Computing Systems.* ACM, 2020. https://doi.org/10.31234/osf.io/z3s5k.
150. Ziv Epstein et al. "Developing an Accuracy-Prompt Toolkit to Reduce COVID-19 Misinformation Online". In: *Harvard Kennedy School Misinformation Review* 2.3 (2021), pp. 1–12. https://doi.org/10.37016/mr-2020-71. URL: https://misinforeview.hks.harvard.edu/?p=7273.
151. Sarah Evanega et al. *Coronavirus Misinformation: Quantifying Sources and Themes in the COVID-19 'Infodemic'.* Tech. rep. Cornell Alliance for Science, 2020.
152. Catharine Evers et al. "Citizen Approval of Nudging Interventions Promoting Healthy Eating: The Role of Intrusiveness and Trustworthiness". In: *BMC public health* 18.1 (2018), pp. 1–10.
153. Frank J. Fabozzi. *Introduction to Securitization.* The Frank J. Fabozzi Series. J. Wiley & Sons, 2008. URL: https://learning.oreilly.com/library/view/-/9780470419571/?ar.
154. Sascha Fahl et al. "On the Ecological Validity of a Password Study". In: *Proceedings of the Ninth Symposium on Usable Privacy and Security.* SOUPS '13. Association for Computing Machinery, 2013. https://doi.org/10.1145/2501604.2501617.
155. Barbara Fasolo et al. "Animation Attracts: The Attraction Effect in an on-Line Shopping Environment". In: *Psychology & Marketing* 23.10 (2006), pp. 799–811. https://doi.org/10.1002/mar.20134. URL: https://onlinelibrary.wiley.com/doi/abs/10.1002/mar.20134.

156. K. J. Kevin Feng et al. "Examining the Impact of Provenance-Enabled Media on Trust and Accuracy Perceptions". In: *Proc. ACM Hum.-Comput. Interact.* 7.CSCW2 (2023), pp. 1–42. https://doi.org/10.1145/3610061.
157. K. J. Kevin Feng et al. "Investigating How University Students in the United States Encounter and Deal With Misinformation in Private WhatsApp Chats During COVID-19". In: *Eighteenth Symposium on Usable Privacy and Security (SOUPS)*. Vol. 18. USENIX Association, 2022, pp. 427–446.
158. Song Feng, Ritwik Banerjee, and Yejin Choi. "Syntactic Stylometry for Deception Detection". In: *Proceedings of the 50th Annual Meeting of the Association for Computational Linguistics (Volume 2: Short Papers)*. Association for Computational Linguistics, 2012, pp. 171–175.
159. Kathrin Figl, Samuel Kiessling, and Ulrich Remus. "Do Symbol and Device Matter? The Effects of Symbol Choice of Fake News Flags and Device on Human Interaction with Fake News on Social Media Platforms". In: *Computers in Human Behavior* 144.107704 (2023), pp. 1–16. https://doi.org/10.1016/j.chb.2023.107704.
160. Sophia Fischer, Katja Soyez, and Sebastian Gurtner. "Adapting Scott and Bruce's General Decision-Making Style Inventory to Patient Decision Making in Provider Choice". In: *Medical Decision Making* 35.4 (2015), pp. 525–532. https://doi.org/10.1177/0272989X15575518. URL: http://journals.sagepub.com/doi/10.1177/0272989X15575518.
161. Uwe Flick. *An Introduction to Qualitative Research*. Edition 5. Sage, 2014. URL: https://uk.sagepub.com/en-gb/eur/an-introduction-to-qualitative-research/book278983.
162. Frans Folkvord et al. "Effect of Source Type and Protective Message on the Critical Evaluation of News Messages on Facebook: Randomized Controlled Trial in the Netherlands". In: *Journal of Medical Internet Research* 24.3 (2022), e27945. https://doi.org/10.2196/27945.
163. Marsha E. Fonteyn, Benjamin Kuipers, and Susan J. Grobe. "A Description of Think Aloud Method and Protocol Analysis". In: *Qualitative Health Research* 3.4 (1993), pp. 430–441. https://doi.org/10.1177/104973239300300403.
164. Anjuli Franz et al. "SoK: Still Plenty of Phish in the Sea—A Taxonomy of User-Oriented Phishing Interventions and Avenues for Future Research". In: *Proceedings of the Seventeenth Symposium on Usable Privacy and Security(SOUPS 2021)*. USENIX, 2021, pp. 339–358. URL: https://www.usenix.org/conference/soups2021/presentation/franz.
165. Diana Freed et al. "Understanding Digital-Safety Experiences of Youth in the U.S." In: *Proceedings of the 2023 CHI Conference on Human Factors in Computing Systems*. ACM, 2023, pp. 1–15. https://doi.org/10.1145/3544548.3581128.
166. Melanie Freeze et al. "Fake Claims of Fake News: Political Misinformation, Warnings, and the Tainted Truth Effect". In: *Political Behavior* 43.4 (2021), pp. 1433–1465. https://doi.org/10.1007/s11109-020-09597-3.
167. Jennifer Fromm et al. "Social Media Data in an Augmented Reality System for Situation Awareness Support in Emergency Control Rooms". In: *Information Systems Frontiers* (2021). https://doi.org/10.1007/s10796-020-10101-9.

168. Norbert Fuhr et al. "An Information Nutritional Label for Online Documents". In: *ACM SIGIR Forum* 51.3 (2018), pp. 46–66. https://doi.org/10.1145/3190580.3190588. URL: https://dl.acm.org/doi/10.1145/3190580.3190588.
169. Steven Furnell, Faisal Alotaibi, and Rawan Esmael. "Aligning Security Practice with Policy : Guiding and Nudging towards Better Behavior". In: *Proceedings of the 52nd Hawaii International Conference on System Sciences* 6 (2019), pp. 5618–5627.
170. Steven Furnell, Adila Jusoh, and Dimitris Katsabas. "The Challenges of Understanding and Using Security: A Survey of End-Users". In: *Computers & Security* 25.1 (2006), pp. 27–35.
171. Tomoya Furuta and Yu Suzuki. "A Fact-checking Assistant System for Textual Documents". In: *2021 IEEE 4th International Conference on Multimedia Information Processing and Retrieval (MIPR)*. IEEE, 2021, pp. 243–246. https://doi.org/10.1109/MIPR51284.2021.00046.
172. Vijaya Gadde and Matt Derella. *An Update on Our Continuity Strategy during COVID-19*. 2020. URL: https://blog.x.com/en_us/topics/company/2020/An-update-on-our-continuity-strategy-during-COVID-19.
173. Mingkun Gao et al. "To Label or Not to Label: The Effect of Stance and Credibility Labels on Readers' Selection and Perception of News Articles". In: *Proceedings of the ACM on Human-Computer Interaction* 2.CSCW (2018), 55:1–55:16. https://doi.org/10.1145/3274324.
174. Diego García-Saiz, Camilo Palazuelos, and Marta Zorrilla. "Data Mining and Social Network Analysis in the Educational Field: An Application for Non-Expert Users". In: *Educational Data Mining: Applications and Trends*. Ed. by Alejandro Peña-Ayala. Springer International Publishing, 2014, pp. 411–439. https://doi.org/10.1007/978-3-319-02738-8_15.
175. Simson Garfinkel and Heather Richter Lipford. "Usable Security: History, Themes, and Challenges". In: *Synthesis Lectures on Information Security, Privacy, and Trust* 5.2 (2014), pp. 1–124.
176. Patrizia Garista and Giancarlo Pocetta. "Digital Resilience: Meanings, Epistemologies and Methodologies for Lifelong Learning". In: 2014. https://doi.org/10.13140/2.1.3552.1605.
177. Anna Gausen, Wayne Luk, and Ce Guo. *Can We Stop Fake News? Using Agent-Based Modelling to Evaluate Countermeasures for Misinformation on Social Media*. 2021.
178. Christine Geeng, Savanna Yee, and Franziska Roesner. "Fake News on Facebook and Twitter: Investigating How People (Don't) Investigate". In: *Proceedings of the 2020 CHI Conference on Human Factors in Computing Systems*. CHI '20. Association for Computing Machinery, 2020, pp. 1–14. https://doi.org/10.1145/3313831.3376784.
179. Anat Gesser-Edelsburg et al. "Correcting Misinformation by Health Organizations during Measles Outbreaks: A Controlled Experiment". In: *PLOS ONE* 13.12 (2018), e0209505. https://doi.org/10.1371/journal.pone.0209505.
180. Marjan Ghazvininejad and Kevin Knight. "How to Memorize a Random 60-Bit String". In: *Proceedings of the 2015 Conference of the North American Chapter of the Association for Computational Linguistics: Human Language Technologies*. Association for Computational Linguistics, 2015, pp. 1569–1575. https://doi.org/10.3115/v1/N15-1180.

181. Fabio Giglietto et al. "Fakes, News and the Election: A New Taxonomy for the Study of Misleading Information within the Hybrid Media System". In: *Current Sociology* 10.1 (2016), p. 40.
182. Henner Gimpel et al. "The Effectiveness of Social Norms in Fighting Fake News on Social Media". In: *Journal of Management Information Systems* 38.1 (2021), pp. 196–221. https://doi.org/10.1080/07421222.2021.1870389.
183. Sean Goggins, Christopher Mascaro, and Stephanie Mascaro. "Relief Work after the 2010 Haiti Earthquake: Leadership in an Online Resource Coordination Network". In: *Proceedings of the ACM 2012 Conference on Computer Supported Cooperative Work*. CSCW '12. Association for Computing Machinery, 2012, pp. 57–66. https://doi.org/10.1145/2145204.2145218.
184. Beth Goldberg. *Can "Inoculation" Build Broad-Based Resistance to Misinformation?* 2021. URL: https://medium.com/jigsaw/can-inoculation-build-broad-based-resistance-to-misinformation-6c67e517e314.
185. Google. *Flutter*. Google. 2020. URL: https://flutter.dev.
186. Wilson Goudalo and Christophe Kolski. "Towards Advanced Enterprise Information Systems Engineering—Solving Resilience, Security and Usability Issues within the Paradigms of Socio-Technical Systems". In: *ICEIS*. 2016.
187. Rebecca Hofstein Grady, Peter Ditto, and Elizabeth Loftus. "Nevertheless, Partisanship Persisted: Fake News Warnings Help Briefly, but Bias Returns with Time". In: *Cognitive Research-Principles and Implications* 6.1 (2021), pp. 1–16. https://doi.org/10.1186/s41235-021-00315-z.
188. Sukeshini Grandhi, Linda Plotnick, and Starr Roxanne Hiltz. "By the Crowd and for the Crowd: Perceived Utility and Willingness to Contribute to Trustworthiness Indicators on Social Media". In: *Proceedings of the ACM on Human-Computer Interaction* 5.GROUP (2021), pp. 1–24. https://doi.org/10.1145/3463930.
189. Paul A Grassi et al. *NIST Special Publication 800 63B Digital Identity Guidelines*. 2016.
190. Shirley Gregor and Alan R. Hevner. "Positioning and Presenting Design Science Research for Maximum Impact". In: *MIS Quarterly* 37.2 (2013), pp. 337–355.
191. Irene Greif. *Computer-Supported Cooperative Work: A Book of Readings*. Morgan Kaufmann, 1988.
192. Guardian. *Arizona Man Dies after Attempting to Take Trump Coronavirus 'Cure'*. 2020. URL: https://www.theguardian.com/world/2020/mar/24/coronavirus-cure-kills-man-after-trump-touts-chloroquine-phosphate.
193. Brian Guay et al. "How to Think about Whether Misinformation Interventions Work". In: *Nature Human Behaviour* 7.8 (2023), pp. 1231–1233. https://doi.org/10.1038/s41562-023-01667-w.
194. Andrew M. Guess et al. "A Digital Media Literacy Intervention Increases Discernment between Mainstream and False News in the United States and India". In: *Proceedings of the Nationale Academy of Science of the United States of America* 117.27 (2020), pp. 15536–15545. https://doi.org/10.1073/pnas.1920498117.
195. Xinning Gui et al. "Managing Uncertainty: Using Social Media for Risk Assessment during a Public Health Crisis". In: *Proceedings of the 2017CHI Conference on Human Factors in Computing Systems*. ACM, 2017, pp. 4520–4533. https://doi.org/10.1145/3025453.3025891.

196. R. Gunning. *The Technique of Clear Writing*. McGraw-Hill, 1952.
197. Hendrik Gunz and Isa Schaller. "Antisemitic Narratives on YouTube and Telegram as Part of Conspiracy Beliefs about COVID-19". In: *Antisemitism on Social Media*. 1st ed. Routledge, 2022, pp. 129–150. https://doi.org/10.4324/9781003200499-9.
198. Bin Guo et al. "The Future of False Information Detection on Social Media: New Perspectives and Trends". In: *ACM Computing Surveys* 53.4 (2021), pp. 1–36. https://doi.org/10.1145/3393880.
199. Chen Guo, Nan Zheng, and Chengqi (John) Guo. "Seeing Is Not Believing: A Nuanced View of Misinformation Warning Efficacy on Video-Sharing Social Media Platforms". In: *Proc. ACM Hum.-Comput. Interact.* 7.CSCW2 (2023), pp. 1–35. https://doi.org/10.1145/3610085.
200. Palmer Haasch. *The Latest Twitter Meme Imitates the 'this Claim Is Disputed' Label the Platform Has Used on Trump's Tweets*. 2021. URL: https://www.insider.com/twitter-trump-censorship-this-claim-is-disputed-meme-election-tweets-2020-11.
201. Lee Hadlington et al. "Perceptions of Fake News, Misinformation, and Disinformation amid the COVID-19 Pandemic: A Qualitative Exploration." In: *Psychology of Popular Media* 12.1 (2023), pp. 40–49. https://doi.org/10.1037/ppm0000387.
202. Christine Hagar. "Using Research to Aid the Design of a Crisis Information Management Course". In: *Association of Library & Information Science Educators SIG*. Vol. 0. ALISE, 2006, pp. 6–10.
203. Loni Hagen et al. "Emoji Use in Twitter White Nationalism Communication". In: *Conference Companion Publication of the 2019 on Computer Supported Cooperative Work and Social Computing*. ACM, 2019, pp. 201–205. https://doi.org/10.1145/3311957.3359495.
204. William Hagman et al. "Public Views on Policies Involving Nudges". In: *Review of philosophy and psychology* 6.3 (2015), pp. 439–453.
205. Michael Hameleers. "Separating Truth from Lies: Comparing the Effects of News Media Literacy Interventions and Fact-Checkers in Response to Political Misinformation in the US and Netherlands". In: *Information,Communication & Society* 25.1 (2022), pp. 110–126. https://doi.org/10.1080/1369118X.2020.1764603.
206. Michael Hameleers and Toni van der Meer. "Striking the Balance between Fake and Real: Under What Conditions Can Media Literacy Messages That Warn about Misinformation Maintain Trust in Accurate Information?" In: *Behaviour & Information Technology* (2023), pp. 1–14. https://doi.org/10.1080/0144929X.2023.2267700.
207. Michael Hameleers et al. "A Picture Paints a Thousand Lies? The Effects and Mechanisms of Multimodal Disinformation and Rebuttals Disseminated via Social Media". In: *Political Communication* 37.2 (2020), pp. 281–301. https://doi.org/10.1080/10584609.2019.1674979.
208. Jeffrey Hancock et al. "On Lying and Being Lied To: A Linguistic Analysis of Deception in Computer-Mediated Communication". In: *Discourse Processes* 45 (2008), pp. 1–23. https://doi.org/10.1080/01638530701739181.
209. Andrew S. Hanks et al. "Healthy Convenience: Nudging Students toward Healthier Choices in the Lunchroom". In: *Journal of Public Health(United Kingdom)* 34.3 (2012), pp. 370–376. https://doi.org/10.1093/pubmed/fds003.

210. Aniko Hannak et al. "Get Back! You Don't Know Me like That: The Social Mediation of Fact-Checking Interventions in Twitter Conversations". In: *Proceedings of the International AAAI Conference on Web and Social Media*. Vol. 8. AAAI, 2014, pp. 187–196.
211. Sven Ove Hansson. "Science Denial as a Form of Pseudoscience". In: *Studies in History and Philosophy of Science Part A* 63 (2017), pp. 39–47. https://doi.org/10.1016/j.shpsa.2017.05.002.
212. Md Mahfuzul Haque et al. "Combating Misinformation in Bangladesh: Roles and Responsibilities as Perceived by Journalists, Fact-checkers, and Users". In: *Proceedings of the ACM on Human-Computer Interaction* 4.CSCW2 (2020), pp. 1–32. https://doi.org/10.1145/3415201. URL: https://dl.acm.org/doi/10.1145/3415201.
213. Marian Harbach et al. "Using Personal Examples to Improve Risk Communication for Security & Privacy Decisions". In: *Proceedings of the SIGCHI Conference on Human Factors in Computing Systems*. CHI '14. Association for Computing Machinery, 2014, pp. 2647–2656. https://doi.org/10.1145/2556288.2556978.
214. Katrin Hartwig, Frederic Doell, and Christian Reuter. "The Landscape of User-centered Misinformation Interventions—A Systematic Literature Review". In: *ACM Computing Surveys (CSUR)* 56.11 (2024), pp. 1–36. https://doi.org/10.1145/3674724.
215. Katrin Hartwig and Christian Reuter. "Fake News technisch begegnen: Detektions- und Behandlungsansätze zur Unterstützung von NutzerInnen". In: *Wahrheit und Fake News im postfaktischen Zeitalter*. Ed. by Peter Klimczak and Thomas Zoglauer. Vol. 3. ars digitalis. Springer Vieweg, 2021, pp. 133–150. URL: https://link.springer.com/chapter/10.1007/978-3-658-32957-0_7.
216. Katrin Hartwig and Christian Reuter. "Fighting Misinformation on Twitter: The Plugin Based Approach TrustyTweet". In: *SCIENCE PEACE SECURITY'19—Proceedings of the Interdisciplinary Conference on Technical Peace and Security Research*. Ed. by Christian Reuter et al. TUprints, 2019, pp. 67–69. URL: https://publications.rwth-aachen.de/record/781544/files/781544.pdf#page=67.
217. Katrin Hartwig and Christian Reuter. "Nudge or Restraint: How Do People Assess Nudging in Cybersecurity—A Representative Study in Germany". In: *Proceedings of the 2021 European Symposium on Usable Security (Euro USEC)*. Association for Computing Machinery, 2021, pp. 141–150. https://doi.org/10.1145/3481357.3481514.
218. Katrin Hartwig and Christian Reuter. "Nudging Users towards Better Security Decisions in Password Creation Using Whitebox-Based Multidimensional Visualisations". In: *Behaviour & Information Technology* 41.7 (2022), pp. 1357–1380. https://doi.org/10.1080/0144929X.2021.1876167.
219. Katrin Hartwig and Christian Reuter. "Nudging Users Towards Better Security Decisions in Password Creation Using Whitebox-based Multidimensional Visualizations". In: *Behaviour & Information Technology (BIT)* 41.7 (2021), pp. 1357–1380. https://doi.org/10.1080/0144929X.2021.1876167.
220. Katrin Hartwig and Christian Reuter. "Transparenz Im Technischen Umgang Mit Fake News". In: *Technik & Mensch* 2 (2021), pp. 9–11. URL: https://tuprints.ulb.tu-darmstadt.de/19267/1/VDI_TuM_2.2021.pdf.
221. Katrin Hartwig and Christian Reuter. "TrustyTweet: An Indicator-Based Browser-Plugin to Assist Users in Dealing with Fake News on Twitter". In: *Proceedings of the*

International Conference on Wirtschaftsinformatik (WI). Vol. 14. AIS, 2019, pp. 1844–1855. URL: https://aisel.aisnet.org/wi2019/specialtrack01/papers/5/.

222. Katrin Hartwig, Ruslan Sandler, and Christian Reuter. "Navigating Misinformation in Voice Messages: Identification of User-Centered Features for Digital Interventions". In: *Risk, Hazards & Crisis in Public Policy* 15.2 (2024), pp. 203–235. https://doi.org/10.1002/rhc3.12296.
223. Katrin Hartwig et al. "Finding Secret Treasure? Improving Memorized Secrets Through Gamification". In: *Proceedings of the 2021 European Symposium on Usable Security (EuroUSEC)*. ACM, 2021, pp. 105–117. https://doi.org/10.1145/3481357.3481509.
224. Katrin Hartwig et al. "From Adolescents' Eyes: Assessing an Indicator-Based Intervention to Combat Misinformation on TikTok". In: *Proceedings of the CHI Conference on Human Factors in Computing Systems*. ACM, 2024, pp. 1–20. https://doi.org/10.1145/3613904.3642264.
225. Katrin Hartwig et al. "Misleading Information in Crises: Exploring Content-Specific Indicators on Twitter from a User Perspective". In: *Behaviour & Information Technology* (2024). https://doi.org/10.1080/0144929X.2024.2373166.
226. Amelia Hassoun et al. "Practicing Information Sensibility: How Gen Z Engages with Online Information". In: *Proceedings of the 2023 CHI Conference on Human Factors in Computing Systems*. CHI '23. Association for Computing Machinery, 2023, pp. 1–17. https://doi.org/10.1145/3544548.3581328.
227. Sarah Hawa et al. "Combating Misinformation Dissemination through Verification and Content Driven Recommendation". In: *2021 Third International Conference on Intelligent Communication Technologies and Virtual Mobile Networks (ICICV)*. Curran Associates, 2021, pp. 917–924. https://doi.org/10.1109/ICICV50876.2021.9388406.
228. Richard Heeks and Angelica V. Ospina. "Conceptualising the Link between Information Systems and Resilience: A Developing Country Field Study". In: *Information Systems Journal* 29 (2018). https://doi.org/10.1111/isj.12177.
229. Marcus Hegner. *Methoden Zur Evaluation von Software*. Vol. 29. IZ-Arbeitsbericht. Informationszentrum Sozialwissenschaften, 2003.
230. Stefan Helmstetter and Heiko Paulheim. "Weakly Supervised Learning for Fake News Detection on Twitter". In: *2018 IEEE/ACM International Conference on Advances in Social Networks Analysis and Mining(ASONAM)*. IEEE, 2018, pp. 274–277. https://doi.org/10.1109/ASONAM.2018.8508520.
231. Aliaksandr Herasimenka et al. "Misinformation and Professional News on Largely Unmoderated Platforms: The Case of Telegram". In: *Journal of Information Technology & Politics* 20.2 (2023), pp. 198–212. https://doi.org/10.1080/19331681.2022.2076272.
232. Paula Herrero-Diz, Jesús Conde-Jiménez, and Salvador Reyes de Cózar. "Teens' Motivations to Spread Fake News on WhatsApp". In: *Social Media + Society* 6.3 (2020), p. 2056305120942879. https://doi.org/10.1177/2056305120942879.
233. Paula Herrero-Diz, Jesús Conde-Jiménez, and Salvador Reyes-de-Cózar. "Spanish Adolescents and Fake News: Level of Awareness and Credibility of Information (Los Adolescentes Españoles Frente a Las Fake News: Nivel de Conciencia y Credibilidad de La Información)". In: *Culture and Education* 33.1 (2021), pp. 1–27. https://doi.org/10.1080/11356405.2020.1859739.

234. Hendrik Heuer and Elena Leah Glassman. "A Comparative Evaluation of Interventions against Misinformation: Augmenting the WHO Checklist". In: *Proceedings of the 2022 CHI Conference on Human Factorsin Computing Systems*. CHI '22. Association for Computing Machinery, 2022, pp. 1–21. https://doi.org/10.1145/3491102.3517717.
235. Alan R. Hevner et al. "Design Science in Information Systems Research". In: *Management Information Systems Quarterly* 28 (2004), pp. 75–.
236. W. Bartley Hildreth et al. *Handbook of Public Administration*. Routledge, 2007.
237. Hyehyun Hong and Hyo Jung Kim. "Antecedents and Consequences of Information Overload in the COVID-19 Pandemic". In: *International Journal of Environmental Research and Public Health* 17.24 (2020). https://doi.org/10.3390/ijerph17249305.
238. Matthew Honnibal et al. *spaCy: Industrial-strength Natural Language Processing in Python*. 2020. https://doi.org/10.5281/zenodo.1212303.
239. Antony Honoré et al. "Some Simple Measures of Richness of Vocabulary". In: *Association for literary and linguistic computing bulletin* 7.2 (1979), pp. 172–177.
240. Benjamin Horne and Sibel Adali. "This Just In: Fake News Packs A Lot In Title, Uses Simpler, Repetitive Content in Text Body, More Similar To Satire Than Real News". In: *Proceedings of the International AAAI Conference on Web and Social Media*. Vol. 11. AAAI, 2017, pp. 759–766. https://doi.org/10.1609/icwsm.v11i1.14976. URL: https://ojs.aaai.org/index.php/ICWSM/article/view/14976.
241. Benjamin Horne, Mauricio Gruppi, and Sibel Adali. "Trustworthy Misinformation Mitigation with Soft Information Nudging". In: *2019First IEEE International Conference on Trust, Privacy and Security in Intelligent Systems and Applications (TPS-ISA)*. IEEE, 2019, pp. 245–254. https://doi.org/10.1109/TPS-ISA48467.2019.00039.
242. Mahmood Hosseini et al. "Four Reference Models for Transparency Requirements in Information Systems". In: *Requirements Engineering* 23 (2017), pp. 251–275.
243. Shin-Yi Huang, Miles M. Yang, and Chien-Hsiung Chen. "When Do Motivational Factors Lead to Negative User Experience on Social Networking Applications?" In: *Australasian Journal of Information Systems* 22 (2018). https://doi.org/10.3127/ajis.v22i0.1533.
244. Y. Linlin Huang et al. "Connected Through Crisis: Emotional Proximity and the Spread of Misinformation Online". In: *Proceedings of the18th ACM Conference on Computer Supported Cooperative Work & Social Computing*. CSCW '15. Association for Computing Machinery, 2015, pp. 969–980. https://doi.org/10.1145/2675133.2675202.
245. Yan Huang and Weirui Wang. "When a Story Contradicts: Correcting Health Misinformation on Social Media through Different Message Formats and Mechanisms". In: *Information Communication & Society* 25.8 (2022), pp. 1192–1209.
246. Hubley, Doug. *A Field Guide to Conspiracy-Theory Rhetoric*. 2019. URL: https://www.bates.edu/news/2019/09/26/stephanie-kelly-romano-guide-to-conspiracy-rhetoric/.
247. Andrew Hutchinson. *TikTok Adds New Video Warning Labels to Stop the Spread of Misinformation*. 2021. URL: https://www.socialmediatoday.com/news/tiktok-adds-new-video-warning-labels-to-stop-the-spread-of-misinformation/594481/.
248. Muhammad Imran, Patrick Meier, and Kees Boersma. "The Use of Social Media for Crisis Management: A Privacy by Design Approach". In: *Big Data, Surveillance and Crisis Management*. Routledge, 2017, pp. 19–37.
249. Instagram. *Combatting Misinformation on Instagram*. 2019. URL: https://about.instagram.com/blog/announcements/combatting-misinformation-on-instagram.

250. Institut für Medienpädagogik und Kommunikation and Wissenschaftsstadt Darmstadt. *Media Check: What Smartphone and Internet Can Mean for You (Tranlated from German).* Tech. rep. Haus der digitalen Medienbildung (HddM), 2023.
251. Dulcie Irving et al. "Correcting Statistical Misinformation about Scientific Findings in the Media: Causation versus Correlation". In: *Journal of Experimental Psychology-Applied* 28.1 (2022), pp. 1–9. https://doi.org/10.1037/xap0000408.
252. MacNeil Isabelle. *WATCH—How to Spot What's Not Real on TikTok CBC Kids News.* 2023. URL: https://www.cbc.ca/kidsnews/post/watch-how-to-spot-whats-not-real-on-tiktok.
253. Md Rafiqul Islam et al. "Deep Learning for Misinformation Detection on Online Social Networks: A Survey and New Perspectives". In: *Social Network Analysis and Mining* 10.1 (2020), p. 82.
254. Rishi Iyengar. *Reddit Takes Action against Groups Spreading Covid Misinformation.* 2021. URL: https://edition.cnn.com/2021/09/01/tech/reddit-covid-misinformation-ban/index.html.
255. Corey Brian Jackson and Yang Wang. "Addressing the Privacy Paradox through Personalized Privacy Notifications". In: *Proceedings of the ACM on interactive, mobile, wearable and ubiquitous technologies* 2.2 (2018), pp. 1–25.
256. Linda A. Jackson and Jin-Liang Wang. "Cultural Differences in Social Networking Site Use: A Comparative Study of China and the United States". In: *Computers in Human Behavior* 29.3 (2013), pp. 910–921. https://doi.org/10.1016/j.chb.2012.11.024. URL: https://www.sciencedirect.com/science/article/pii/S0747563212003299.
257. Matthew O. Jackson, Suraj Malladi, and David McAdams. "Learning through the Grapevine and the Impact of the Breadth and Depth of Social Networks". In: *Proceedings of the National Academy of Sciences* 119.34 (2022), e2205549119. https://doi.org/10.1073/pnas.2205549119.
258. Nicole J. Jackson. "The Canadian Government's Response to Foreign Disinformation: Rhetoric, Stated Policy Intentions, and Practices". In: *International Journal* 76.4 (2021), pp. 544–563. https://doi.org/10.1177/00207020221076402.
259. Farnaz Jahanbakhsh and David R Karger. "A Browser Extension for In-Place Signaling and Assessment of Misinformation". In: *Proceedings of the CHI Conference on Human Factors in Computing Systems.* ACM, 2024, pp. 1–21. https://doi.org/10.1145/3613904.3642473.
260. Farnaz Jahanbakhsh, Amy X. Zhang, and David R. Karger. "Leveraging Structured Trusted-Peer Assessments to Combat Misinformation". In: *Proceedings of the ACM on Human-Computer Interaction 6.CSCW2* (2022), 524:1–524:40. https://doi.org/10.1145/3555637.
261. Farnaz Jahanbakhsh et al. "Exploring Lightweight Interventions at Posting Time to Reduce the Sharing of Misinformation on Social Media". In: *Proceedings of the ACM on Human-Computer Interaction* 5.CSCW1 (2021), 18:1–18:42. https://doi.org/10.1145/3449092.
262. Farnaz Jahanbakhsh et al. "Exploring the Use of Personalized AI for Identifying Misinformation on Social Media". In: *Proceedings of the 2023 CHI Conference on Human Factors in Computing Systems.* CHI '23. Association for Computing Machinery, 2023, pp. 1–27. https://doi.org/10.1145/3544548.3581219.

263. Farnaz Jahanbakhsh et al. "Our Browser Extension Lets Readers Change the Headlines on News Articles, and You Won't Believe What They Did!" In: *Proc. ACM Hum.-Comput. Interact.* 6.CSCW2 (2022), pp. 1–33. https://doi.org/10.1145/3555643.
264. M. Rosie Jahng, Elizabeth Stoycheff, and Annisa Rochadiat. "They Said It's "Fake": Effects of Discounting Cues in Online Comments on Information Quality Judgments and Information Authentication". In: *Mass Communication and Society* 24.4 (2021), pp. 527–552. https://doi.org/10.1080/15205436.2020.1870143.
265. Kamila Janmohamed et al. "Interventions to Mitigate COVID-19 Misinformation: A Systematic Review and Meta-Analysis". In: *Journal of Health Communication* 26.12 (2021), pp. 846–857. https://doi.org/10.1080/10810730.2021.2021460.
266. Jay Jennings and Natalie Jomini Stroud. "Asymmetric Adjustment: Partisanship and Correcting Misinformation on Facebook". In: *New Media & Society* 0.0 (2021), p. 146144482110217. https://doi.org/10.1177/14614448211021720.
267. Youngseung Jeon et al. "HearHere: Mitigating Echo Chambers in News Consumption through an AI-based Web System". In: *Proceedings of the ACM on Human-Computer Interaction* 8.CSCW1 (2024), pp. 1–34. https://doi.org/10.1145/3637340.
268. Se-Hoon Jeong, Hyunyi Cho, and Yoori Hwang. "Media Literacy Interventions: A Meta-Analytic Review". In: *Journal of Communication* 62.3 (2012), pp. 454–472. https://doi.org/10.1111/j.1460-2466.2012.01643.x.
269. Jennifer Jerit and Yangzi Zhao. "Political Misinformation". In: *Annual Review of Political Science.* 1st ser. 23.1 (2020), pp. 77–94. https://doi.org/10.1146/annurev-polisci-050718-032814.
270. Debora Jeske, Lynne Coventry, and Pam Briggs. "Nudging Whom How : IT Proficiency , Impulse Control and Secure Behaviour". In: *Proceedings of the CHI Workshop on Personalizing Behavior Change Technologies* April (2014), pp. 1–4.
271. Chenyan Jia et al. "Understanding Effects of Algorithmic vs. Community Label on Perceived Accuracy of Hyper-partisan Misinformation". In: *Proceedings of the ACM on Human-Computer Interaction* 6.CSCW2 (2022), 371:1–371:27. https://doi.org/10.1145/3555096.
272. Shan Jiang, Ronald E. Robertson, and Christo Wilson. "Bias Misperceived:The Role of Partisanship and Misinformation in YouTube Comment Moderation". In: *Proceedings of the International AAAI Conference on Web and Social Media* 13 (2019), pp. 278–289. https://doi.org/10.1609/icwsm.v13i01.3229.
273. Hyerim Jo, Fan Yang, and Qing Yan. "Spreaders vs Victims: The Nuanced Relationship between Age and Misinformation via FoMO and Digital Literacy in Different Cultures". In: *New Media & Society* 11.1 (2022), p. 146144482211304. https://doi.org/10.1177/14614448221130476.
274. Wonkwang Jo et al. "Online Information Exchange and Anxiety Spread in the Early Stage of the Novel Coronavirus (COVID-19) Outbreak in South Korea: Structural Topic Model and Network Analysis". In: *Journal of Medical Internet Research* 22.6 (2020). https://doi.org/10.2196/19455.
275. Joireman et al. "Two-Factor Consideration of Future Consequences-14 Scale Promotion Orientation Explains Why Future-Oriented People Exercise and Eat Healthy : Evidence From The". In: 38.10 (2012), pp. 1272–1287. URL: https://www.semanticscholar.org/paper/Two-Factor-Consideration-of-Future-Consequences-14-Joireman-Shaffer/02bdaef4d8121bb5aef31f5886bb7abb44a97fea.

276. Gargi Joshi et al. "Explainable Misinformation Detection across Multiple Social Media Platforms". In: *IEEE access : practical innovations,open solutions* 11 (2023), pp. 23634–23646. https://doi.org/10.1109/ACCESS.2023.3251892.
277. Janice Y. Jung and Barbara A. Mellers. "American Attitudes toward Nudges." In: *Judgment & Decision Making* 11.1 (2016).
278. Astrid Junghans, Tracy Cheung, and Denise De Ridder. "Under Consumers' Scrutiny-an Investigation into Consumers' Attitudes and Concerns about Nudging in the Realm of Health Behavior". In: *BMC public health* 15.1 (2015), p. 336.
279. Beverly Kahn, Diane Strong, and Richard Wang. "Information Quality Benchmarks: Product and Service Performance". In: *Communications of The Acm* 45 (2002), pp. 184–192. https://doi.org/10.1145/505999.506007.
280. Joseph Kahne and Benjamin Bowyer. "Educating for Democracy in a Partisan Age: Confronting the Challenges of Motivated Reasoning and Misinformation". In: *American Educational Research Journal* 54.1 (2017), pp. 3–34. https://doi.org/10.3102/0002831216679817.
281. Daniel Kahneman. *Thinking, Fast and Slow*. Macmillan, 2011.
282. Jeffrey P. Kaleta, Jong Seok Lee, and Sungjin Yoo. "Nudging with Construal Level Theory to Improve Online Password Use and Intended Password Choice". In: *Information Technology & People* (2019).
283. Bente Kalsnes. "Fake News". In: *Oxford Research Encyclopedia of Communication*. Oxford University Press, 2018.
284. Shipi Kankane, Carlina Dirusso, and Christen Buckley. "Can We Nudge Users toward Better Password Management ? An Initial Study". In: *Extended Abstracts of the 2018 CHI Conference on Human Factors in Computing System*. ACM (2018), pp. 1–6.
285. Eleni Kapantai et al. "A Systematic Literature Review on Disinformation: Toward a Unified Taxonomical Framework". In: *New Media & Society* 23.5 (2021), pp. 1301–1326. https://doi.org/10.1177/1461444820959296.
286. Alireza Karduni et al. "Vulnerable to Misinformation? Verifi!" In: *Proceedings of the 24th International Conference on Intelligent User Interfaces*. IUI '19. Association for Computing Machinery, 2019, pp. 312–323. https://doi.org/10.1145/3301275.3302320.
287. Kianoosh Karimi and Javad Nickpayam. "Gamification from the Viewpoint of Motivational Theory". In: *Emerging Science Journal* 1.1 (2017), pp. 34–42. https://doi.org/10.28991/esj-2017-01114.
288. Marc-André Kaufhold et al. "Avoiding Chaotic Use of Social Media before, during, and after Emergencies: Design and Evaluation of Citizens' Guidelines". In: *Journal of Contingencies and Crisis Management* 27.3 (2019), pp. 198–213. https://doi.org/10.1111/1468-5973.12249.
289. Marc-André Kaufhold et al. "Mitigating Information Overload in Social Media during Conflicts and Crises: Design and Evaluation of a Cross-Platform Alerting System". In: *Behaviour & Information Technology* 39.3 (2020), pp. 319–342.
290. Kulvinder Kaur and Pawan Kumar. "Social Media: A Blessing or a Curse? Voice of Owners in the Beauty and Wellness Industry". In: *The TQM Journal* 34.5 (2022), pp. 1039–1056. https://doi.org/10.1108/TQM-03-2021-0074.
291. Sabrina Heike Kessler and Eva Bachmann. "Debunking Health Myths on the Internet: The Persuasive Effect of (Visual) Online Communication". In: *Journal of Public*

Health-Heidelberg 30.8 (2022), pp. 1823–1835. https://doi.org/10.1007/s10389-022-01694-3.

292. Yash Khivasara, Yash Khare, and Tejas Bhadane. "Fake News Detection System Using Web-Extension". In: *2020 IEEE Pune Section International Conference (PuneCon)*. IEEE, 2020, pp. 119–123. https://doi.org/10.1109/PuneCon50868.2020.9362384.
293. Johannes Kiesel, Benno Stein, and Stefan Lucks. "A Large-scale Analysis of the Mnemonic Password Advice." In: *NDSS*. 2017.
294. Antino Kim, Patricia L. Moravec, and Alan R. Dennis. "Combating Fake News on Social Media with Source Ratings: The Effects of User and Expert Reputation Ratings". In: *Journal of Managment Information Systems* 36.3 (2019), pp. 931–968. https://doi.org/10.1080/07421222.2019.1628921.
295. Paul Kim et al. "Controversy Score Calculation for News Articles". In: *2019 First International Conference on Transdisciplinary AI (TransAI)*. IEEE, 2019, pp. 56–63. https://doi.org/10.1109/TransAI46475.2019.00018.
296. Sojung Claire Kim, Emily K. Vraga, and John Cook. "An Eye Tracking Approach to Understanding Misinformation and Correction Strategies on Social Media: The Mediating Role of Attention and Credibility to Reduce HPV Vaccine Misperceptions". In: *Health Communication* 36.13 (2021), pp. 1687–1696. https://doi.org/10.1080/10410236.2020.1787933.
297. Yeongdae Kim et al. "Exoskeleton for the Mind: Exploring Strategies against Misinformation with a Metacognitive Agent". In: *Proceedings of the Augmented Humans International Conference 2023*. AHs '23. Association for Computing Machinery, 2023, pp. 209–220. https://doi.org/10.1145/3582700.3582725.
298. Peter Kincaid et al. *Derivation of New Readability Formulas (Automated Readability Index, Fog Count and Flesch Reading Ease Formula) for Navy Enlisted Personnel:* tech. rep. 8-75. Institute for Simulation and Trainings, 1975.
299. Jan Kirchner and Christian Reuter. "Countering Fake News: A Comparison of Possible Solutions Regarding User Acceptance and Effectiveness". In: *Proceedings of the ACM: Human Computer Interaction (PACM): Computer-Supported Cooperative Work and Social Computing* 4.CSCW2 (2020), pp. 1–27. https://doi.org/10.1145/3415211.
300. Bart Knijnenburg. "Privacy? I Can't Even! Making a Case for User-Tailored Privacy". In: *IEEE Security and Privacy* 15.4 (2017), pp. 62–67. https://doi.org/10.1109/MSP.2017.3151331.
301. Timo K. Koch, Lena Frischlich, and Eva Lermer. "Effects of Fact-Checking Warning Labels and Social Endorsement Cues on Climate Change Fake News Credibility and Engagement on Social Media". In: *Journal of Applied Social Psychology* 53.6 (2023), pp. 495–507. https://doi.org/10.1111/jasp.12959.
302. Jonna Koivisto and Juho Hamari. "The Rise of Motivational Information Systems: A Review of Gamification Research". In: *International Journal of Information Management* 45 (2019), pp. 191–210.
303. Tibor Koltay. "Information Architecture, Information Overload, and the Literacies". In: *Journal of Information Architecture*. 4 (2012), pp. 41–56.
304. Saranga Komanduri et al. "Of Passwords and People: Measuring the Effect of Password-Composition Policies". In: *Proceedings of the Sigchi Conference on Human Factors in Computing Systems*. 2011, pp. 2595–2604.

305. Nadejda Komendantova, Dmitry Erokhin, and Teresa Albano. "Misinformation and Its Impact on Contested Policy Issues: The Example of Migration Discourses". In: *Societies* 13.7 (2023), p. 168. https://doi.org/10.3390/soc13070168.
306. Nadejda Komendantova et al. "A Value-Driven Approach to Addressing Misinformation in Social Media". In: *Humanities and Social Sciences Communications* 8.1 (2021), pp. 1–12. https://doi.org/10.1057/s41599-020-00702-9.
307. Loukas Konstantinou and Evangelos Karapanos. "Nudging for Online Misinformation: A Design Inquiry". In: *Computer Supported Cooperative Work and Social Computing*. ACM, 2023, pp. 69–75. https://doi.org/10.1145/3584931.3607015. URL: https://dl.acm.org/doi/10.1145/3584931.3607015.
308. Dimosthenis Kotsopoulos et al. "User-Centered Gamification: The Case of IoT-Enabled Energy Conservation at Work". In: *International Journal of E-Services and Mobile Applications (IJESMA)* 12.2 (2020), pp. 15–39.
309. Nicole M. Krause et al. "Fact-Checking as Risk Communication: The Multi-Layered Risk of Misinformation in Times of COVID-19". In: *Journal of Risk Research* (2020), pp. 1–8. https://doi.org/10.1080/13669877.2020.1756385.
310. Sarah E. Kreps and Douglas L. Kriner. "The COVID-19 Infodemic and the Efficacy of Interventions Intended to Reduce Misinformation". In: *Public Opinion Quarterly* 86.1 (2022), pp. 162–175. https://doi.org/10.1093/poq/nfab075.
311. Hye-Chung Kum et al. "Enhancing Privacy through an Interactive On-Demand Incremental Information Disclosure Interface: Applying Privacy-by-Design to Record Link age". In: *Fifteenth Symposium on Usable Privacy and Security (SOUPS 2019)*. 2019.
312. Markus Langer et al. "Trust in Artificial Intelligence: Comparing Trust Processes Between Human and Automated Trustees in Light of Unfair Bias". In: *Journal of Business and Psychology* 38.3 (2023), pp. 493–508. https://doi.org/10.1007/s10869-022-09829-9.
313. Matthias Laschke et al. "Overcoming Procrastination with ReMind". In: *Proceedings of the 6th International Conference on Designing Pleasurable Products and Interfaces*. ACM, 2013, pp. 77–85. https://doi.org/10.1145/2513506.2513515. URL: https://dl.acm.org/doi/10.1145/2513506.2513515.
314. Jonathan Lazar, Jinjuan Heidi Feng, and Harry Hochheiser. "Chapter 8—Interviews and Focus Groups". In: *Research Methods in Human Computer Interaction (Second Edition)*. Ed. by Jonathan Lazar, Jinjuan Heidi Feng, and Harry Hochheiser. Morgan Kaufmann, 2017, pp. 187–228. https://doi.org/10.1016/B978-0-12-805390-4.00008-X.
315. Alex Leavitt and Joshua A. Clark. "Upvoting Hurricane Sandy: Event-Based News Production Processes on a Social News Site". In: *Proceedings of the SIGCHI Conference on Human Factors in Computing Systems*. CHI '14. Association for Computing Machinery, 2014, pp. 1495–1504. https://doi.org/10.1145/2556288.2557140.
316. Alex Leavitt and John J. Robinson. "The Role of Information Visibility in Network Gatekeeping: Information Aggregation on Reddit during Crisis Events". In: *Proceedings of the 2017 ACM Conference on Computer Supported Cooperative Work and Social Computing*. CSCW '17. Association for Computing Machinery, 2017, pp. 1246–1261. https://doi.org/10.1145/2998181.2998299.

317. Ulrike Lechner. "Future Security: Processes or Properties?—Research Directions in Cybersecurity". In: *Models, Mindsets, Meta: The What, the How, and the Why Not?* Springer, 2019, pp. 235–246.
318. Jiyoung Lee. "The Effect of Web Add-on Correction and Narrative Correction on Belief in Misinformation Depending on Motivations for Using Social Media". In: *Behaviour & Information Technology* 41.3 (2022), pp. 629–643.
319. Jiyoung Lee and Kim Bissell. "Correcting Vaccine Misinformation on Social Media: The Inadvertent Effects of Repeating Misinformation within Such Corrections on COVID-19 Vaccine Misperceptions". In: *Current Psychology* (2024), pp. 1–13. https://doi.org/10.1007/s12144-024-05651-z.
320. Jiyoung Lee and Kim Bissell. "User Agency-Based versus Machine Agency-Based Misinformation Interventions: The Effects of Commenting and AI Fact-Checking Labeling on Attitudes toward the COVID-19 Vaccination". In: *New Media & Society* (2023), pp. 1–21. https://doi.org/10.1177/14614448231163228.
321. Kate Sangwon Lee. "Explicit Disaster Response Features in Social Media: Safety Check and Community Help Usage on Facebook during Typhoon Mangkhut". In: *Proceedings of the 21st International Conference on Human-Computer Interaction with Mobile Devices and Services.* 2019, pp. 1–12.
322. Seongmin Lee et al. "MisVis: Explaining Web Misinformation Connections via Visual Summary". In: *Extended Abstracts of the 2022CHI Conference on Human Factors in Computing Systems.* CHI EA '22. Association for Computing Machinery, 2022, pp. 1–6. https://doi.org/10.1145/3491101.3519711.
323. Yang W. Lee et al. "AIMQ: A Methodology for Information Quality Assessment". In: *Information & Management* 40.2 (2002), pp. 133–146. https://doi.org/10.1016/S0378-7206(02)00043-5.
324. Amanda Lehman and Sophie Jo Miller. "A Theoretical Conversation about Responses to Information Overload". In: *Information-an International Interdisciplinary Journal* 11.8 (2020). https://doi.org/10.3390/info11080379.
325. Matthew S. Levendusky. "Why Do Partisan Media Polarize Viewers?" In: *American Journal of Political Science* 57.3 (2013), pp. 611–623. https://doi.org/10.1111/ajps.12008.
326. Sarah Levitan et al. "Combining Acoustic-Prosodic, Lexical, and Phonotactic Features for Automatic Deception Detection". In: *Proceedings of Interspeech.* ISCA, 2016, pp. 2006–2010. https://doi.org/10.21437/Interspeech.2016-1519.
327. Stephan Lewandowsky and Sander van der Linden. "Countering Misinformation and Fake News Through Inoculation and Prebunking". In: *European Review of Social Psychology* 32.2 (2021), pp. 348–384. https://doi.org/10.1080/10463283.2021.1876983.
328. Stephan Lewandowsky et al. "Misinformation and Its Correction: Continued Influence and Successful Debiasing". In: *Psychological Science in the Public Interest* 13.3 (2012), pp. 106–131. https://doi.org/10.1177/1529100612451018.
329. Clayton Lewis. *Using the Thinking-Aloud Method in Cognitive Interface Design.* IBM TJ Watson Research Center, 1982.
330. Yang-Jun Li et al. "Health Misinformation on Social Media: A Systematic Literature Review and Future Research Directions". In: *AIS Transactions on Human-Computer Interaction* 14.2 (2022), pp. 116–149. https://doi.org/10.17705/1thci.00164.

331. Raymond Liaw et al. "Maater: Crowdsourcing to Improve Online Journalism". In: *CHI '13 Extended Abstracts on Human Factors in Computing Systems on - CHI EA '13*. ACM Press, 2013, p. 2549. https://doi.org/10.1145/2468356.2468828.
332. Torrin M. Liddell and John K. Kruschke. "Analyzing Ordinal Data with Metric Models: What Could Possibly Go Wrong?" In: *Journal of Experimental Social Psychology* 79 (2018), pp. 328–348. URL: https://www.sciencedirect.com/science/article/pii/S0022103117307746.
333. Helen M Lillie et al. "Using Narratives to Correct Politically Charged Health Misinformation and Address Affective Belief Echoes". In: *Journal of Public Health* (2024), pp. 1–7. https://doi.org/10.1093/pubmed/fdae050.
334. Gionnieve Lim and Simon T. Perrault. "Effects of Automated Misinformation Warning Labels on the Intents to like, Comment and Share Posts". In: *Proceedings of the 11th International Conference on Human-Agent Interaction*. HAI '23. Association for Computing Machinery, 2023, pp. 299–305. https://doi.org/10.1145/3623809.3623856.
335. Gionnieve Lim and Simon T. Perrault. "Intents and Motivations to Like, Comment and Share Posts With Warnings of Misinformation". In: *Proceedings of the 35th Australian Computer-Human Interaction Conference*. ACM, 2023, pp. 108–113. https://doi.org/10.1145/3638380.3638390.
336. Gionnieve Lim and Simon Tangi Perrault. "Local Perceptions and Practices of News Sharing and Fake News". In: *Companion Publication of the 2021 Conference on Computer Supported Cooperative Work and Social Computing*. ACM, 2021, pp. 117–120. https://doi.org/10.1145/3462204.3481767.
337. Yiling Lin, Magda Osman, and Richard Ashcroft. "Nudge: Concept, Effectiveness, and Ethics". In: *Basic and Applied Social Psychology* 39.6 (2017), pp. 293–306. https://doi.org/10.1080/01973533.2017.1356304.
338. Chen Ling, Krishna P. Gummadi, and Savvas Zannettou. ""Learn the Facts about COVID-19": Analyzing the Use of Warning Labels on TikTok Videos". In: *Proceedings of the International AAAI Conference on Web and Social Media* 17 (2023), pp. 554–565. https://doi.org/10.1609/icwsm.v17i1.22168.
339. Bin Liu et al. "Follow My Recommendations: A Personalized Privacy Assistant for Mobile App Permissions". In: *Twelfth Symposium on Usable Privacy and Security (SOUPS 2016)*. 2016, pp. 27–41.
340. Xingyu Liu et al. "Checking the Fact-Checkers: The Role of Source Type, Perceived Credibility, and Individual Differences in Fact-Checking Effectiveness". In: *Communication Research* (2023), pp. 1–28. https://doi.org/10.1177/00936502231206419.
341. Kuan-Chieh Lo et al. "All the Wiser: Fake News Intervention Using User Reading Preferences". In: *Proceedings of the 14th ACM International Conference on Web Search and Data Mining*. ACM, 2021, pp. 1069–1072. https://doi.org/10.1145/3437963.3441696.
342. Eugène Loos, Loredana Ivan, and Donald Leu. ""Save the Pacific Northwest Tree Octopus": A Hoax Revisited. Or: How Vulnerable Are School Children to Fake News?" In: *Information and Learning Science* 119.9/10 (2018), pp. 514–528. https://doi.org/10.1108/ILS-04-2018-0031.
343. Jennifer S. Love, Adam Blumenberg, and Zane Horowitz. "The Parallel Pandemic: Medical Misinformation and COVID-19". In: *Journal of General Internal Medicine* 35.8 (2020), pp. 2435–2436. https://doi.org/10.1007/s11606-020-05897-w.

344. Zhuoran Lu et al. "The Effects of AI-based Credibility Indicators on the Detection and Spread of Misinformation under Social Influence". In: *Proceedings of the ACM on Human-Computer Interaction* 6.CSCW2 (2022), pp. 1–27. https://doi.org/10.1145/3555562.
345. Nathan Malkin. "Waiting Makes the Heart Grow Fonder and the Password Grow Stronger". In: *Symposium on Usable Privacy and Security(SOUPS)-Posters. USENIX Association, Newcastle, UK.* 2013, pp. 1–2.
346. H. B. Mann and D. R. Whitney. "On a Test of Whether One of Two Random Variables Is Stochastically Larger than the Other". In: *The Annals of Mathematical Statistics* 18.1 (1947), pp. 50–60. URL: https://www.jstor.org/stable/2236101.
347. Alexandre Maros et al. "A Study of Misinformation in Audio Messages Shared in WhatsApp Groups". In: *Disinformation in Open Online Media.* Vol. 12887. Springer International Publishing, 2021, pp. 85–100. https://doi.org/10.1007/978-3-030-87031-7_6.
348. Alexandre Maros et al. "Analyzing the Use of Audio Messages in WhatsApp Groups". In: *Proceedings of The Web Conference 2020.* ACM, 2020, pp. 3005–3011. https://doi.org/10.1145/3366423.3380070.
349. Cameron Martel, Mohsen Mosleh, and David Rand. "You're Definitely Wrong, Maybe: Correction Style Has Minimal Effect on Corrections of Misinformation Online". In: *Media and Communication* 9.1 (2021), pp. 120–133. https://doi.org/10.17645/mac.v9i1.3519.
350. Giovanni Da San Martino et al. "Prta: A System to Support the Analysis of Propaganda Techniques in the News". In: *Proceedings of the 58th Annual Meeting of the Association for Computational Linguistics: System Demonstrations.* Association for Computational Linguistics, 2020, pp. 287–293.
351. Azza El-Masri, Martin J. Riedl, and Samuel Woolley. "Audio Misinformation on WhatsApp: A Case Study from Lebanon". In: *Harvard Kennedy School Misinformation Review* 3.4 (2022), pp. 1–13. https://doi.org/10.37016/mr-2020-102.
352. R. McAlpine. *Global English: For Global Business.* Pearson Education New Zealand Limited, 1997.
353. Julian McDougall. "Media Literacy versus Fake News: Critical Thinking, Resilience and Civic Engagement". In: *Media Studies* 10.19 (2019), pp. 29–45. https://doi.org/10.20901/ms.10.19.2.
354. G. Mclaughlin. "SMOG Grading—A New Readability Formula." In: *The Journal of Reading* 12.8 (1969), pp. 639–646.
355. Ben Medeiros and Pawan Singh. "Addressing Misinformation on WhatsApp in India Through Intermediary Liability Policy, Platform Design Modification, and Media Literacy". In: *Journal of Information Policy* 10 (2020), pp. 276–298.
356. Shahan Ali Memon and Kathleen M Carley. "Characterizing Covid-19 Misinformation Communities Using a Novel Twitter Dataset". In: *arXivpreprint* arXiv:2008.00791 (2020).
357. Paul Mena. "Cleaning Up Social Media: The Effect of Warning Labels on Likelihood of Sharing False News on Facebook". In: *Policy and Internet* 12.2 (2020), pp. 165–183. https://doi.org/10.1002/poi3.214.

358. Martin Mende et al. "Fighting Infodemics: Labels as Antidotes to Mis- and Disinformation?!" In: *Journal of Public Policy and Marketing* 43.1, SI (2024), pp. 31–52. https://doi.org/10.1177/07439156231184816.
359. Marcelo Mendoza, Barbara Poblete, and Carlos Castillo. "Twitter under Crisis: Can We Trust What We RT?" In: *Proceedings of the First Workshop on Social Media Analytics*. ACM, 2010, pp. 71–79. https://doi.org/10.1145/1964858.1964869. URL: https://dl.acm.org/doi/10.1145/1964858.1964869.
360. Meta. *About Fact-Checking on Facebook*. 2022. URL: https://www.facebook.com/business/help/2593586717571940?id=673052479947730.
361. Miriam J. Metzger and Andrew J. Flanagin. "Credibility and Trust of Information in Online Environments: The Use of Cognitive Heuristics". In: *Journal of Pragmatics*. Biases and Constraints in Communication: Argumentation, Persuasion and Manipulation 59 (2013), pp. 210–220. https://doi.org/10.1016/j.pragma.2013.07.012.
362. Nicholas Micallef and Nalin Asanka Gamagedara Arachchilage. "A Serious Game Design : Nudging Users' Memorability of Security Questions". In: *Australasian Conference on Information Systems*. 2017, pp. 1–11.
363. Nicholas Micallef et al. "Fakey: A Game Intervention to Improve News Literacy on Social Media". In: *Proc. ACM Hum.-Comput. Interact.* 5.CSCW1 (2021). https://doi.org/10.1145/3449080.
364. Nicholas Micallef et al. "Stop Annoying Me! An Empirical Investigation of the Usability of App Privacy Notifications". In: *Proceedings of the 29thAustralian Conference on Computer-Human Interaction*. 2017, pp. 371–375.
365. Paul Mihailidis and Samantha Viotty. "Spreadable Spectacle in Digital Culture: Civic Expression, Fake News, and the Role of Media Literacies in "Post-Fact" Society". In: *American Behavioral Scientist* 61.4 (2017), pp. 441–454. https://doi.org/10.1177/0002764217701217.
366. Elena Milani, Emma Weitkamp, and Peter Webb. "The Visual Vaccine Debate on Twitter: A Social Network Analysis". In: *Media and Communication* 8.2 (2020), pp. 364–375. https://doi.org/10.17645/mac.v8i2.2847.
367. Thomas Mildner et al. "About Engaging and Governing Strategies: A Thematic Analysis of Dark Patterns in Social Networking Services". In: *Proceedings of the 2023 CHI Conference on Human Factors in Computing Systems*. ACM, 2023, pp. 1–15. https://doi.org/10.1145/3544548.3580695.
368. Stuart Mills. "Personalized Nudging". In: *Behavioural Public Policy* 6.1 (2022), pp. 150–159. https://doi.org/10.1017/bpp.2020.7. URL: https://www.cambridge.org/core/journals/behavioural-public-policy/article/personalized-nudging/E854A04226DEA94B623ECA2ACF64C8D0.
369. Ashlee Milton et al. ""I See Me Here": Mental Health Content, Community, and Algorithmic Curation on TikTok". In: *Proceedings of the2023 CHI Conference on Human Factors in Computing Systems*. CHI '23. Association for Computing Machinery, 2023, pp. 1–17. https://doi.org/10.1145/3544548.3581489.
370. Yisroel Mirsky and Wenke Lee. "The Creation and Detection of Deepfakes: A Survey". In: *ACM Computing Surveys* 54.1 (2022), pp. 1–41. https://doi.org/10.1145/3425780.
371. Shalini Misra, Patrick Roberts, and Matthew Rhodes. "Information Overload, Stress, and Emergency Managerial Thinking". In: *International Journal of Disaster Risk Reduction* 51 (2020), p. 101762. https://doi.org/10.1016/j.ijdrr.2020.101762.

372. Maria D. Molina et al. " "Fake News" Is Not Simply False Information: A Concept Explication and Taxonomy of Online Content". In: *American Behavioral Scientist* 65.2 (2021), pp. 180–212. https://doi.org/10.1177/0002764219878224. URL: http://journals.sagepub.com/doi/10.1177/0002764219878224.

373. Won-Ki Moon, Myojung Chung, and Mo Jones-Jang. "How Can We Fight Partisan Biases in the COVID-19 Pandemic? AI Source Labels on Fact-checking Messages Reduce Motivated Reasoning". In: *Mass Communication and Society* 26.4 (2023), pp. 646–670. https://doi.org/10.1080/15205436.2022.2097926.

374. Patricia L. Moravec, Antino Kim, and Alan R. Dennis. "Appealing to Sense and Sensibility: System 1 and System 2 Interventions for Fake News on Social Media". In: *Information Systems Research* 31.3 (2020), pp. 987–1006. https://doi.org/10.1287/isre.2020.0927.

375. Meredith Ringel Morris et al. "Tweeting Is Believing? Understanding Microblog Credibility Perceptions". In: *Proceedings of the ACM2012 Conference on Computer Supported Cooperative Work*. CSCW '12. Association for Computing Machinery, 2012, pp. 441–450. https://doi.org/10.1145/2145204.2145274.

376. Fred Morstatter et al. "From Alt-Right to Alt-Rechts: Twitter Analysis of the 2017 German Federal Election". In: *Companion of the The Web Conference 2018 on The Web Conference 2018—WWW '18*. ACM Press, 2018, pp. 621–628. https://doi.org/10.1145/3184558.3188733. URL: http://dl.acm.org/citation.cfm?doid=3184558.3188733.

377. Dr Philipp Müller and Nora Denner. *Was tun gegen Fake News? Eine Analyse anhand der Entstehungsbedingungen und Wirkweisen gezielter Falschmeldungen im Internet : Kurzgutachten im Auftrag der Friedrich-Naumann-Stiftung für die Freiheit*. Tech. rep. Friedrich-Naumann-Stiftung für die Freiheit, 2019, pp. 1–32.

378. Günter Müller, Thomas G. Koslowski, and Rafael Accorsi. "Resilience - A New Research Field in Business Information Systems?" In: *Business Information Systems Workshops*. Ed. by Witold Abramowicz. Springer Berlin Heidelberg, 2013, pp. 3–14.

379. Randall Munroe. *XKCD #936: "Password Strength"*. 2011. URL: https://www.explainxkcd.com/wiki/index.php/936:_Password_Strength.

380. Muhammad Naeem and Ozuem Wilson. "Understanding Misinformation and Rumors That Generated Panic Buying as a Social Practice during COVID-19 Pandemic: Evidence from Twitter, YouTube and Focus Group Interviews". In: *Information Technology & People* 35.7 (2022), pp. 2140–2166.

381. Anja Naumann et al. "Intuitive Use of User Interfaces: Defining a Vague Concept". In: *Engineering Psychology and Cognitive Ergonomics*. Ed. by Don Harris. Springer Berlin Heidelberg, 2007, pp. 128–136.

382. Elmie Nekmat. "Nudge Effect of Fact-Check Alerts: Source Influence and Media Skepticism on Sharing of News Misinformation in Social Media". In: *Social Media + Society* 6.1 (2020), pp. 8–14. https://doi.org/10.1177/2056305119897322.

383. Adina Nerghes, Peter Kerkhof, and Iina Hellsten. "Early Public Responses to the Zika-Virus on YouTube: Prevalence of and Differences Between Conspiracy Theory and Informational Videos". In: *Proceedings of the 10th ACM Conference on Web Science*. WebSci '18. Association for Computing Machinery, 2018, pp. 127–134. https://doi.org/10.1145/3201064.3201086.

384. Lynnette Hui Xian Ng and Jia Yuan Loke. "Analyzing Public Opinion and Misinformation in a COVID-19 Telegram Group Chat". In: *IEEE Internet Computing* 25.2 (2021), pp. 84–91. https://doi.org/10.1109/MIC.2020.3040516.
385. James Nicholson, Lynne Coventry, and Pam Briggs. "Can We Fight Social Engineering Attacks by Social Means? Assessing Social Salience as a Means to Improve Phish Detection". In: *Thirteenth Symposium on Usable Privacy and Security (SOUPS 2017)*. 2017, pp. 285–298.
386. Conrad Nied et al. "Alternative Narratives of Crisis Events: Communities and Social Botnets Engaged on Social Media". In: *CSCW2017—Companion of the 2017 ACM Conference on Computer Supported Cooperative Work and Social Computing* (2017), pp. 263–266. https://doi.org/10.1145/3022198.3026307.
387. Shuo Niu et al. "Building Credibility, Trust, and Safety on Video-Sharing Platforms". In: *Extended Abstracts of the 2023 CHI Conference on Human Factors in Computing Systems*. CHI EA '23. Association for Computing Machinery, 2023, pp. 1–7. https://doi.org/10.1145/3544549.3573809.
388. Michael I. Norton, Daniel Mochon, and Dan Ariely. "The IKEA Effect: When Labor Leads to Love". In: *Journal of Consumer Psychology* 22.3 (2012), pp. 453–460. https://doi.org/10.1016/j.jcps.2011.08.002.
389. Felix Nwobi and Felix Akanno. "Power Comparison of ANOVA and Kruskal-Wallis Tests When Error Assumptions Are Violated". In: *Advances in Methodology and Statistics* 18.2 (2021). https://doi.org/10.51936/ltgt2135. URL: https://mz.mf.uni-lj.si/article/view/214.
390. Thomas Nygren and Mona Guath. "Swedish Teenagers' Difficulties and Abilities to Determine Digital News Credibility". In: *Nordicom Review* 40.1 (2019), pp. 23–42. https://doi.org/10.2478/nor-2019-0002.
391. Brendan Nyhan and Jason Reifler. "When Corrections Fail: The Persistence of Political Misperceptions". In: *Political Behavior* 32.2 (2010), pp. 303–330. https://doi.org/10.1007/s11109-010-9112-2.
392. Niall J. O'Sullivan et al. "The Unintentional Spread of Misinformation on 'TikTok'; A Paediatric Urological Perspective". In: *Journal of Pediatric Urology* 18.3 (2022), pp. 371–375. https://doi.org/10.1016/j.jpurol.2022.03.001.
393. Katherine O'Toole. "Collaborative Creativity in TikTok Music Duets". In: *Proceedings of the 2023 CHI Conference on Human Factors in Computing Systems*. CHI '23. Association for Computing Machinery, 2023, pp. 1–16. https://doi.org/10.1145/3544548.3581380.
394. Jacques Ophoff and Frauke Dietz. "Using Gamification to Improve Information Security Behavior: A Password Strength Experiment". In: *IFIP Advances in Information and Communication Technology* 557 (2019), pp. 157–169. https://doi.org/10.1007/978-3-030-23451-5_12.
395. Mathias Osmundsen et al. "Partisan Polarization Is the Primary Psychological Motivation behind Political Fake News Sharing on Twitter". In: *American Political Science Review* 115.3 (2021), pp. 999–1015. https://doi.org/10.1017/S0003055421000290.
396. Eva Ostertagová, Oskar Ostertag, and Jozef Kováč. "Methodology and Application of the Kruskal-Wallis Test". In: *Applied Mechanics and Materials* 611 (2014), pp. 115–120. https://doi.org/10.4028/www.scientific.net/AMM.611.115. URL: https://www.scientific.net/AMM.611.115.

397. Cristina Laura Oyarzun et al. "Automatic Segmentation of the Structures in the Nasal Cavity and the Ethmoidal Sinus for the Quantification of Nasal Septal Deviations". In: *Medical Imaging 2021:Computer-Aided Diagnosis*. Vol. 11597. SPIE, 2021, pp. 582–591. https://doi.org/10.1117/12.2582251.
398. Cristina Laura Oyarzun et al. "Web-Based Prostate Visualization Tool". In: *Current Directions in Biomedical Engineering* 6.3 (2020), pp. 563–566. https://doi.org/10.1515/cdbme-2020-3144.
399. Elise Özalp, Katrin Hartwig, and Christian Reuter. "Trends in Explainable Artificial Intelligence for Non-Experts". In: *Artificial Intelligence—Limits and Prospects*. Ed. by Peter Klimczak and Christer Petersen. 4. Transcript Verlag, 2023, pp. 223–243. URL: https://doi.org/10.14361/9783839457320.
400. Pinar Ozturk, Huaye Li, and Yasuaki Sakamoto. "Combating Rumor Spread on Social Media: The Effectiveness of Refutation and Warning". In: *2015 48th Hawaii International Conference on System Sciences*. IEEE, 2015, pp. 2406–2414. https://doi.org/10.1109/HICSS.2015.288.
401. Marinella Paciello, Giuseppe Corbelli, and Francesca D'Errico. "The Role of Self-Efficacy Beliefs in Dealing with Misinformation among Adolescents". In: *Frontiers in Psychology* 14 (2023), pp. 1–7.
402. Rosemary Pacini and Seymour Epstein. "The Relation of Rational and Experiential Information Processing Styles to Personality, Basic Beliefs, and the Ratio-Bias Phenomenon." In: *Journal of personality and social psychology* 76.6 (1999), p. 972.
403. Matthew J. Page et al. "The PRISMA 2020 Statement: An Updated Guideline for Reporting Systematic Reviews". In: *Systematic Reviews* 10.1 (2021), p. 89. https://doi.org/10.1186/s13643-021-01626-4.
404. Leysia Palen and Kenneth M. Anderson. "Crisis Informatics—New Data for Extraordinary Times". In: *Science* 353.6296 (2016), pp. 224–225. https://doi.org/10.1126/science.aag2579. URL: https://www.science.org/doi/full/10.1126/science.aag2579.
405. Leysia Palen et al. "Crisis in a Networked World: Features of Computer-Mediated Communication in the April 16, 2007, Virginia Tech Event". In: *Social Science Computer Review* 27.4 (2009), pp. 467–480. https://doi.org/10.1177/0894439309332302.
406. Olga Papadopoulou et al. "MeVer NetworkX: Network Analysis and Visualization for Tracing Disinformation". In: *Future Internet* 14.5 (2022), p. 147. https://doi.org/10.3390/fi14050147.
407. Orestis Papakyriakopoulos and Ellen Goodman. "The Impact of Twitter Labels on Misinformation Spread and User Engagement: Lessons from Trump's Election Tweets". In: *Proceedings of the ACM Web Conference 2022*. WWW '22. Association for Computing Machinery, 2022, pp. 2541–2551. https://doi.org/10.1145/3485447.3512126.
408. Concetta Papapicco, Isabella Lamanna, and Francesca D'Errico. "Adolescents' Vulnerability to Fake News and to Racial Hoaxes: A Qualitative Analysis on Italian Sample". In: *Multimodal Technologies and Interaction* 6.3 (2022), p. 20. https://doi.org/10.3390/mti6030020.
409. Saumya Pareek and Jorge Goncalves. "Peer-Supplied Credibility Labels as an Online Misinformation Intervention". In: *International Journal of Human-Computer Studies* 188 (2024), pp. 1–17. https://doi.org/10.1016/j.ijhcs.2024.103276.
410. Sungkyu Park et al. "An Experimental Study to Understand User Experience and Perception Bias Occurred by Fact-checking Messages". In: *Proceedings of the Web*

Conference 2021. ACM, 2021, pp. 2769–2780. https://doi.org/10.1145/3442381.3450121.
411. Demetris Paschalides et al. *Check-It: A Plugin for Detecting and Reducing the Spread of Fake News and Misinformation on the Web*. 2019.
412. Irene V. Pasquetto et al. "Social Debunking of Misinformation on WhatsApp: The Case for Strong and In-group Ties". In: *Proceedings of the ACM on Human-Computer Interaction* 6.CSCW1 (2022), pp. 1–35. https://doi.org/10.1145/3512964.
413. Eyal Pe'er et al. "Do Minorities like Nudges? The Role of Group Norms in Attitudes towards Behavioral Policy". In: *Judgment and Decision Making* 14.1 (2019), p. 40.
414. Eyal Pe'er et al. "Nudge Me Right: Personalizing Online Nudges to People's Decision-Making Styles". In: *SSRN Electronic Journal* (2019). https://doi.org/10.2139/ssrn.3324907.
415. Sarah Pearman et al. "Why People (Don't) Use Password Managers Effectively". In: *Fifteenth Symposium on Usable Privacy and Security(SOUPS)* (2019).
416. Eyal Peer et al. "Beyond the Turk: Alternative Platforms for Crowdsourcing Behavioral Research". In: *Journal of Experimental Social Psychology* 70 (2017), pp. 153–163. https://doi.org/10.1016/j.jesp.2017.01.006. URL: https://www.sciencedirect.com/science/article/pii/S0022103116303201.
417. Ken Peffers et al. "The Design Science Research Process: A Model for Producing and Presenting Information Systems Research". In: *Proceedings of the First International Conference on Design Science Research in Information Systems and Technology (DESRIST 2006)*, Claremont, CA, USA. 2006, pp. 83–106.
418. Gordon Pennycook and David G. Rand. "Accuracy Prompts Are a Replicable and Generalizable Approach for Reducing the Spread of Misinformation". In: *Nature Communications* 13.2333 (2022), pp. 1–12. https://doi.org/10.1038/s41467-022-30073-5.
419. Gordon Pennycook and David G. Rand. "Lazy, Not Biased: Susceptibility to Partisan Fake News Is Better Explained by Lack of Reasoning than by Motivated Reasoning". In: *Cognition* 188 (2019), pp. 39–50. https://doi.org/10.1016/j.cognition.2018.06.011.
420. Gordon Pennycook and David G. Rand. "Who Falls for Fake News? The Roles of Bullshit Receptivity, Overclaiming, Familiarity, and Analytic Thinking". In: *Journal of Personality* 88.2 (2020), pp. 185–200. https://doi.org/10.1111/jopy.12476.
421. Gordon Pennycook et al. "A Practical Guide to Doing Behavioral Research on Fake News and Misinformation". In: *Collabra: Psychology* 7.1 (2021), p. 25293. https://doi.org/10.1525/collabra.25293.
422. Gordon Pennycook et al. "Fighting COVID-19 Misinformation on Social Media: Experimental Evidence for a Scalable Accuracy-Nudge Intervention". In: *Psychological Science* 31.7 (2020), pp. 770–780. https://doi.org/10.1177/0956797620939054.
423. Gordon Pennycook et al. "Shifting Attention to Accuracy Can Reduce Misinformation Online". In: *Nature* 592.7855 (2021), pp. 590–595. https://doi.org/10.1038/s41586-021-03344-2.
424. Gordon Pennycook et al. "The Implied Truth Effect: Attaching Warnings to a Subset of Fake News Headlines Increases Perceived Accuracy of Headlines Without Warnings". In: *Management Science* 66.11 (2020), pp. 4944–4957. https://doi.org/10.1287/mnsc.2019.3478.

425. Ana Pérez et al. "Fostering Teenagers' Assessment of Information Reliability: Effects of a Classroom Intervention Focused on Critical Source Dimensions". In: *Learning and Instruction* 58 (2018), pp. 53–64. https://doi.org/10.1016/j.learninstruc.2018.04.006.
426. Kimberly Perzel and David Kane. "Usability Patterns for Applications on the World Wide Web". In: *Proceedings of the Pattern Languages of Programming Conference*. Vol. 99. 1999.
427. Raunak M. Pillai and Lisa K. Fazio. "Explaining Why Headlines Are True or False Reduces Intentions to Share False Information". In: *Collabra-Psychology* 9.87617 (2023), pp. 1–11. https://doi.org/10.1525/collabra.87617.
428. Linda Plotnick et al. *Real or Fake? User Behavior and Attitudes Related to Determining the Veracity of Social Media Posts*. 2019. https://doi.org/10.48550/arXiv.1904.03989.
429. Sara Pluviano, Caroline Watt, and Sergio Della Sala. "Misinformation Lingers in Memory: Failure of Three pro-Vaccination Strategies". In: *PLOS ONE* 12.7 (2017), e0181640. https://doi.org/10.1371/journal.pone.0181640.
430. Erez Porat, Ina Blau, and Azy Barak. "Measuring Digital Literacies: Junior High-School Students' Perceived Competencies versus Actual Performance". In: *Computers & Education* 126 (2018), pp. 23–36. https://doi.org/10.1016/j.compedu.2018.06.030.
431. Ethan Porter, Yamil Velez, and Thomas J. Wood. "Factual Corrections Eliminate False Beliefs about COVID-19 Vaccines". In: *Public Opinion Quarterly* 86.3 (2022), pp. 762–773. https://doi.org/10.1093/poq/nfac034.
432. Ethan Porter and Thomas J. Wood. "Political Misinformation and Factual Corrections on the Facebook News Feed: Experimental Evidence". In: *Journal of Politics* 84.3 (2022), pp. 1812–1817. https://doi.org/10.1086/719271.
433. Martin Potthast et al. "A Stylometric Inquiry into Hyperpartisan and Fake News". In: *Proceedings of the 56th Annual Meeting of the Association for Computational Linguistics (Volume 1: Long Papers)*. Association for Computational Linguistics, 2018, pp. 231–240. https://doi.org/10.18653/v1/P18-1022.
434. Martin Potthast et al. "Clickbait Detection". In: *Proceedings of Advances in Information Retrieval*. Ed. by Nicola Ferro et al. Lecture Notes in Computer Science. Springer International Publishing, 2016, pp. 810–817. https://doi.org/10.1007/978-3-319-30671-1_72.
435. Pardis Pourghomi et al. "How to Stop Spread of Misinformation on Social Media: Facebook Plans vs. Right-Click Authenticate Approach". In: *2017 International Conference on Engineering & MIS (ICEMIS)*. IEEE, 2017, pp. 1–8. https://doi.org/10.1109/ICEMIS.2017.8272957.
436. Jenny Preece et al. *Human-Computer Interaction*. Addison-Wesley Longman, 1994. https://doi.org/10.5555/561701. URL: https://dl.acm.org/doi/abs/10.5555/561701.
437. Clara Pretus et al. "The Misleading Count: An Identity-Based Intervention to Counter Partisan Misinformation Sharing". In: *Philisophical Transactions of the Royal Society B- Biological Sciences* 379.20230040 (2024), pp. 1–9. https://doi.org/10.1098/rstb.2023.0040.
438. Toby Prike, Lucy H. Butler, and Ullrich K. H. Ecker. "Source-Credibility Information and Social Norms Improve Truth Discernment and Reduce Engagement with Misinformation Online". In: *Scientific Reports* 14.1 (2024), pp. 1–11. https://doi.org/10.1038/s41598-024-57560-7.

439. Toby Prike and Ullrich K. H. Ecker. "Effective Correction of Misinformation". In: *Current Opinion in Psychology* 54.101712 (2023), pp. 1–6. https://doi.org/10.1016/j.copsyc.2023.101712.
440. Jeffrey T. Prince and Scott Wallsten. "How Much Is Privacy Worth around the World and across Platforms?" In: *Journal of Economics & Management Strategy* 31.4 (2022), pp. 841–861. https://doi.org/10.1111/jems.12481. URL: https://onlinelibrary.wiley.com/doi/abs/10.1111/jems.12481.
441. Aditya Kumar Purohit and Agnes Helme-Guizon. "The Power of Digital Nudge: Moving Towards Public Transportation and Debunking False Beliefs. A Conceptual Framework: Structured Abstract". In: *22nd Academy of Marketing Science World Marketing Congress*. 2019. URL: https://libra.unine.ch/handle/123456789/26892.
442. Cynthia Putnam et al. "Adaptation of the System Usability Scale for User Testing with Children". In: *Extended Abstracts of the 2020 CHI Conference on Human Factors in Computing Systems*. ACM, 2020, pp. 1–7. https://doi.org/10.1145/3334480.3382840.
443. Python-Software-Foundation. *Python 3.9.1*. 2020. URL: https://www.python.org/.
444. Sijia Qian, Cuihua Shen, and Jingwen Zhang. "Fighting Cheapfakes: Using a Digital Media Literacy Intervention to Motivate Reverse Search of out-of-Context Visual Misinformation". In: *Journal of Computer-Mediated Communication* 28.1 (2023), pp. 1–12. https://doi.org/10.1093/jcmc/zmac024.
445. Muireann Quigley. "Nudging for Health: On Public Policy and Designing Choice Architecture". In: *Medical Law Review* 21.4 (2013), pp. 588–621. https://doi.org/10.1093/medlaw/fwt022.
446. Alec Radford et al. *Robust Speech Recognition via Large-Scale Weak Supervision*. 2022. https://doi.org/10.48550/arXiv.2212.04356.
447. Dipti P. Rana, Isha Agarwal, and Anjali More. "A Review of Techniques to Combat The Peril of Fake News". In: *Proceedings of the 2018 4thInternational Conference on Computing Communication and Automation(ICCCA)*. 2018, pp. 1–7. https://doi.org/10.1109/CCAA.2018.8777676.
448. George Raptis et al. "Better, Funner, Stronger: A Gameful Approach to Nudge People into Making Less Predictable Graphical Password Choices". In: *Proceedings of the 2021 CHI Conference on Human Factors in Computing Systems*. 2021, pp. 1–17.
449. Steve Rathje et al. "Accuracy and Social Motivations Shape Judgements of (Mis)Information". In: *Nature Human Behaviour* 7.6 (2023), pp. 892–903. https://doi.org/10.1038/s41562-023-01540-w. URL: https://www.nature.com/articles/s41562-023-01540-w.
450. Shubhadip Ray et al. "Dynamics in Data Privacy and Sharing Economics". In: *IEEE Transactions on Technology and Society* 2.3 (2021), pp. 114–115. https://doi.org/10.1109/TTS.2021.3077534. URL: https://ieeexplore.ieee.org/document/9422747.
451. Christopher G. Reddick. *Public Administration and Information Technology*. Jones & Bartlett Learning, 2012.
452. Thomas Reinhold and Christian Reuter. "Zur Debatte über die Einhegung eines Cyberwars: Analyse militärischer Cyberaktivitäten im Krieg Russlands gegen die Ukraine". In: *Zeitschrift für Friedens -und Konfliktforschung* 12.1 (2023), pp. 135–149. https://doi.org/10.1007/s42597-023-00094-y.
453. Lucia A. Reisch and Cass R. Sunstein. "Do Europeans like Nudges?" In: *Judgment and Decision making* 11.4 (2016), pp. 310–325.

454. Karen Renaud and Verena Zimmerman. "Enriched Nudges Lead to Stronger Password Replacements ... but Implement Mindfully". In: *2017 Information Security for South Africa (ISSA)*. IEEE, 2017, pp. 1–9. https://doi.org/10.1109/ISSA.2017.8251779. URL: https://ieeexplore.ieee.org/document/8251779/.
455. Karen Renaud and Verena Zimmermann. "Ethical Guidelines for Nudging in Information Security & Privacy". In: *International Journal of Human-Computer Studies* 120 (2018), pp. 22–35.
456. Karen Renaud and Verena Zimmermann. "Nudging Folks towards Stronger Password Choices: Providing Certainty Is the Key". In: *Behavioural Public Policy* 3.02 (2018), pp. 228–258. https://doi.org/10.1017/bpp.2018.3.
457. Karen Renaud et al. "Lessons Learned from Evaluating Eight Password Nudges in the Wild". In: *The LASER Workshop: Learning from Authoritative Security Experiment Results (LASER 2017)*. 2017, pp. 25–37.
458. Gustavo Resende et al. "(Mis)Information Dissemination in WhatsApp: Gathering, Analyzing and Countermeasures". In: *The World Wide Web Conference*. ACM, 2019, pp. 818–828. https://doi.org/10.1145/3308558.3313688.
459. Christian Reuter. *A European Perspective on Crisis Informatics: Citizens' and Authorities' Attitudes Towards Social Media for Public Safety and Security*. The Radboud University Thesis Repository, 2022.
460. Christian Reuter and Marc-André Kaufhold. "Fifteen Years of Social Media in Emergencies: A Retrospective Review and Future Directions for Crisis Informatics". In: *Journal of Contingencies and Crisis Management* 26.1 (2018), pp. 41–57. https://doi.org/10.1111/1468-5973.12196.
461. Christian Reuter, Thomas Ludwig, and Volkmar Pipek. "Ad Hoc Participation in Situation Assessment: Supporting Mobile Collaboration in Emergencies". In: *ACM Trans. Comput.-Hum. Interact.* 21.5 (2014). https://doi.org/10.1145/2651365.
462. Christian Reuter et al. "Fake News Perception in Germany: A Representative Study of People's Attitudes and Approaches to Counteract Disinformation". In: *Proceedings of the International Conference on Wirtschaftsinformatik (WI)*. AIS, 2019, pp. 1–15. URL: https://aisel.aisnet.org/wi2019/track09/papers/5/.
463. Christian Reuter et al. "Social Media in Emergencies: A Representative Study on Citizens' Perception in Germany". In: *Proc. ACM Hum.-Comput. Interact.* 1.CSCW (2017). https://doi.org/10.1145/3134725.
464. Patrick R. Rich and Maria S. Zaragoza. "Correcting Misinformation in News Stories: An Investigation of Correction Timing and Correction Durability". In: *Journal of Applied Research in Memory and Cognition* 9.3 (2020), pp. 310–322. https://doi.org/10.1016/j.jarmac.2020.04.001.
465. Adam S. Richards and John A. Banas. "Inoculating Against Reactance to Persuasive Health Messages". In: *Health Communication* 30.5 (2015), pp. 451–460. https://doi.org/10.1080/10410236.2013.867005.
466. Elisabeth Roberts, John Farrington, and Sarah Skerratt. "Evaluating New Digital Technologies Through a Framework of Resilience". In: *Scottish Geographical Journal* 131.3-4 (2015), pp. 253–264. https://doi.org/10.1080/14702541.2015.1068947.
467. Jon Roozenbeek, Eileen Culloty, and Jane Suiter. "Countering Misinformation Evidence, Knowledge Gaps, and Implications of Current Interventions". In: *European Psychologist* 28.3 (2023), pp. 189–205. https://doi.org/10.1027/1016-9040/a000492.

468. Jon Roozenbeek and Sander van der Linden. "Fake News Game Confers Psychological Resistance against Online Misinformation". In: *Palgrave Communications* 5.1 (2019), p. 65. https://doi.org/10.1057/s41599-019-0279-9.
469. Jon Roozenbeek and Sander van der Linden. "The Fake News Game: Actively Inoculating against the Risk of Misinformation". In: *Journal of Risk Research* 22 (2018), pp. 570–580. https://doi.org/10.1080/13669877.2018.1443491.
470. Victoria L. Rubin and Tatiana Lukoianova. "Truth and Deception at the Rhetorical Structure Level". In: *Journal of the Association for Information Science and Technology* 66.5 (2015), pp. 905–917. https://doi.org/10.1002/asi.23216.
471. Eva Ruegenhagen and Bernard Rummel. "System Usability Scale–Jetzt Auch Auf Deutsch". In: *SAP Global Design Enablement Team* (2013).
472. Margie Ruffin, Gang Wang, and Kirill Levchenko. "Explaining Why Fake Photos Are Fake: Does It Work?" In: *Proc. ACM Hum.-Comput. Interact.* 7.GROUP (2022), pp. 1–22. https://doi.org/10.1145/3567558.
473. Hilda Ruokolainen and Gunilla Widén. "Conceptualising Misinformation in the Context of Asylum Seekers". In: *Information Processing & Management* 57.3 (2020), p. 102127. https://doi.org/10.1016/j.ipm.2019.102127.
474. Morva Saaty et al. *Integrating Social Media into the Design Process*. 2022.
475. McKenzie Sadeghi, Jack Brewster, and Macrina Wang. *X's Unchecked Propaganda: Engagement Soared by 70% for Russian, Chinese and Iranian Disinformation Sources Following a Change by Elon Musk - Misinformation Monitor: September 2023*. 2023. URL: https://www.newsguardtech.com/misinformation-monitor/september-2023.
476. Fadi Safieddine, Wassim Masri, and Pardis Pourghomi. "Corporate Responsibility in Combating Online Misinformation". In: *International Journal of Advanced Computer Science and Applications* 7.2 (2016), pp. 126–132. https://doi.org/10.14569/IJACSA.2016.070217.
477. Nina Sakhnini and Debaleena Chattopadhyay. "A Review of Smartphone Fact-Checking Apps and Their (Non) Use Among Older Adults". In: *Adjunct Publication of the 24th International Conference on Human-Computer Interaction with Mobile Devices and Services*. MobileHCI '22. Association for Computing Machinery, 2022, pp. 1–8. https://doi.org/10.1145/3528575.3551448.
478. Joni Salminen et al. "Personas and Analytics: A Comparative User Study of Efficiency and Effectiveness for a User Identification Task". In: *Proceedings of the 2020 CHI Conference on Human Factors in Computing Systems*. ACM, 2020, pp. 1–13. https://doi.org/10.1145/3313831.3376770.
479. Emily Saltz, Claire R Leibowicz, and Claire Wardle. "Encounters with Visual Misinformation and Labels Across Platforms: An Interview and Diary Study to Inform Ecosystem Approaches to Misinformation Interventions". In: *Extended Abstracts of the 2021 CHI Conference on Human Factors in Computing Systems*. ACM, 2021, pp. 1–6. https://doi.org/10.1145/3411763.3451807.
480. Emily Saltz et al. "Misinformation Interventions Are Common, Divisive, and Poorly Understood". In: *Harvard Kennedy School Misinformation Review* 2.5 (2021), pp. 1–25. https://doi.org/10.37016/mr-2020-81.
481. Carmen Sanchez and David Dunning. "Overconfidence among Beginners: Is a Little Learning a Dangerous Thing?" In: *Journal of Personality and Social Psychology* 114.1 (2018), pp. 10–28. https://doi.org/10.1037/pspa0000102.

482. Jasmyne A. Sanderson et al. "Listening to Misinformation While Driving: Cognitive Load and the Effectiveness of (Repeated) Corrections". In: *Journal of Applied Research in Memory and Cognition* 12.3 (2023), pp. 325–334. https://doi.org/10.1037/mac0000057.
483. Angeline Sangalang, Yotam Ophir, and Joseph N. Cappella. "The Potential for Narrative Correctives to Combat Misinformation". In: *Journal of Communication* 69.3 (2019), pp. 298–319. https://doi.org/10.1093/joc/jqz014.
484. M. A. Sasse and I. Flechais. "Usable Security: Why Do We Need It? How Do We Get It?" In: *Security and Usability: Designing Secure Systems That People Can Use*. Ed. by L. F. Cranor and S. Garfinkel. O'Reilly, 2005, pp. 13–30.
485. Florian Saurwein and Charlotte Spencer-Smith. "Combating Disinformation on Social Media: Multilevel Governance and Distributed Accountability in Europe". In: *Digital Journalism* 8.6 (2020), pp. 820–841. https://doi.org/10.1080/21670811.2020.1765401.
486. Rosa Scardigno and Giuseppe Mininni. "The Rhetoric Side of Fake News: A New Weapon for Anti-Politics?" In: *World Futures* 76.2 (2020), pp. 81–101. https://doi.org/10.1080/02604027.2019.1703158. URL: https://www.tandfonline.com/doi/full/10.1080/02604027.2019.1703158.
487. Anastasia Schaadhardt et al. ""Laughing so I Don't Cry": How TikTok Users Employ Humor and Compassion to Connect around Psychiatric Hospitalization". In: *Proceedings of the 2023 CHI Conference on Human Factors in Computing Systems*. CHI '23. Association for Computing Machinery, 2023, pp. 1–13. https://doi.org/10.1145/3544548.3581559.
488. Leonie Schaewitz and Nicole Kramer. "Combating Disinformation: Effects of Timing and Correction Format on Factual Knowledge and Personal Beliefs". In: *Multidisciplinary International Symposium on Disinformation in Open Online Media*. Ed. by Max VanDuijn et al. Vol. 12259. Springer, 2020, pp. 233–245. https://doi.org/10.1007/978-3-030-61841-4_16.
489. Lisa Scharrer, Vanessa Pape, and Marc Stadtler. "Watch Out: Fake! How Warning Labels Affect Laypeople's Evaluation of Simplified Scientific Misinformation". In: *Discourse Processes* 59.8 (2022), pp. 575–590. https://doi.org/10.1080/0163853X.2022.2096364.
490. Philipp Schmid and Cornelia Betsch. "Benefits and Pitfalls of Debunking Interventions to Counter mRNA Vaccination Misinformation during the COVID-19 Pandemic". In: *Science Communication* 44.5 (2022), pp. 531–558. https://doi.org/10.1177/10755470221129608.
491. Stefka Schmid et al. "Digital Resilience in Dealing with Misinformation on Social Media during COVID-19 A Web Application to Assist Users in Crises". In: *Information Systems Frontiers* 26.2 (2024), pp. 477–499. https://doi.org/10.1007/s10796-022-10347-5.
492. Hannah Schmid-Petri et al. "The Multilingual Twitter Discourse on Vaccination in Germany During the Covid-19 Pandemic". In: *Media and Communication* 11.1 (2023), pp. 293–305. https://doi.org/10.17645/mac.v11i1.6058. URL: https://www.proquest.com/docview/2800736452/abstract/27D66BFFC6E64DE1PQ/1.
493. Andreas T. Schmidt and Bart Engelen. "The Ethics of Nudging: An Overview". In: *Philosophy Compass* 15.4 (2020). https://doi.org/10.1111/phc3.12658.

494. Verena Schoenmueller, Simon J. Blanchard, and Gita V. Johar. *Empowering Fake-News Mitigation: Insights from Sharers' Social Media Post-Histories*. 2023. URL: http://arxiv.org/abs/2203.10560.
495. Sam Scholefield and Lynsay A Shepherd. "Gamification Techniques for Raising Cyber Security Awareness". In: *International Conference on HCI for Cybersecurity, Privacy and Trust (HCI-CPT)*. 2019.
496. Charlotte Schöning, Christian Matt, and Thomas Hess. "Personalised Nudging for More Data Disclosure? On the Adaption of Data Usage Policies Format to Cognitive Styles". In: *Proceedings of the 52nd Hawaii International Conference on System Sciences*. IEEE Computer Society, Conference Publishing Services, 2019. URL: https://scholarspace.manoa.hawaii.edu/handle/10125/59877.
497. Guido Schryen et al. "A Knowledge Development Perspective on Literature Reviews: Validation of a New Typology in the IS Field". In: *Communications of the Association for Information Systems* 46.7 (2020), pp. 134–186. https://doi.org/10.17705/1CAIS.04607.
498. Stephanie Ricker Schulte and Victor Pickard. "Fixing Fake News: Self-Regulation and Technological Solutionism". In: *Fake News: Understanding Media and Misinformation in the Digital Age*. MIT Press, 2020.
499. Wolfgang Schweiger et al. *Informationskompetenz: Erkennen, Was Wahr Und Richtig Ist*. *Tech.* rep. Landesmedienzentrum Baden-Württemberg, 2020.
500. Susanne G Scott and Reginald A Bruce. "Decision-Making Style: The Development and Assessment of a New Measure". In: *Educational and psychological measurement* 55.5 (1995), pp. 818–831.
501. Katie Seaborn, Mark Chignell, and Jacek Gwizdka. "Psychological Resilience during COVID-19: A Meta-Review Protocol". In: *BMJ Open* 11 (2021), e051417. https://doi.org/10.1136/bmjopen-2021-051417.
502. Tara Sell, Divya Hosangadi, and Marc Trotochaud. "Misinformation and the US Ebola Communication Crisis: Analyzing the Veracity and Content of Social Media Messages Related to a Fear-Inducing Infectious Disease Outbreak". In: *BMC Public Health* 20 (2020), p. 550. https://doi.org/10.1186/s12889-020-08697-3.
503. Haeseung Seo, Aiping Xiong, and Dongwon Lee. "Trust It or Not: Effects of Machine-Learning Warnings in Helping Individuals Mitigate Misinformation". In: *Proceedings of the 10th ACM Conference on Web Science*. ACM, 2019, pp. 265–274. https://doi.org/10.1145/3292522.3326012.
504. Lanyu Shang et al. "A Multimodal Misinformation Detector for COVID-19 Short Videos on TikTok". In: *2021 IEEE International Conference on Big Data (Big Data)*. IEEE, 2021, pp. 899–908. https://doi.org/10.1109/BigData52589.2021.9671928.
505. Chengcheng Shao et al. "Anatomy of an Online Misinformation Network". In: *PLOS ONE* 13 (2018). https://doi.org/10.1371/journal.pone.0196087.
506. Filipo Sharevski and Donald Gover. "Two Truths and a Lie: Exploring Soft Moderation of COVID-19 Misinformation with Amazon Alexa". In: *The 16th International Conference on Availability, Reliability and Security*. ACM, 2021, pp. 1–9. https://doi.org/10.1145/3465481.3470017.
507. Filipo Sharevski and Aziz N Zeidieh. " "I Just Didn't Notice It:" Experiences with Misinformation Warnings on Social Media amongst Users Who Are Low Vision or Blind".

In: *Proceedings of the 2023 New Security Paradigms Workshop*. NSPW '23. Association for Computing Machinery, 2023, pp. 17–33. https://doi.org/10.1145/3633500.3633502.
508. Filipo Sharevski et al. "Meaningful Context, a Red Flag, or Both? Preferences for Enhanced Misinformation Warnings among US Twitter Users". In: *Proceedings of the 2022 European Symposium on Usable Security*. EuroUSEC '22. Association for Computing Machinery, 2022, pp. 189–201. https://doi.org/10.1145/3549015.3555671.
509. Karishma Sharma et al. "Coronavirus on Social Media: Analyzing Misinformation in Twitter Conversations". In: *arXiv preprintarXiv:2003.12309* (2020).
510. Karishma Sharma et al. *COVID-19 on Social Media: Analyzing Misinformation in Twitter Conversations*. 2020. https://doi.org/10.48550/arXiv.2003.12309. URL: http://arxiv.org/abs/2003.12309.
511. Zien Sheikh Ali et al. "Tahaqqaq: A Real-Time System for Assisting Twitter Users in Arabic Claim Verification". In: *Proceedings of the 46th International ACM SIGIR Conference on Research and Development in Information Retrieval*. SIGIR '23. Association for Computing Machinery, 2023, pp. 3170–3174. https://doi.org/10.1145/3539618.3591815.
512. Steve Sheng et al. "Anti-Phishing Phil: The Design and Evaluation of a Game That Teaches People Not to Fall for Phish". In: *Proceedings of the 3rd Symposium on Usable Privacy and Security (SOUPS '07)*. Vol. 229. 2007, pp. 88–99. https://doi.org/10.1145/1280680.1280692.
513. Imani N. Sherman, Jack W. Stokes, and Elissa M. Redmiles. "Designing Media Provenance Indicators to Combat Fake Media". In: *24th International Symposium on Research in Attacks, Intrusions and Defenses*. ACM, 2021, pp. 324–339. https://doi.org/10.1145/3471621.3471860.
514. Swapneel Sheth, Gail Kaiser, and Walid Maalej. "Us and Them: A Study of Privacy Requirements across North America, Asia, and Europe". In: *Proceedings of the 36th International Conference on Software Engineering*. ACM, 2014, pp. 859–870. https://doi.org/10.1145/2568225.2568244. URL: https://dl.acm.org/doi/10.1145/2568225.2568244.
515. Anu Shrestha and Francesca Spezzano. "Online Misinformation: From the Deceiver to the Victim". In: *Proceedings of the 2019 IEEE/ACM International Conference on Advances in Social Networks Analysis and Mining*. ASONAM '19. Association for Computing Machinery, 2020, pp. 847–850. https://doi.org/10.1145/3341161.3343536.
516. Anup Shrestha, Aileen Cater-Steel, and Mark Toleman. "How to Communicate Evaluation Work in Design Science Research? An Exemplar Case Study". In: *25th Australasian Conference on Information Systems* (2014).
517. Kai Shu, H Russell Bernard, and Huan Liu. "Studying Fake News via Network Analysis: Detection and Mitigation". In: *Emerging Research Challenges and Opportunities in Computational Social Network Analysis and Mining*. 2019, pp. 43–65. https://doi.org/10.1007/978-3-319-94105-9_3.
518. Kai Shu et al. "Fake News Detection on Social Media: A Data Mining Perspective". In: *ACM SIGKDD explorations newsletter* 19.1 (2017), p. 15.
519. H. S. Sichel. "On a Distribution Law for Word Frequencies". In: *Journal of the American Statistical Association* 70.351 (1975), pp. 542–547. https://doi.org/10.2307/2285930.

520. Safat Siddiqui. "Personalized Interaction-Focused Interventions for Mitigating Misinformation". PhD thesis. 2022. URL: https://www.proquest.com/docview/2723183824/abstract/9F3C77A85DAF4747PQ/1?sourcetype=Dissertations%20&%20Theses.
521. Mario Silic and Paul Benjamin Lowry. "Using Design-Science Based Gamification to Improve Organizational Security Training and Compliance". In: *Journal of Management Information Systems (JMIS)* (2019).
522. E. H. Simpson. "Measurement of Diversity". In: *Nature* 163.4148 (1949), pp. 688–688. https://doi.org/10.1038/163688a0.
523. H. Colleen Sinclair. *10 Ways to Spot Online Misinformation*. 2020. URL: https://theconversation.com/10-ways-to-spot-online-misinformation-132246.
524. Yvonne Skipper, Daniel Jolley, and Joseph Reddington. "'But Wait, That Isn't Real': A Proof-of-Concept Study Evaluating 'Project Real', a Co-Created Intervention That Helps Young People to Spot Fake News Online". In: *British Journal of Development Psychology* 41.4 (2023), pp. 371–384. https://doi.org/10.1111/bjdp.12456.
525. Ciarra N. Smith and Holly H. Seitz. "Correcting Misinformation About Neuroscience via Social Media". In: *Science Communication* 41.6 (2019), pp. 790–819. https://doi.org/10.1177/1075547019890073.
526. Edgar A. Smith and J. Peter Kincaid. "Derivation and Validation of the Automated Readability Index for Use with Technical Materials". In: *Human Factors: The Journal of the Human Factors and Ergonomics Society* 12.5 (1970), pp. 457–564. https://doi.org/10.1177/001872087001200505.
527. Robert Soden and Leysia Palen. "Informating Crisis: Expanding Critical Perspectives in Crisis Informatics". In: *Proceedings of the ACM on Human-Computer Interaction* 2 (2018), pp. 1–22. https://doi.org/10.1145/3274431.
528. Yunya Song, Sai Wang, and Qian Xu. "Fighting Misinformation on Social Media: Effects of Evidence Type and Presentation Mode". In: *Health Education Research* 37.3 (2022), pp. 185–198. https://doi.org/10.1093/her/cyac011.
529. Catherine Sotirakou, Theodoros Paraskevas, and Constantinos Mourlas. "Toward the Design of a Gamification Framework for Enhancing Motivation Among Journalists, Experts, and the Public to Combat Disinformation: The Case of CALYPSO Platform". In: *HCI in Games*. Ed. by Xiaowen Fang. Lecture Notes in Computer Science. Springer International Publishing, 2022, pp. 542–554. https://doi.org/10.1007/978-3-031-05637-6_35.
530. Francesca Spezzano. "Using Service-Learning in Graduate Curriculum to Address Teenagers' Vulnerability to Web Misinformation". In: *Proceedings of the 26th ACM Conference on Innovation and Technology in Computer Science Education V. 2. ACM*, 2021, pp. 637–638. https://doi.org/10.1145/3456565.3460039.
531. Kate Starbird. "Crowd Computation: Organizing Information during Mass Disruption Events". In: *Proceedings of the ACM 2012 Conference on Computer Supported Cooperative Work Companion*. ACM, 2012, pp. 339–342. https://doi.org/10.1145/2141512.2141615. URL: https://dl.acm.org/doi/10.1145/2141512.2141615.
532. Kate Starbird. "Examining the Alternative Media Ecosystem Through the Production of Alternative Narratives of Mass Shooting Events on Twitter". In: *Proceedings of the Eleventh International Conference on Web and Social Media, ICWSM 2017, Montr*éal, Québec, Canada. 2017, pp. 230–239.

533. Kate Starbird, Ahmer Arif, and Tom Wilson. "Disinformation as Collaborative Work: Surfacing the Participatory Nature of Strategic Information Operations". In: *Proc. ACM Hum.-Comput. Interact.* 3.CSCW (2019), pp. 1–26. https://doi.org/10.1145/3359229.
534. Kate Starbird and Leysia Palen. " "Voluntweeters": Self-Organizing by Digital Volunteers in Times of Crisis". In: *Proceedings of the International Conference on Human Factors in Computing Systems* (2011).
535. Kate Starbird et al. "Engage Early, Correct More: How Journalists Participate in False Rumors Online during Crisis Events". In: *Proceedings of the 2018 CHI Conference on Human Factors in Computing Systems (CHI 2018).* ACM, 2018, pp. 1–12. https://doi.org/10.1145/3173574.3173679.
536. Kate Starbird et al. "Rumors, False Flags, and Digital Vigilantes: Misinformation on Twitter after the 2013 Boston Marathon Bombing". In: *IConference* (2014). https://doi.org/10.9776/14308.
537. Kate Starbird et al. "Social Media, Public Participation, and the 2010 BP Deepwater Horizon Oil Spill". In: *Human and Ecological Risk Assessment: An International Journal* 21.3 (2015), pp. 605–630. https://doi.org/10.1080/10807039.2014.947866. URL: http://www.tandfonline.com/doi/abs/10.1080/10807039.2014.947866.
538. Statista. *Ranking of the Biggest Social Networks and Messengers Regarding Number of Users in January 2023 (in German).* 2023. URL: https://de.statista.com/statistik/daten/studie/181086/umfrage/die-weltweit-groessten-social-networks-nach-anzahl-der-user/.
539. Martin Steinebach et al., eds. *Desinformation aufdecken und bekämpfen: Interdisziplinäre Ansätze gegen Desinformationskampagnen und für Meinungspluralität.* Nomos Verlagsgesellschaft mbH & Co. KG, 2020. https://doi.org/10.5771/9783748904816.
540. Karin Stoegner. " 'We Are the New Jews!' And 'The Jewish Lobby' – Antisemitism and the Construction of a National Identity by the Austrian Freedom Party". In: *Nations and Nationalism* 22.3 (2016), pp. 484–504. https://doi.org/10.1111/nana.12165. URL: https://onlinelibrary.wiley.com/doi/abs/10.1111/nana.12165.
541. Peter Story et al. "From Intent to Action: Nudging Users Towards Secure Mobile Payments". In: *Sixteenth Symposium on Usable Privacy and Security (SOUPS 2020).* 2020, pp. 379–415.
542. Alan Strathman et al. "The Consideration of Future Consequences: Weighing Immediate and Distant Outcomes of Behavior". In: *Journal of Personality and Social Psychology* 66.4 (1994), pp. 742–752. https://doi.org/10.1037/0022-3514.66.4.742.
543. Stefan StrauB. "Deep Automation Bias: How to Tackle a Wicked Problem of AI?" In: *Big Data and Cognitive Computing* 5.2 (2021), p. 18. https://doi.org/10.3390/bdcc5020018.
544. M. Connor Sullivan. "Leveraging Library Trust to Combat Misinformation on Social Media". In: *Library & Information Science Research* 41.1 (2019), pp. 2–10. https://doi.org/10.1016/j.lisr.2019.02.004.
545. Sharifa Sultana and Susan R. Fussell. "Dissemination, Situated Fact-checking, and Social Effects of Misinformation among Rural Bangladeshi Villagers During the COVID-19 Pandemic". In: *Proceedings of the ACM on Human-Computer Interaction* 5.CSCW2 (2021), 436:1–436:34. https://doi.org/10.1145/3479580.

546. Cass R. Sunstein, Lucia A. Reisch, and Julius Rauber. “Behavioral Insights All over the World? Public Attitudes toward Nudging in a Multi-Country Study”. In: *Public Attitudes Toward Nudging in a Multi-Country Study (February 21, 2017)* (2017).
547. Tagesschau. *Entsetzen Über Eskalation Am Reichstagsgebäude: Protest Gegen Corona-Politik*. 2020.
548. Tetsuji Takada and Yumeji Hattori. “Giving Motivation for Using Secure Credentials through User Authentication by Game”. In: *Proceedings of the International Conference on Advanced Visual Interfaces*. ACM, 2020, pp. 1–3. https://doi.org/10.1145/3399715.3399950.
549. Damian Tambini. *Fake News: Public Policy Responses*. Tech. rep. Media Policy Project, London School of Economics and Political Science, 2017.
550. Yuko Tanaka and Rumi Hirayama. “Exposure to Countering Messages Online: Alleviating or Strengthening False Belief?” In: *Cyberpsychology,Behavior, and Social Networking* 22.11 (2019), pp. 742–746. https://doi.org/10.1089/cyber.2019.0227.
551. Yuko Tanaka, Yasuaki Sakamoto, and Toshihiko Matsuka. “Toward a Social-Technological System That Inactivates False Rumors through the Critical Thinking of Crowds”. In: *2013 46th Hawaii International Conference on System Sciences*. IEEE, 2013, pp. 649–658. https://doi.org/10.1109/HICSS.2013.557.
552. Ran Tao et al. “Hope over Fear: The Interplay between Threat Information and Hope Appeal Corrections in Debunking Early COVID-19 Misinformation”. In: *Social Science & Medicine* 333.116132 (2023). https://doi.org/10.1016/j.socscimed.2023.116132.
553. Jack E. Taylor et al. “Rating Norms Should Be Calculated from Cumulative Link Mixed Effects Models”. In: *Behavior Research Methods* 55.5 (2022), pp. 2175–2196. URL: https://link.springer.com/10.3758/s13428-022-01814-7.
554. Thaler, Richard H. and Sunstein, Cass R. *Nudge: Improving Decisions about Health, Wealth, and Happiness*. Penguin Books, 2009.
555. Saravanan Thirumuruganathan, Michael Simpson, and Laks V.S. Lakshmanan. “To Intervene or Not To Intervene: Cost Based Intervention for Combating Fake News”. In: *Proceedings of the2021 International Conference on Management of Data*. ACM, 2021, pp. 2300–2309.
556. Kurt Thomas et al. “ “It’s Common and a Part of Being a Content Creator”: Understanding How Creators Experience and Cope with Hate and Harassment Online”. In: *Proceedings of the 2022 CHI Conference on Human Factors in Computing Systems*. CHI ’22. Association for Computing Machinery, 2022, pp. 1–15. https://doi.org/10.1145/3491102.3501879.
557. Calum Thornhill et al. “A Digital Nudge to Counter Confirmation Bias”. In: *Frontiers in Big Data* 2 (2019), p. 11. https://doi.org/10.3389/fdata.2019.00011.
558. Nguyen Hoang Thuan, Andreas Drechsler, and Pedro Antunes. “Construction of Design Science Research Questions”. In: *Communications of the Association for Information Systems* 44 (2019), pp. 332–363. https://doi.org/10.17705/1CAIS.04420.
559. Thi Tran et al. “An Investigation of Misinformation Harms Related to Social Media during Two Humanitarian Crises”. In: *Information Systems Frontiers* (2020), pp. 1–9. https://doi.org/10.1007/s10796-020-10088-3.
560. Yu-Chia Tseng et al. “Investigating Correction Effects of Different Modalities for Misinformation about COVID-19”. In: *Companion Publication of the 2022 Conference on*

Computer Supported Cooperative Work and Social Computing. CSCW'22 Companion. Association for Computing Machinery, 2022, pp. 54–58. https://doi.org/10.1145/3500868.3559455.
561. Gleb Tsipursky, Fabio Votta, and Kathryn M Roose. "Fighting Fake News and Post-Truth Politics with Behavioral Science: The Pro-Truth Pledge". In: *Behavior and Social Issues* 27.1 (2018), pp. 47–70.
562. Marina Tulin et al. "Beyond Belief Correction: Effects of the Truth Sandwich on Perceptions of Fact-Checkers and Verification Intentions". In: *Journalism Practice* (2024), pp. 1–21. https://doi.org/10.1080/17512786.2024.2311311.
563. Melissa Tully, Leticia Bode, and Emily K. Vraga. "Mobilizing Users: Does Exposure to Misinformation and Its Correction Affect Users' Responses to a Health Misinformation Post?" In: *Social Media + Society* 6.4 (2020), pp. 1–12. https://doi.org/10.1177/2056305120978377.
564. Melissa Tully, Emily K. Vraga, and Leticia Bode. "Designing and Testing News Literacy Messages for Social Media". In: *Mass Communication and Society* 23.1 (2020), pp. 22–46. https://doi.org/10.1080/15205436.2019.1604970.
565. Iis Tussyadiah, Shujun Li, and Graham Miller. "Privacy Protection in Tourism: Where We Are and Where We Should Be Heading For". In: *Information and Communication Technologies in Tourism* 2019. Springer, 2019, pp. 278–290.
566. Fiona J. Tweedie and R. Harald Baayen. "How Variable May a Constant Be? Measures of Lexical Richness in Perspective". In: *Computers and the Humanities* 32.5 (1998), pp. 323–352.
567. Twitter. *Introducing Birdwatch, a Community-Based Approach to Misinformation*. 2021. URL: https://blog.x.com/en_us/topics/product/2021/introducing-birdwatch-a-community-based-approach-to-misinformation.
568. Twitter. *Updating Our Approach to Misleading Information*. 2020. URL: https://blog.x.com/en_us/topics/product/2020/updating-our-approach-to-misleading-information.
569. Blase Ur et al. "Design and Evaluation of a Data-Driven Password Meter". In: *Proceedings of the 2017 CHI Conference on Human Factors in Computing Systems*. 2017, pp. 3775–3786.
570. Blase Ur et al. "How Does Your Password Measure up? The Effect of Strength Meters on Password Creation". In: *Presented as Part of the 21st USENIX Security Symposium (USENIX Security 12)*. 2012, pp. 65–80.
571. Marvin van Bekkum and Frederik Zuiderveen Borgesius. "Using Sensitive Data to Prevent Discrimination by Artificial Intelligence: Does the GDPR Need a New Exception?" In: *Computer Law & Security Review* 48 (2023), p. 105770. https://doi.org/10.1016/j.clsr.2022.105770. URL: https://www.sciencedirect.com/science/article/pii/S0267364922001133.
572. Toni G. L. A. van der Meer, Michael Hameleers, and Jakob Ohme. "Can Fighting Misinformation Have a Negative Spillover Effect? How Warnings for the Threat of Misinformation Can Decrease General News Credibility". In: *Journalism Studies* 24.6 (2023), pp. 803–823. https://doi.org/10.1080/1461670X.2023.2187652.
573. Toni G. L. A. van der Meer and Yan Jin. "Seeking Formula for Misinformation Treatment in Public Health Crises: The Effects of Corrective Information Type and Source". In: *Health Communication* 35.5 (2020), pp. 560–575. https://doi.org/10.1080/10410236.2019.1573295.

574. Jan-Willem van Prooijen and Karen M. Douglas. "Belief in Conspiracy Theories: Basic Principles of an Emerging Research Domain". In: *European Journal of Social Psychology* 48.7 (2018), pp. 897–908. https://doi.org/10.1002/ejsp.2530.
575. Rama Adithya Varanasi, Joyojeet Pal, and Aditya Vashistha. "Accost, Accede, or Amplify: Attitudes towards COVID-19 Misinformation on WhatsApp in India". In: *Proceedings of the 2022 CHI Conference on Human Factors in Computing Systems*. CHI '22. Association for Computing Machinery, 2022, pp. 1–17. https://doi.org/10.1145/3491102.3517588.
576. Abigail T. Velasco et al. "Factit: A Fact-Checking Browser Extension". In: *2023 IEEE 12th International Conference on Educational and Information Technology (ICEIT)*. IEEE, 2023, pp. 342–347. https://doi.org/10.1109/ICEIT57125.2023.10107833.
577. Yamil R. Velez, Ethan Porter, and Thomas J. Wood. "Latino-Targeted Misinformation and the Power of Factual Corrections". In: *Journal of Politics* 85.2 (2023). https://doi.org/10.1086/722345.
578. Arnold P. O. S. Vermeeren et al. "User Experience Evaluation Methods: Current State and Development Needs". In: *Proceedings of the 6thNordic Conference on Human-Computer Interaction: Extending Boundaries*. NordiCHI '10. ACM, 2010, pp. 521–530. https://doi.org/10.1145/1868914.1868973.
579. Sarah Vieweg. "Twitter Communications in Mass Emergency: Contributions to Situational Awareness". In: *Proceedings of the ACM 2012 Conference on Computer Supported Cooperative Work Companion*. ACM, 2012, pp. 227–230.
580. Ingrid Volkmer. *Social Media and COVID-19: A Global Study of Digital Crisis Interaction among Gen Z and Millennials*. Tech. rep. University of Melbourne, 2022.
581. Christian von der Weth, Jithin Vachery, and Mohan Kankanhalli. "Nudging Users to Slow Down the Spread of Fake News in Social Media". In: *2020 IEEE International Conference on Multimedia & Expo Workshops (ICMEW)*. IEEE, 2020, pp. 1–6. https://doi.org/10.1109/ICMEW46912.2020.9106003.
582. Emily K. Vraga and Leticia Bode. "I Do Not Believe You: How Providing a Source Corrects Health Misperceptions across Social Media Platforms". In: *Information, Communication & Society* 21.10 (2018), pp. 1337–1353. https://doi.org/10.1080/1369118X.2017.1313883.
583. Emily K. Vraga and Leticia Bode. "Using Expert Sources to Correct Health Misinformation in Social Media". In: *Science Communication* 39.5 (2017), pp. 621–645. https://doi.org/10.1177/1075547017731776.
584. Emily K. Vraga, Leticia Bode, and Melissa Tully. "Creating News Literacy Messages to Enhance Expert Corrections of Misinformation on Twitter". In: *Communication Research and Practice* 49.2 (2022), pp. 245–267.
585. Emily K. Vraga, Leticia Bode, and Melissa Tully. "The Effects of a News Literacy Video and Real-Time Corrections to Video Misinformation Related to Sunscreen and Skin Cancer". In: *Health Communication* 37.13 (2021), pp. 1622–1630.
586. Emily K. Vraga, Sojung Claire Kim, and John Cook. "Testing Logic-based and Humor-based Corrections for Science, Health, and Political Misinformation on Social Media". In: *Jourbal of Broadcasting & Electronic Media* 63.3 (2019), pp. 393–414. https://doi.org/10.1080/08838151.2019.1653102.
587. Emily K. Vraga, Melissa Tully, and Leticia Bode. "Assessing the Relative Merits of News Literacy and Corrections in Responding to Misinformation on Twitter". In: *New*

Media & Society 24.10 (2021), pp. 2354–2371. https://doi.org/10.1177/146144482199 8691.

588. Emily K. Vraga et al. "Testing the Effectiveness of Correction Placement and Type on Instagram". In: *The International Journal of Press/Politics* 25.4 (2020), pp. 632–652. https://doi.org/10.1177/1940161220919082.
589. Christopher N. Wahlheim, Timothy R. Alexander, and Carson D. Peske. "Reminders of Everyday Misinformation Statements Can Enhance Memory for and Beliefs in Corrections of Those Statements in the Short Term". In: *Psychological Science* 31.10 (2020), pp. 1325–1339. https://doi.org/10.1177/0956797620952797.
590. Sarenne Wallbridge, Peter Bell, and Catherine Lai. *It's Not What You Said, It's How You Said It: Discriminative Perception of Speech as a Multichannel Communication System*. 2021. https://doi.org/10.48550/arXiv.2105.00260.
591. Thomas Waltemate et al. "The Impact of Avatar Personalization and Immersion on Virtual Body Ownership, Presence, and Emotional Response". In: *IEEE transactions on visualization and computer graphics* 24.4 (2018), pp. 1643–1652.
592. Franz Waltenberger, Simon Höferlin, and Michael Froehlich. "Reddit Insights: Improving Online Discussion Culture by Contextualizing User Profiles". In: *Extended Abstracts of the 2023 CHI Conference on Human Factors in Computing Systems*. CHI EA '23. Association for Computing Machinery, 2023, pp. 1–6. https://doi.org/10.1145/3544549.3585671.
593. Nayer Wanas et al. "Automatic Scoring of Online Discussion Posts". In: *Proceedings of the 2nd ACM Workshop on Information Credibility on the Web*. Association for Computing Machinery, 2008, pp. 19–26.
594. Austin Horng-En Wang. "PM Me the Truth? The Conditional Effectiveness of Fact-Checks across Social Media Sites". In: *Social Media+ Society* 8.098347 (2022), pp. 1–16. https://doi.org/10.1177/20563051221098347.
595. Ran Wang et al. "RumorLens: Interactive Analysis and Validation of Suspected Rumors on Social Media". In: *Extended Abstracts of the 2022CHI Conference on Human Factors in Computing Systems*. CHI EA '22. Association for Computing Machinery, 2022, pp. 1–7. https://doi.org/10.1145/3491101.3519712.
596. Weirui Wang and Yan Huang. "Countering the "Harmless E-Cigarette" Myth: The Interplay of Message Format, Message Sidedness, and Prior Experience With E-Cigarette Use in Misinformation Correction". In: *Science Communication* 43.2 (2021), pp. 170–198. https://doi.org/10.1177/1075547020974384.
597. Yuxi Wang et al. "Systematic Literature Review on the Spread of Health-related Misinformation on Social Media". In: *Social Science & Medicine* 240 (2019), p. 112552. https://doi.org/10.1016/j.socscimed.2019.112552.
598. Colin Ware. *Information Visualization: Perception for Design*. 4th ed. Elsevier, 2012.
599. Herman Wasserman and Dani Madrid-Morales. "An Exploratory Study of "Fake News" and Media Trust in Kenya, Nigeria and South Africa". In: *African Journalism Studies* 40.1 (2019), pp. 107–123. https://doi.org/10.1080/23743670.2019.1627230. URL: https://doi.org/10.1080/23743670.2019.1627230.
600. Theresa Webb and Kathryn Martin. "Evaluation of a Us School-Based Media Literacy Violence Prevention Curriculum on Changes in Knowledge and Critical Thinking Among Adolescents". In: *Journal of Children and Media* 6.4 (2012), pp. 430–449. https://doi.org/10.1080/17482798.2012.724591.

601. Victoria Westbrook, Duane T. Wegener, and Mark W. Susmann. "Mechanisms in Continued Influence: The Impact of Misinformation Corrections on Source Perceptions". In: *Memory & Cognition* 51.6 (2023), pp. 1317–1330. https://doi.org/10.3758/s13421-023-01402-w.

602. WhatsApp. *Introducing Voice Messages*. 2013. URL: https://blog.whatsapp.com/introducing-voice-messages.

603. WhatsApp. *We're Making Voice Messages Even Better*. 2022. URL: https://blog.whatsapp.com/making-voice-messages-better?lang=af.

604. Daniel Lowe Wheeler. "Zxcvbn: Low-budget Password Strength Estimation". In: *25th USENIX Security Symposium (USENIX Security 16)*. 2016, pp. 157–173.

605. Winnifred Wijnker et al. "Debunking Strategies for Misleading Bar Charts". In: *JCOM—Journal of Science Communication* 21.7 (2022), pp. 1–27. https://doi.org/10.22323/2.21070207.

606. Tom Wilson, Kaitlyn Zhou, and Kate Starbird. "Assembling Strategic Narratives: Information Operations as Collaborative Work within an Online Community". In: *Proceedings of the ACM on Human-Computer Interaction* 2.CSCW (2018), pp. 1–26. https://doi.org/10.1145/3274452.

607. Don Winiecki, Francesca Spezzano, and Chandler Underwood. "Understanding Teenagers' Real and Fake News Sharing on Social Media". In: *Proceedings of the 22nd Annual ACM Interaction Design and Children Conference*. ACM, 2023, pp. 598–602. https://doi.org/10.1145/3585088.3593864.

608. Bang Wong. "Points of View: Color Blindness". In: *Nature* 8.441 (2011). https://doi.org/10.1038/nmeth.1618.

609. Reed M. Wood et al. "Promoting COVID-19 Vaccine Confidence through Public Responses to Misinformation: The Joint Influence of Message Source and Message Content". In: *Social Science & Medicine* 324.115863 (2023), pp. 1–11. https://doi.org/10.1016/j.socscimed.2023.115863.

610. Thomas Wood and Ethan Porter. "The Elusive Backfire Effect: Mass Attitudes' Steadfast Factual Adherence". In: *Political Behavior* 41.1 (2019), pp. 135–163. https://doi.org/10.1007/s11109-018-9443-y.

611. World Health Organization. *Novel Coronavirus (2019-nCoV) Situation Report, 13. Clarification, Clinical and Presentation, Clinical and Factors, Risk*. Tech. rep. February. 2020.

612. Luke Wroblewski. *Web Form Design: Filling in the Blanks*. Rosenfeld Media, 2008.

613. Liang Wu and Huan Liu. "Tracing Fake-News Footprints: Characterizing Social Media Messages by How They Propagate". In: *Proceedings of the Eleventh ACM International Conference on Web Search and Data Mining*. ACM, 2018, pp. 637–645. https://doi.org/10.1145/3159652.3159677.

614. Volker Wulf et al. "Engaging with Practices: Design Case Studies as a Research Framework in CSCW". In: *Proceedings of the ACM Conference on Computer Supported Cooperative Work, CSCW*. 2011, pp. 505–512. https://doi.org/10.1145/1958824.1958902.

615. Alex J. Xu et al. "TikTok and Prostate Cancer: Misinformation and Quality of Information Using Validated Questionnaires". In: *BJU International* 128.4 (2021), pp. 435–437. https://doi.org/10.1111/bju.15403.

616. Fan Xu, Victor S. Sheng, and Mingwen Wang. "A Unified Perspective for Disinformation Detection and Truth Discovery in Social Sensing: A Survey". In: *ACM Computing Surveys* 55.1 (2021), 6:1–6:33. https://doi.org/10.1145/3477138.
617. Kamya Yadav. *Platform Interventions: How Social Media Counters Influence Operations*. 2021.
618. Rui Yan, Mirella Lapata, and Xiaoming Li. "Tweet Recommendation with Graph Co-Ranking". In: *Proceedings of the 50th Annual Meeting of the Association for Computational Linguistics*. 2012, pp. 516–525.
619. Jeongwon Yang and Sunhee Lee. "Framing the MERS Information Crisis: An Analysis on Online News Media's Rumour Coverage". In: *Journal of Contingencies and Crisis Management* (2020), pp. 1–13. https://doi.org/10.1111/1468-5973.12292.
620. Waheeb Yaqub et al. "Effects of Credibility Indicators on Social Media News Sharing Intent". In: *Proceedings of the 2020 CHI Conference on Human Factors in Computing Systems*. CHI '20. Association for Computing Machinery, 2020, pp. 1–14. https://doi.org/10.1145/3313831.3376213. URL: https://doi.org/10.1145/3313831.3376213.
621. Wan Yit Yong, Rajesh Jaiswal, and Fernando Perez Tellez. "Explainability in NLP Model: Detection of Covid-19 Twitter Fake News". In: *Proceedings of the 2023 Conference on Human Centered Artificial Intelligence: Education and Practice*. HCAIep '23. Association for Computing Machinery, 2023, pp. 1–7. https://doi.org/10.1145/3633083.3633212.
622. Yule, G.U. *The Statistical Study of Literary Vocabulary*. Cambridge University Press, 1944.
623. Himanshu Zade et al. "Tweet Trajectory and AMPS-based Contextual Cues Can Help Users Identify Misinformation". In: *Proc. ACM Hum.-Comput. Interact.* 7.CSCW1 (2023), pp. 1–27. https://doi.org/10.1145/3579536.
624. Jing Zeng and Chung-hong Chan. "A Cross-National Diagnosis of Infodemics: Comparing the Topical and Temporal Features of Misinformation around COVID-19 in China, India, the US, Germany and France". In: *Online Information Review* 45.4 (2021), pp. 709–728. https://doi.org/10.1108/OIR-09-2020-0417. URL: https://www.emerald.com/insight/content/doi/10.1108/OIR-09-2020-0417/full/html.
625. L. Zeng, K. Starbird, and E. S. Spiro. "Rumors at the Speed of Light? Modeling the Rate of Rumor Transmission During Crisis". In: *2016 49th Hawaii International Conference on System Sciences (HICSS)*. 2016, pp. 1969–1978. https://doi.org/10.1109/HICSS.2016.248.
626. Amy X. Zhang et al. "A Structured Response to Misinformation: Defining and Annotating Credibility Indicators in News Articles". In: *Companion of the The Web Conference 2018 on The Web Conference 2018 -WWW '18*. ACM Press, 2018, pp. 603–612. https://doi.org/10.1145/3184558.3188731.
627. Yuqi Zhang et al. "Investigation of the Determinants for Misinformation Correction Effectiveness on Social Media during COVID-19 Pandemic". In: *Information Processing & Management* 59.102935 (2022), pp. 1–16. https://doi.org/10.1016/j.ipm.2022.102935.
628. Wenqing Zhao. "Misinformation Correction across Social Media Platforms". In: *2019 International Conference on Computational Science and Computational Intelligence (CSCI)*. IEEE, 2019, pp. 1371–1376. https://doi.org/10.1109/CSCI49370.2019.00256.

629. Chengbo Zheng and Xiaojuan Ma. "Evaluating the Effect of Enhanced Text-Visualization Integration on Combating Misinformation in Data Story". In: *2022 IEEE 15th Pacific Visualization Symposium (Pacific Vis)*. IEEE, 2022, pp. 141–150. https://doi.org/10.1109/PacificVis53943.2022.00023.
630. Rong Zheng et al. "A Framework for Authorship Identification of Online Messages: Writing-style Features and Classification Techniques". In: *Journal of the American Society for Information Science and Technology* 57.3 (2006), pp. 378–393. https://doi.org/10.1002/asi.20316.
631. Xinyi Zhou and Reza Zafarani. "A Survey of Fake News: Fundamental Theories, Detection Methods, and Opportunities". In: *ACM Computing Surveys* 53.5 (2020), pp. 1–40. https://doi.org/10.1145/3395046.
632. Verena Zimmermann and Nina Gerber. "The Password Is Dead, Long Live the Password—A Laboratory Study on User Perceptions of Authentication Schemes". In: *International Journal of Human-Computer Studies* 133 (2020), pp. 26–44.
633. Verena Zimmermann, Karola Marky, and Karen Renaud. "Hybrid Password Meters for More Secure Passwords—a Comprehensive Study of Password Meters Including Nudges and Password Information". In: *Behaviour & Information Technology* 42.6 (2023), pp. 700–743. https://doi.org/10.1080/0144929X.2022.2042384. URL: https://www.tandfonline.com/doi/full/10.1080/0144929X.2022.2042384.
634. Verena Zimmermann and Karen Renaud. "The Nudge Puzzle: Matching Nudge Interventions to Cybersecurity Decisions". In: *ACM Transactions on Computer-Human Interaction (TOCHI)* 28.1 (2021), pp. 1–45.
635. Luisa Dolores Zozaya-Durazo, Charo Sádaba-Chalezquer, and Beatriz Feijoo-Fernández. ""Fake or Not, I'm Sharing It": Teen Perception about Disinformation in Social Networks". In: *Young Consumers* 0.0 (2023), pp. 1–18. https://doi.org/10.1108/YC-06-2022-1552.
636. Arkaitz Zubiaga et al. "Detection and Resolution of Rumours in Social Media: A Survey". In: *ACM Computing Surveys* 51.2 (2019), pp. 1–36. https://doi.org/10.1145/3161603.

The manufacturer's authorised representative in the EU is Springer
Nature Customer Service Centre GmbH, Europaplatz 3, 69115 Heidelberg,
Germany. If you have any concerns regarding our products, please
contact ProductSafety@springernature.com

Printed and bound by CPI Group (UK) Ltd, Croydon, CR0 4YY
30/09/2026
02225566-0006